SOIL AND WATER
CONSERVATION ENGINEERING

SOIL AND WATER CONSERVATION ENGINEERING

Third Edition

Glenn O. Schwab
Professor of Agricultural Engineering
The Ohio State University
and
Ohio Agricultural Research and Development
Center, Columbus, Ohio

Richard K. Frevert
Professor of Agricultural Engineering
The University of Arizona
Tucson, Arizona

Talcott W. Edminster
Late Administrator, Agricultural Research,
Science Education Administration
United States Department of Agriculture
Washington, D.C.

Kenneth K. Barnes
Late Professor of Agricultural Engineering
and Head of Soils, Water, and Engineering
The University of Arizona
Tucson, Arizona

JOHN WILEY & SONS

New York • Chichester • Brisbane • Toronto • Singapore

Library of Congress Cataloging in Publication Data:

Main entry under title:

Soil and Water conservation engineering.

  Previous 2 editions published in 1955 and 1966
entered under R. K. Frevert.
  Includes index.
  1. Soil conservations.   2. Water conservation.
3. Agricultural engineering.   I. Schwab, Clenn Orville,
1914–   . II. Frevert, Richard K. Soil and water
conservation engineering.   III. Title.
S623.F68   1981   631.4   80-27961
ISBN 0-471-03078-3

Printed in the United States of America

10 9 8 7 6 5 4 3 2

PREFACE

We were most gratified by the general acceptance of the first two editions of *Soil and Water Conservation Engineering*. In this edition changes have been made to give greater emphasis to engineering design of soil and water conservation practices. For this reason, as well as to reduce the length, the chapter on soil, water, and plant relationships and the chapter on subsurface drainage principles have been omitted or incorporated into other design chapters. The chapter on legal aspects has also been assimilated into other chapters where appropriate. All chapters have been updated to reflect the latest design practices.

Many suggestions from instructors and other users have been included. The major change has been the conversion from English to System International (SI) units. Examples and problems at the end of chapters are in SI units with English units in parentheses. In many cases the SI units are rounded for ease in computation, which may result in slightly different answers with the two systems of units. Most graphs and illustrations show both systems of units to aid in making the transition. Because of the many conversions, the authors realize there may be some errors, for which we ask your kind indulgence.

As in the two previous editions, the purpose of this book is to provide a professional text for agricultural engineering students. This book includes subject matter on all the engineering phases of soil and water conservation as well as a limited section on hydrology. The first chapter covers the general aspects of the book; Chapters 2 through 4, hydrology; Chapters 5 through 21, erosion and its control, earth dams, flood control, drainage, and irrigation.

We have again assumed in preparing this revision that the student has taken such basic courses as surveying, mechanics, hydraulics, and soils. However, a knowledge of these subjects is not essential for understanding many portions of the text. In presenting the subject, we have attempted to emphasize the analytical approach supplemented with sufficient field data to point out practical applications. Although stressing principles rather than tables, charts, and diagrams, the book may provide considerable basic data for practicing engineers as well. Class problems and many examples have been included to emphasize design principles and to facilitate an understanding of the subject matter.

G. O. Schwab
R. K. Frevert

v

ACKNOWLEDGMENTS

We are deeply indebted to many individuals and organizations for the use of material. We are especially grateful to The Ferguson Foundation, Detroit, Mich., for making the first edition possible by defraying the cost of its development. Harold E. Pinches, formerly with The Ferguson Foundation, was instrumental in promoting this project. We are grateful to Massey-Ferguson Limited, Toronto, Canada, for providing funds to prepare the second edition. E. L. Barger of Massey-Ferguson Limited offered helpful advice and encouragement throughout the preparation of earlier editions of this book. The following individuals have reviewed portions of the manuscript and made valuable suggestions: H. J. Braud, Jr., S. T. Chu, D. D. Fangmeier, M. E. Jensen, C. L. Larson, W. D. Lembke, L. Lyles, G. E. Merva, B. H. Nolte, D. Nir, and M. L. Palmer. A. D. Fenemor and Mrs. Kenneth K. Barnes reviewed a majority of the chapters and made helpful comments.

Special appreciation goes to the many friends and colleagues at The University of Arizona, The Ohio State University, the Soil Conservation Service, and the United States Department of Agriculture, Science Education Administration, Agricultural Research, who have contributed in many ways through frequent contacts.

We wish also to express appreciation to our wives and families for their sympathetic understanding during the preparation of this edition.

We are dedicating this book to the memory of our colleagues Kenneth K. Barnes and Talcott W. Edminster, who contributed much to the earlier editions and participated in the planning of the third edition.

G. O. S.
R. K. F.

CONTENTS

ABBREVIATIONS

Agron.	Agronomy		hp	horsepower
ARS	Agriculture Research Service (SEA in 1978)		L.F.	load factor
			mimeo.	mineographed
ASAE	American Society of Agricultural Engineers		mo.	month
			NOAA	National Oceanic and Atmospheric Administration
ASCE	American Society of Civil Engineers		P.C.	point of curvature
			P.I.	point of intersection
ASTM	American Society for Testing Materials		P.T.	point of tangency
bu/ac	bushels per acre		publ.	publication
cons.	conservation		res.	research
dhp	drawbar horsepower		rpm	revolution per minute
dia.	diameter		serv.	service
ESSA	Environmental Science Service Administration		SCS	Soil Conservation Service
			SEA	Science and Education Administration
expt.	experiment		soc.	society
geophys.	geophysical		USBR	U.S. Bureau Reclamation
GPO	Government Printing Office		USDA	U.S. Department Agriculture
			V.I.	vertical interval

ENGLISH UNITS

ac-ft	acre feet		gpm	gallons per minute
bu/ac	bushels per acre		iph	inches per hour
cfm	cubic feet per minute		ipd	inches per day
cfd	cubic feet per day		mph	miles per hour
cfs	cubic feet per second		mpd	miles per day
fpm	feet per minute		pcf	pounds per cubic foot
fps	feet per second		ppm	parts per million
fpd	feet per day		t/a	tons per acre

SYSTEM INTERNATIONAL UNITS (SI)

cm	centimeter	kW	kilowatt
d	day	L	liter
g	grams	m	meter
g/L	grams per liter	Mg	megagrams (1 metric ton)
h	hour	mL	milliliter
ha	hectare (10^4 m^2)	m/s	meters per second
ha-m	hectare-meter	mm	millimeter
kg	kilogram (10^3 g)	N	newton
kg/m^3	kilogram per cubic meter	Pa	pascal (N/m^2)
km	kilometer	s	second
kN	kilonewton	W	watts

SIGNS AND SYMBOLS

a cross-sectional area; constant
A watershed area in acres; energy; soil loss
b constant; width
b_n width of notch
b_w width of waterway
B_c outside diameter
B_d width of trench
c cut; chord length; crop coefficient
C coefficient; cropping-management factor; energy
C_v coefficient variation
C_s coefficient skew
C_u uniformity coefficient
d diameter; depth; dry density; distance; day
d_c critical depth
d_i inside diameter
d_w wet density
D diameter; depth; runoff; degree of curvature
D_c drainage coefficient
D_f length of exposure of water surface
e void ratio; distance; deflection angle; vapor pressure
E efficiency; specific energy head; degree of erosion factor; evaporation
EC electrical conductivity
ET evapotranspiration
f infiltration rate; hydraulic friction; depth; monthly evapotranspiration factor; fill
F total infiltration; fertility factor; Froude number; soil erodibility
g acceleration of gravity
G specific gravity of solids; wind erosion climatic factor; sensible heat
G_a apparent specific gravity
G_f specific gravity of fluids
h head; wave height; height; hour; depth of channel
H total head; height; total head loss including friction
H_e specific energy head
H_f friction head loss
H_{rp} riser pipe height
i rainfall intensity; inflow rate
I total rainfall; angle of intersection
I_a initial rainfall extraction

I_c impact coefficient
k constant; permeability; time conversion factor; capillary conductivity; von Karman's constant
K constant; permeability; evapotranspiration coefficient; soil-erodibility factor; conductivity
K_c head loss coefficient for pipe and square conduits
K_e entrance head loss coefficient
K_s Scobey's coefficient of retardation
L length; liter; slope length factor
LR leaching requirement
m exponent; moisture content; water table height; meters
M watershed area in square miles
N revolutions per minute
n roughness coefficient; porosity; number of values
o outflow rate
p wetted perimeter; percent daytime hours of year
P power; pressure; peak runoff rate; rainfall; conservation practice factor
q seepage rate; sprinkler discharge rate; runoff rate; flow rate
Q discharge rate; runoff volume
r scale ratio (prototype to model); radius; reflectance
R hydraulic radius; radius; rainfall and runoff erosivity index; rotation factor; radiation
s slope gradient; rate of storage; distance; standard deviation; second
S slope in percent; storage; settlement; sprinkler spacing; slope erosion factor; energy
t mean monthly temperature; thickness; time; width
T conversion time interval; concentrated surface load; tangent distance; width; return period; tractive force; dimensionless time
T_a absolute temperature (°C + 273)
T_c time of concentration
T_L time of lag; recession time lag
T_p time of peak
T_r time of recession
u monthly evapotranspiration; volume conversion factor; wind velocity
U seasonal evapotranspiration
v velocity; rate of capillary movement; rate of soil moisture movement
v_t threshold velocity
V volume; vegetative cover factor
w unit weight of soil; flow conversion factor
W weight; top width of dam; watershed characteristics; field width; water quantity; width of slope
W_c soil load on conduits
W_d dry weight of soil; water delivered
W_i volume of water input
W_s volume of water stored in the root zone
W_t load on conduits due to a concentrated surface load

W_w wet weight of soil
X constant for geographical location
Y constant for intake rate and cover
y depth
z side slope ratio (horizontal to vertical); depth; height above soil surface
Z vertical distance
θ side slope angle
ρ density
μ dynamic viscosity
ϕ soil moisture potential (capillary)
ψ gravitational potential
Φ potential
γ constant
σ Stefan-Boltzmann constant
τ shear stress

CHAPTER 1

Introduction

Soil and water conservation engineering is the application of engineering principles to the solution of soil and water management problems. The conservation of these vital resources implies *utilization without waste* so as to make possible a high level of production that can be continued indefinitely.

The engineering problems involved in soil and water conservation may be divided into the six following phases: erosion control, drainage, irrigation, flood control, moisture conservation, and water resource development. Although soil erosion takes place even under virgin conditions, the problems to be considered are caused principally by man's removal of the protective cover of natural vegetation. Drainage is the removal of excess water from wet land; irrigation is the application of water to land having a deficiency of moisture for optimum crop growth. Flood control consists of the prevention of overflow on low land and the reduction of flow in streams during and after heavy storms.

Moisture conservation entails application of modified tillage and crop management practices including natural and artificial mulching techniques, level bench terracing, contouring, pitting, ponds, and other physical means of retaining precipitation on the land and reducing evaporative losses from the soil surface. Water resource development involves the collection and storage of surface water as well as the recharge and orderly development of ground water supplies.

The two principal ways of increasing crop production are to develop new land not now in production and to improve the productivity of present cropland. The development of new land is brought about primarily by drainage, irrigation, and removal of shrubs, trees, and rocks. The engineering phases mentioned apply primarily to those measures that will increase efficiency of production on present arable land. A major challenge is to develop systems for greater precision in water and plant nutrient control so as to increase use efficiency of soil, water, and energy resources and to improve the environment for humans.

Population, crop production, energy, and pollution problems in many countries are much more serious than in the United States. Compared to developed countries, such as the United States, underdeveloped countries of the world

1

have a higher population (1967), about twice the population growth rate, a much lower economic growth rate per capita, and a much higher need for an increase in food production.

The distribution of tillable and pasture land, which must produce most of our foods, is shown in Tables 1.1 and 1.2. Since only 10.6 percent of the world's land is tillable, it will require ever increasing soil and water conservation measures and more intensive land use to meet the future food needs of the world. In 1967 the United States had about 12.6 percent of the world's tillable land.

1.1. Agricultural Engineers in Soil and Water Conservation. Sound soil and water conservation is based on the full integration of engineering, atmospheric, plant, and soil sciences. The agricultural engineer,* because of his training in soils, plants, and other basic agricultural subjects, in addition to his engineering background, is well suited to carry out the integration of these sciences. To carry out this plan the engineer must have a knowledge of the soil including its physical and chemical characteristics as well as a sound over-all viewpoint. The agricultural engineer has a unique role because his efforts are directed toward the creation of the proper environment for the optimum production of plants and animals. All professional groups should have an appreciation of each other's problems and should cooperate to the fullest extent since few problems can be solved within the limits of any one profession.

To be fully effective in applying technical training, the agricultural engineer must also acquaint himself with the social and economic backgrounds that relate to soil and water conservation. He must have a full understanding of the various governmental structures and mechanisms that have been developed to implement sound soil and water conservation programs. He should also become familiar with mapping and classifying land for its use in accordance with its capabilities.

The historical, social, and economic backgrounds that have had a major influence on the conservation movement have been reviewed by Bennett (1939), Clawson et al. (1960), Golze (1952), and Highsmith et al. (1963).

1.2. Land Resource Regions. For purposes of making recommendations for various conservation practices, it is helpful to subdivide the United States into a series of major land resource regions, as shown in Fig. 1.1. These regions have been delineated on the basis of broad physiographic characteristics and serve as a basis for much of the current study and planning of conservation needs in both research and action programs.

1.3. Investment in Soil and Water Conservation Practices. The net investment in soil and water practices is shown in Fig. 1.2. These 5-year average

---

* The impersonal he, him, or his appears occasionally in this book for reasons of style and accepted English usage and is intended to refer to both males and females.

Table 1.1 World Land Use, 1966

	Land Use in Millions of km²			
	Tilled	*Pasture*	*Forest & Other*	*Total*
Europe	1.5	0.9	2.5	4.9
U.S.S.R.	2.3	3.7	16.4	22.4
Asia	4.5	4.5	18.9	27.8
Africa	2.3	7.0	21.0	30.2
N. America	2.6	3.7	16.1	22.4
S. America	0.8	4.1	12.9	17.8
Oceania	0.4	4.6	3.5	8.5
TOTAL	14.3	23.6	91.4	134.2
Percentage	10.6%	21.3%	68.1%	100%

Source: From Borgstrom (1969).

curves moderate year to year changes and reflect only excess investment over depreciation.

SOIL EROSION CONTROL

The control of soil erosion caused by water and by wind is of great importance in the maintenance of crop yields. It is estimated from available measurements that at least 3 billion Mg (metric tons) of soil are washed out of the fields and pastures of the nation every year. Assuming a weight of 1280 kg/m³, this quantity of soil represents a volume equivalent to a depth of about 1 m on 6400 32.4 ha (80 ac) farms. In addition to these losses by water, there are also large losses due to wind erosion. Not only is soil lost in the erosion process but also a proportionally higher percentage of plant nutrients, organic matter, and fine soil particles in the removed material is lost then in the original soil.

Table 1.2 United States Land Use (nonfederal land excluding Alaska)

	Land Use in Millions of km²				
Year	*Tilled*	*Pasture*	*Forest*	*Urban & Others*	*Total*
1950	1.9	1.6	1.6	0.4	5.5[a]
1958	1.8	2.0	1.8	0.5	6.1
1967	1.8	2.0	1.8	0.5	6.1
1977	1.7	2.2	1.5	0.7	6.1
1977	27.4%	36.0%	24.5%	12.1%	100.0%

[a] Excludes Hawaii and Caribbean area.
Source: From USDA inventories reported by Dideriksen et al. (1978).

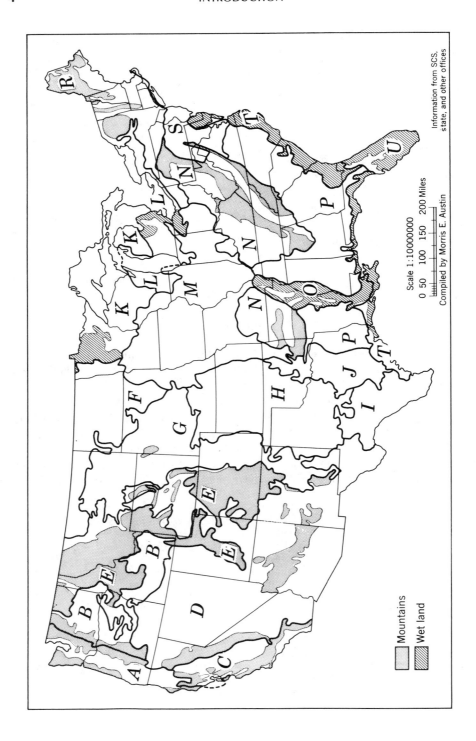

Scale 1:10000000

0 50 100 150 200 Miles

Compiled by Morris E. Austin

Information from SCS,
state, and other offices

Mountains

Wet land

Erosion control is essential to maintain the crop productivity of the soil as well as to control sedimentation in streams and lakes. Federal legislation in 1978 directed all states to develop plans for controlling sediment pollution from nonpoint sources, which includes farm land. Much effort and funding is expected in this program.

Several types of erosion are shown in Fig. 1.3. Sheet and rill erosion from cropland are in excess of 22.4 Mg/ha (10 tons/ac) in Missouri, Tennessee, and Mississippi (Dideriksen et al., 1978) compared to the average of 9 Mg/ha for all cropland in the nation. Government reports of changes in national erosion trends with time are not consistent, but all studies show that erosion is a major problem. Wind erosion from cropland is greater than 22.4 Mg/ha in Texas, New Mexico, and Arizona.

The relative degree of erosion and its distribution in the United States are indicated in Fig. 1.4. This map shows areas having slight, moderate, or severe erosion and does not differentiate between that caused by water and that caused by wind. Many small areas where severe erosion may occur locally cannot be shown on a map of this scale.

DRAINAGE

In the 1977 SCS inventory, Dideriksen et al. (1978) reported that excess water is a major problem on 25 percent or 42 million hectares (104 million acres) of the total cropland in the United States. The 1969 Agricultural Census indicated that organized drainage projects provided drainage to over 36 million hectares (90 million acres) of land. The magnitude of these drainage operations is seen in the total cost of drainage project works and services for 1971-72 of about 57 million dollars. This amount does not include the investments made for drainage on farms outside of organized districts.

Fig. 1.1. Major land resource regions of the United States. (A) Northwestern forest, forage, and specialty crop region. (B) Northwestern wheat and range region. (C) California subtropical fruit, truck, and specialty crop region. (D) Western range and irrigated region. (E) Rocky Mountain range and forest region. (F) Northern Great Plains spring wheat region. (G) Western Great Plains range and irrigated region. (H) Central Great Plains winter wheat and range region. (I) Southwestern plateaus and plains, range and cotton region. (J) Southwestern prairies, cotton and forage region. (K) Northern lake states forest and forage region. (L) Lake states fruit, truck, and dairy region. (M) Central feed grains and livestock region. (N) East and Central general farming and forest region. (O) Mississippi Delta cotton and feed grains region. (P) South Atlantic and Gulf Slope cash crop, forest, and livestock region. (R) Northeastern forage and forest region. (S) Northern Atlantic Slope truck, fruit, and poultry region. (T) Atlantic and Gulf Coast lowlands, forest and truck crop region. (U) Florida subtropical fruit, truck crop and range region. (Courtesy Soil Conservation Service.)

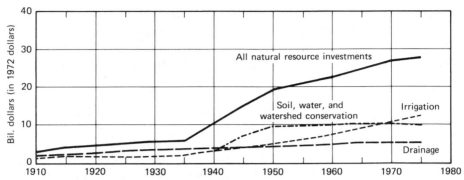

Fig. 1.2. Net investments, both public and private, in soil and water conservation practices (5-year averages). Investments are on a net or depreciating basis as of the year shown. (U.S. Department Agriculture, Agr. Hbk. 524, 1977).

Fig. 1.3. Sheet, rill, and gully erosion together with wind erosion on cropland. (Courtesy Soil Conservation Service.)

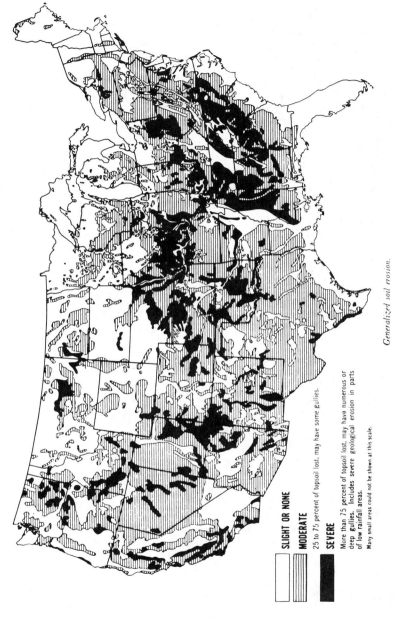

SLIGHT OR NONE

MODERATE

25 to 75 percent of topsoil lost, may have some gullies.

SEVERE

More than 75 percent of topsoil lost, may have numerous or deep gullies. Includes severe geological erosion in parts of low rainfall areas.

Many small areas could not be shown at this scale.

Generalized soil erosion.

Fig. 1.4. Distribution of soil erosion in the United States. (Courtesy Soil Conservation Service.)

An estimated 22 million hectares (54 million acres) could yet be drained for improved agricultural production. Of the land now drained, 75 percent will need redraining because of poor design or lack of maintenance. Drainage installation tends to drop in dry years and expand during wet periods. For example, from 1974 to 1975 pipe drainage increased 40 percent (SCS), and this percentage was even greater in some Corn Belt states.

The distribution of drained land in the United States is shown in Fig. 1.5. About two-thirds of our wetland needing drainage is located in the South, and about one-sixth is in Michigan, Wisconsin, and Minnesota. An area needing drainage is shown in Fig. 1.6.

In removing excess water from the land in humid areas it is usually necessary to use either surface ditches or pipe drains, or a combination of both. Wetland is usually flat, has high fertility, and does not have serious erosion problems. Where two or more landowners are involved, organized drainage districts may be formed to obtain outlets for drainage systems. Such drainage enterprises have been developed principally in (1) the prairie and level uplands of the Midwest, (2) the bottom lands of the Mississippi Valley, (3) the bottom lands in the Piedmont and hill areas of the South, (4) the coastal plains of the East and South, and (5) the irrigated areas of the West.

Drainage in humid areas often precedes land development, while in arid

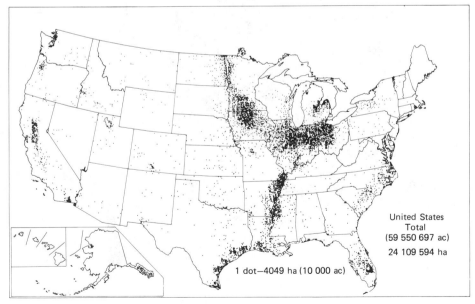

United States
Total
(59 550 697 ac)
24 109 594 ha

1 dot—4049 ha (10 000 ac)

Fig. 1.5. Distribution of drained wetland (Class 1-5 farms) in the United States. (Courtesy U.S. Census: 1969).

Fig. 1.6. Inadequate surface and subsurface drainage result in flooding following heavy rain. (Courtesy Soil Conservation Service.)

regions it normally accompanies irrigation. The principle purposes of drainage in irrigated regions are to reclaim saline and sodic soils by leaching and to prevent salinity problems by maintaining a low water table. Where salinity problems exist, land should not be developed for irrigation unless drainage facilities can be provided.

IRRIGATION

Irrigation provides one of the greatest possibilities for increasing potential production. Although it is most extensive in the West, more and more irrigation is being carried out in the eastern states. Where the annual rainfall is less than 250 mm, irrigation is a necessity; where rainfall is from 250 to 500 mm, crop production is limited unless the land is irrigated; and where rainfall is more than 500 mm, irrigation is often required for maximum production.

Whether one is intensifying crop production in the humid East with a modern corner center-pivot sprinkler irrigation system, as shown in Fig. 1.7a, or converting desert land to lush productive irrigated valleys as in Fig. 1.7b, the basic needs are the same—that is, good soils, good drainage, and a reliable supply of good quality water.

Reclamation of arid land often requires the removal of brush, smoothing of the surface, and provisions for draining the land at the time of development or at some future date.

Fig. 1.7. Irrigation has wide application from (a) supplementing rainfall in the East to (b) developing productive valleys from desert land. (Courtesy Soil Conservation Service and Valmont Industries, Inc.)

Irrigation is limited largely by the available water supply. Relatively large quantities of water are required to satisfy the needs of the crop and to supply conveyance, evaporation, and seepage losses.

In the 18 western states, irrigation needs varied from 350 to 1500 mm depth. The conveyance losses, exclusive of losses caused by evaporation and other causes, were approximately 2 million ha-m. Additional losses from evaporation, seepage, and inefficient irrigation management at the farm level may reach as high as 30 percent of the water delivered at the gate.

The problems affecting the efficient use of water diverted for irrigation are illustrated in Fig. 1.8. The agricultural engineer has a major responsibility in helping to meet these problems through development and application of practices that will reduce the unnecessary losses in storage, transmission, and application; and thus provide a higher percentage of the water for actual use by the irrigated crop.

According to the 1969 Census of Agriculture, 257 147 farms are irrigated in the 50 states with a total irrigated area of 15.8 million hectares (39.1 million acres) distributed as shown in Fig. 1.9. The 17 western states and Louisiana accounted for 91 percent of the irrigated land. Census data for the 30 humid states showed that 18.2 percent of 46 800 farmers irrigated 9.3 percent of the total national irrigated land. The rate of increase (1976) is about 0.4 million hectares (1 million acres) per year, nationally. Although only about 8 percent of the total cropland is irrigated, it produces about 25 percent of the total value of farm crops. Sprinkler and trickle irrigation is continuing to expand on much of the irrigated land in the East and West.

SOIL MOISTURE CONSERVATION

A major critical problem of agriculture in much of the United States is the recurring deficiency of soil moisture for crops and range production. The dry farming areas of the Great Plains, the Pacific Northwest, and many of the

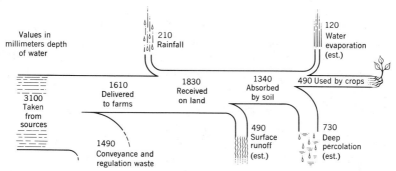

Fig. 1.8. Disposal of water diverted for irrigation where advanced irrigation practices have not yet been applied. (Courtesy U.S. Department Agriculture, 1955 Yearbook.)

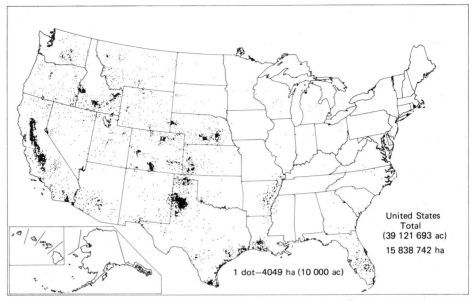

United States
Total
(39 121 693 ac)
15 838 742 ha

1 dot—4049 ha (10 000 ac)

Fig. 1.9.   Irrigated land in the United States. (Courtesy U.S. Census: 1969.)

intermountain valleys are particularly affected. Because of nonuniformity of precipitation patterns, many humid region areas are also influenced by critical moisture shortages during certain periods in the growing season. The enormity of the problem is evident in the estimates that two-thirds of the rainfall in the Great Plains states is lost by evaporation alone. Studies have shown that if this evaporation from the soil surface in the ten Great Plains states alone could be reduced by the equivalent of 76 mm of precipitation, these states would have an additional 37 million hectare-meters of water—enough to fill Lake Mead. Figure 1.10 shows a typical balance sheet on moisture supply in such an area.

   The agricultural engineer has a major responsibility in developing new practices that will permit the entrapment and storage in the soil profile of a greater percentage of the available precipitation. Design of effective terrace systems with special water catchment areas, improved snow trapping techniques (as shown in Fig. 1.11), tillage practices that modify the soil surface configuration so as to retain precipitation and reduce the total evaporation potential, and surface evaporation control through use of mulches and films (as shown in Fig. 1.12) are all challenges for the agricultural engineer.

WATER RESOURCE DEVELOPMENT

   Agriculture, being the greatest user of water, is directly concerned with the increasing demand for water. Irrigation alone in 1980 withdrew about 31 percent of all the developed water supplies in the United States. As industrial,

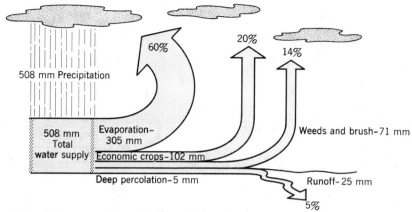

Fig. 1.10. Water budget for the Great Plains. (Courtesy Agricultural Research Service.)

municipal, recreational, and other nonagricultural uses of water increase their demand for a share of the limited water supplies of the nation, it will become increasingly important for agriculture to improve the efficiency with which it uses its share of the nation's water supply. Increasing emphasis must be placed on more fully developing the water resources that are directly available for agricultural use, in lieu of diversion and transmission of water resources that could be more effectively used by other components of our society. Of the 19

Fig. 1.11. Drifting snow trapped by rows of sorghum stubble spaced 12–24 m apart added 50 mm of soil moisture for the succeeding crop by holding most of the 80 mm of snow that fell on the field. (Courtesy Agricultural Research Service.)

Fig. 1.12.   Plastic film mulches reduce evaporation from the soil surface, thereby providing effective moisture conservation. (Courtesy Agricultural Research Service.)

million ha-m (152 million ac-ft) of water estimated to be withdrawn for irrigation in 1980, 7.5 million ha-m (61 million ac-ft) or 40 percent (Todd, 1970) will not be available to the crop. This loss includes conveyance losses, seepage, water that percolates below the root zone, evaporation, and transpiration by phreatophytes. Elimination of these losses is equivalent to new water supplies that could be developed.

Recharge of ground water supplies by water spreading and recharge wells, replenishment irrigation with drainage water, and similar practices provide other means of water resource development.

In areas of limited water supply the development of small quantities of water is essential for livestock, for farmstead use, and for specialty crop irrigation. Ground covers and sealants that provide a stable, waterproof surface to maximize runoff into a suitable storage structure is referred to as *water harvest*. A practical example of a water harvest system is shown in Fig. 1.13.

FLOOD CONTROL

The agricultural engineer is primarily concerned with floods that occur in headwater areas of less than 2590 km² (1000 square miles). Since downstream floods on major tributaries are more spectacular and damages are more evident,

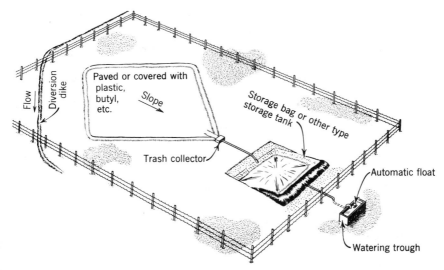

Flow

Diversion dike

Paved or covered with plastic, butyl, etc.

Slope

Storage bag or other type storage tank

Trash collector

Automatic float

Watering trough

Fig. 1.13. Example of "water harvesting" as a means of making maximum use of limited precipitation. (Courtesy Agricultural Research Service.)

floods in headwaters have too often been neglected. The total flood losses increase with the size of the drainage area, but the losses per unit area decrease. Damage from floods in headwater areas is primarily on agricultural land; downstream floods cause major damage to metropolitan areas. Flood damage to agricultural land is shown in Fig. 1.14. Damaging floods of greater or lesser degree occur on some streams every year.

Fig. 1.14. Flood damage to agricultural land. (Courtesy Soil Conservation Service.)

The principal headwater flood control measures include proper watershed management and the storage of water in small reservoirs. Proper watershed control measures reduce runoff, and they also result in a corresponding decrease in soil loss. Headwater flood control programs are also concerned with such related activities as drainage, irrigation, gully and streambank erosion control, and land clearing.

REFERENCES

Beasley, R. P. (1972). *Erosion and Sediment Pollution Control.* Iowa State University Press, Iowa.

Bennett, H. H. (1939). *Soil Conservation.* McGraw-Hill, New York.

Borgstrom, G. (1969). *Too Many.* Macmillan, New York.

Clawson, M., R. B. Held, and C. H. Stoddard (1960). *Land for the Future.* Johns Hopkins Press, Baltimore, Md.

Dideriksen, R. I., A. R. Hildebaugh, and K. O. Schmude (1978). Erosion Inventory-Sheet and Rill Erosion. ASAE Paper 78-2514, presented at ASAE meeting in Dec. 1978.

Golze, A. R. (1952). *Reclamation in the United States.* McGraw-Hill, New York.

Hudson, N. (1971). *Soil Conservation.* Cornell University Press, Ithaca, New York.

Highsmith, R. M., Jr., J. G. Jensen, and R. D. Rudd (1963). *Conservation in the United States.* Rand McNally, Chicago, Ill.

Israelsen, O. W., and V. E. Hansen (1962). *Irrigation Principles and Practices* (3rd ed.). John Wiley, New York.

Kohnke, H., and A. R. Bertrand (1959). *Soil Conservation.* McGraw-Hill, New York.

Todd, D. K. (1970). *The Water Encyclopedia.* Water Information Center, Port Washington, New York.

U.S. Department Agriculture (1962). *Basic Statistics of the National Inventory of Soil and Water Conservation Needs.* Statistical Bull. 317.

——— (1977). *Handbook of Agriculture Charts.* Agr. Hbk. 524.

——— (1958). "Land." *Yearbook of Agriculture.* U.S. GPO.

——— (1965). *Land Resource Regions and Major Land Resource Areas of the United States.* Agr. Hbk. 296.

——— (1957). "Soil." *Yearbook of Agriculture.* U.S. GPO.

——— (1955). "Water." *Yearbook of Agriculture.* U.S. GPO.

U.S. Census: 1969 (1973). *Irrigation,* Vol. IV. 1969 Census of Agriculture, U.S. GPO.

——— (1973). *Drainage of Agricultural Lands,* Vol. VI. 1969 Census of Agriculture, U.S. GPO.

CHAPTER 2

Precipitation

Precipitation, along with the atmospheric phenomena of heat, moisture, and air movement, is a part of the science of meteorology. This science of weather is of particular interest to those concerned with the effective use of soil and water. The weather is often the controlling factor in problems of preventing excessive movement of soil, or retaining needed moisture, of increasing the intake of surface water, of adding needed water by irrigation, and of removing excess water by drainage. Moisture, whether too much, too little, or poorly distributed, is one of the major limitations in agricultural production.

2.1. The Hydrologic Cycle. The science of meteorology is a part of the much broader field of hydrology, which includes the study of water as it occurs in the atmosphere as well as on and below the surface of the earth. One representation of the hydrologic cycle is given in Fig. 2.1. It shows the formation of precipitation, which may occur as rain, snow, sleet, or hail. Some of this precipitation evaporates partially or completely before reaching the ground. Precipitation reaching the earth's surface may be intercepted by vegetation, it may infiltrate the surface of the ground, it may evaporate, or it may run off the surface. Evaporation may be from the surface of the ground, from free water surfaces, or from the leaves of plants through transpiration. A portion of the total rainfall moves over the earth's surface as runoff while another portion moves into the soil surface, is used by vegetation, becomes part of the deep ground water supply, or seeps slowly to streams and to the ocean.

Figure 2.1 shows also the measurements commonly made of those portions of the hydrologic cycle of special interest to agricultural engineers. These include the measurements of precipitation by rain or snow gages, the measurement of accumulated snow by snow surveys over established ranges, the measurement of runoff by gaging stream channels, and the measurement of ground water levels. These ground water levels may be measurements either of the deep water tables as indicated by the height the water rises in wells or of the shallower perched water tables of particular interest in analyzing drainage problems.

2.2. Forms of Precipitation. Precipitation may occur in any of a number of forms and may change from one form to another during its descent. The forms

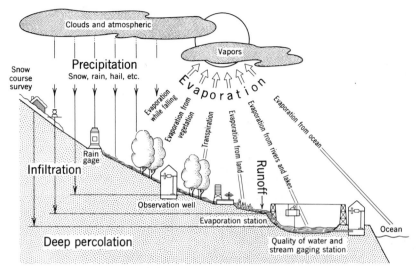

Fig. 2.1. The hydrologic cycle.

of precipitation consisting of falling water droplets may be classified as drizzle or rain. Drizzle consists of quite uniform precipitation with drops less than 0.5 mm in diameter. Rain consists of generally larger particles.

Precipitation may also occur as frozen water particles including snow, sleet, and hail. Snow is composed of a grouping of small ice crystals known as snowflakes. Sleet forms when raindrops are falling through air having a temperature below freezing; a hail stone is an accumulation of many thin layers of ice over a snow pellet. Of the forms of precipitation, rain and snow make the greatest contribution to our water supply.

Moisture at the soil surface is also made available by direct condensation and absorption from the atmosphere, commonly referred to as dew. Studies of dew formation by Brawand and Kohnke (1958) showed that 30 mm per year condensed on bare soil, but only 25 mm on a grass cover. About 15 mm was collected on corn leaves during the summer, and 33 mm were condensed on soybean leaves. Although dew is normally evaporated by noon, it is effective in reducing the rate of soil moisture depletion.

2.3. Characteristics of Raindrops. Since by far the largest portion of precipitation occurs as rain, and since rainfall directly affects soil erosion, the characteristics of raindrops are of interest. Raindrops were found by Laws and Parsons (1943) to include water particles as large as 7 mm in diameter. They found that the size distribution in any one storm covered a considerable range and that this size distribution varied with the rainfall intensity. Figure 2.2 gives

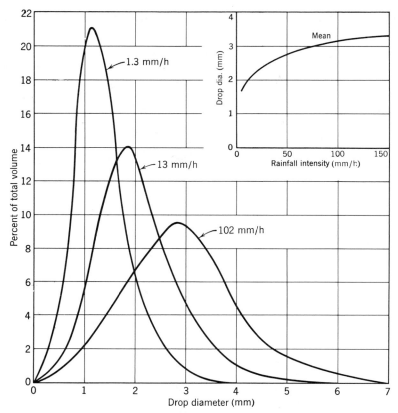

Fig. 2.2. Effect of rainfall intensity on raindrop size and their contribution to total rainfall. (Redrawn from Laws and Parsons, 1943; and Wischmeier and Smith 1958.)

the raindrop diameter for three of the intensities studied. Not only does the higher-intensity storm have more large-diameter raindrops, but it also has a wider range of raindrop diameters.

Raindrops are not necessarily spherical, or even streamlined. Falling raindrops are deformed from spherical shape by unequal pressures, due to air resistance, developing over their surfaces. Large raindrops divide in the air—drops over 5 mm in diameter being generally unstable.

In studies of soil erosion the velocities of raindrops are important. Laws (1941) found that the velocity of fall depended on the size of the particle, and that large drops fell more rapidly. As the height of fall was increased, the velocity increased only to a height of about 11 m; the drops then approached a terminal velocity, which varied from about 5 m per second for a 1-mm drop to about 9 m per second for a 5-mm drop.

2.4.   Weather Maps.   The weather picture is commonly depicted by weather
maps showing the position of the isobars, the ground position of the fronts, and
the areas of precipitation. Such weather maps are shown in Figs. 2.3a and 2.4.
Official maps show air temperature, dew point, wind direction and velocity,
barometric pressure in millibars, and pressure change during the last three
hours. The function of the weather forecaster is to prognosticate the movement
of these frontal areas, their development, and the probable precipitation. As
weather maps are now commonly included in daily newspapers, the individual
is provided with an opportunity to practice forecasting, and to compare his
predictions with those of the Weather Bureau. The cloud picture shown in Fig.

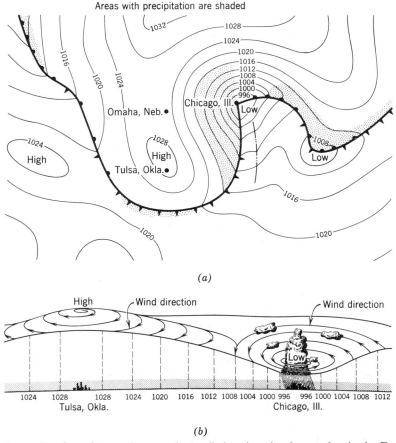

Fig. 2.3.   (a) Portion of a weather map in April showing cloudy weather in the East, rain
in the Middle West, and clear skies in the Southwest. (b) Wind circulation around a high
pressure center at Tulsa and a low center at Chicago.

2.4*a* shows the possibility of using meteorological satellites for weather map analysis in sparse data regions.

MEASUREMENT OF PRECIPITATION

Since most estimates of runoff rates are based on precipitation data, information regarding the amounts and intensity of precipitation is of great importance.

2.5. Gaging Rainfall. The purpose of the rain gage is to measure the depth and intensity of rain falling on a flat surface. The many problems of measurements with gages include effects of topography and nearby vegetation as well as the design of the gage itself. Rain gages generally used in the United States are vertical, cylindrical containers with top openings 203 mm (8 in.) in diameter. A funnel-shaped hood is inserted to minimize evaporation losses.

Rain gages may be classified as recording or nonrecording. Nonrecording rain gages, as shown in Fig. 2.5*a*, are economical, require servicing only after rains, and are relatively free of maintenance. The gage illustrated here is the Weather Bureau type. The water is funneled into an inner cylinder one-tenth of the cross section of the catch area. This provides a magnification of ten times the depth of the water and makes it possible to measure to the nearest 0.25 mm (0.01 in.).

Recording rain gages may be of several types. The type shown in Fig. 2.5*b* is the Fergusson weighing rain gage. Water is caught in a bucket placed above the recording mechanism. The weight of the water places a tension on the spring. The amount of displacement is recorded through an appropriate linkage on a chart placed on a clock-driven drum. The recording mechanism shown, which allows the pen to traverse the chart three times, gives a large vertical scale and makes possible a more accurate reading of the chart. Some weighing gages do not have the reversing mechanism. There are other types of recording rain gages, most of them using either the tipping bucket or the float-and-siphon principle.

For inaccessible locations, rain gages may be equipped with telemetry equipment. Radar measurement of cloud density and rainfall is also possible.

2.6. Measuring Snowfall. Since the water content of freshly fallen snow varies from less than 40 mm to over 400 mm of water per meter of snow, snowfall is much more difficult to measure than rainfall. While this wide variation in density makes it hazardous to indicate the amount of snow by simple depth measurements, a water equivalent depth of 10 percent of snow depth is a commonly accepted mean. Water content of compacted snow, however, is often 30 to 50 percent of snow depth.

Snowfall measurements are often made with regular rain gages, the evapora-

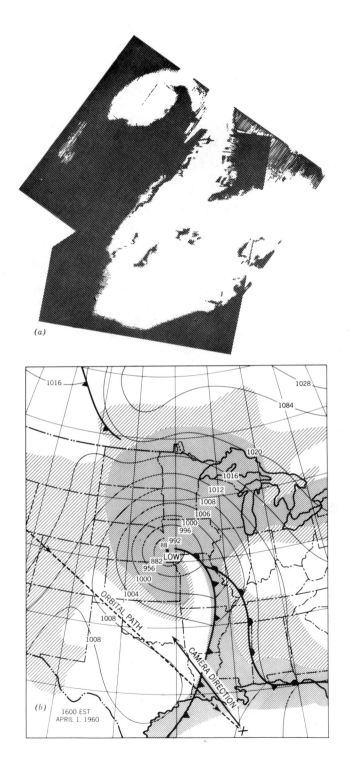

(a)

(b) 1600 EST
APRIL 1. 1960

22

tion hood having been removed. A measured quantity of some noncorrosive, nonevaporative, antifreeze material is generally placed in the rain gage to cause the snow to melt upon entrance. Errors due to wind are more serious in measuring snowfall than in measuring rain. Snow may also be measured by sampling the depth on a level surface with a metal sampling tube or with the top of the rain gage.

Another method of measuring snowfall is by determining the depth of snow by a snow survey. Such surveys are particularly useful in mountainous areas. These snow courses consist of ranges that are sampled at specified intervals. The sampling equipment consists of specially designed tubes that take a sample of the complete depth of the snow. The sample is then weighed and the equivalent depth of water recorded. Liquid-filled plastic snow pillows may also be placed on the soil surface before the snow season. Water equivalent is determined from a manometer gage that records the snow pressure.

By measuring these snow courses for a period of years and comparing the equivalent water depth with the observed runoff from the snow field, one can make predictions of the amount of runoff. Aerial snow surveys can be made by photographing depth gages or by picking up radio transmitted signals from gamma-ray depth-measuring equipment. Such devices are set up on snow ranges at suitable locations. These predictions are of particular value in planning for the most effective use of the quantities of irrigation water available during the following summer, as well as in forecasting the probability of spring floods.

2.7. Errors in Measurement. Many errors in measurement result from carelessness in handling the equipment and in analyzing data. Errors characteristic of the nonrecording rain gage of the Weather Bureau type include the water creeping up on the measuring stick, evaporation, leaks in the funnel or can, and the denting of the cans. The volume of water displaced by the measuring stick is about 2 percent and may be taken as the correction for evaporation.

Another class of errors is due to obstructions such as trees, buildings, and uneven topography. These errors can be minimized by proper location of the rain gages. The gages are normally placed with the opening about 760 mm above the surface of the ground. They should be located so as to minimize turbulence in the wind passing across the gage. A practical rule is to have a clearance of 45 degrees from the vertical center line through the gage, but a safer rule is to be sure that the distance from the obstruction to the gage is equal to at least 2 times the height of the obstruction.

Fig. 2.4. (a) Photograph of cloud cover (white area) taken by TIROS I, the first photographic meteorological satellite. (b) Corresponding synoptic chart of the frontal system with cross hatching to indicate cloud cover. (Courtesy U.S. Weather Bureau.)

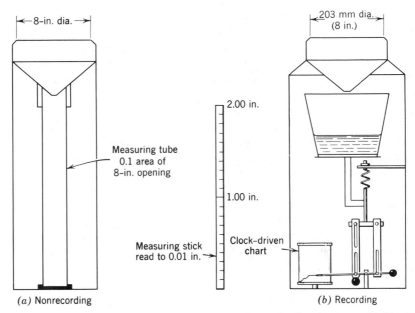

Fig. 2.5. (a) U.S. Weather Bureau nonrecording (standard) rain gage. (b) Fergusson weighing (recording) rain gage.

The wind velocity also affects the amount of water caught. A wind of 16 km/h would cause a deficit catch of about 17 percent, but at 48 km/h the deficit is increased to about 60 percent. Whenever possible, the gage should be located on level ground as the upward or downward wind movement often found on uneven topography may easily affect the amount of precipitation caught.

2.8. The Gaging Network in the United States. Precipitation records have been kept in this country ever since it has been settled. However, only since about 1890 have recording rain gages, giving the intensity of precipitation, been used. Rain gages have steadily increased in number until the gaging network in the United States consists of about 11,000 nonrecording and 3500 recording instruments. Many of the nonrecording gages are serviced by volunteer personnel; most of the recording equipment is either connected with local, state, or federal installations. The results of these extensive gaging activities are given in the various publications of the Environmental Science Service Administration, ESSA (formerly U.S. Weather Bureau) and in reports of other federal and state agencies.

The ESSA maintains a central data processing center at Asheville, North Carolina. Precipitation as well as other climatological data are placed on punch cards or computer tapes for rapid processing and evaluation.

ANALYSIS OF PRECIPITATION DATA

Rainfall data are of interest both in a specific locality and over considerable areas. Since a rain gage gives the precipitation at a given point, it is easier to make a point rainfall analysis than to study rainfall over an area.

2.9. Intensity, Duration, and Frequency of Rainfall. One of the most important rainfall characteristics is rainfall intensity, usually expressed in millimeters per hour. Very intense storms are not necessarily more frequent in areas having a high total annual rainfall. Storms of high intensity generally last for fairly short periods and cover small areas. Storms covering large areas are seldom of high intensity but may last for several days. The infrequent combination of relatively high intensity and long duration gives large total amounts of rainfall. These storms do much erosion damage and may cause devastating floods. These unusually heavy storms are generally associated with warm-front precipitation. They are most apt to occur when the rate of frontal movement has decreased, when other fronts may pass by at close intervals, when stationary fronts persist in an area for a considerable period, or when tropical cyclones move into the area.

Intense rainstorms of varying duration occur from time to time over almost all portions of the United States. However, the probability of these heavy rainfalls varies with the locality. The first step in designing a water-control facility is to determine the probable recurrence of storms of different intensity and duration so that an economical size of structure can be provided. For most purposes it is not feasible to provide a structure that will withstand the greatest rainfall that has ever occurred. It is often more economical to have a periodic failure than to design for a very intense storm. Where human life is endangered, however, the design should handle runoff from storms even greater than have been recorded. For these purposes, data providing return periods of storms of various intensities and durations are essential. This return period, sometimes called recurrence interval, is defined as the average period of time within which the depth of rainfall for a given duration will be equaled or exceeded once on the average.

A general expression for rainfall intensity is given by

$$i = \frac{KT^x}{t^n} \tag{2.1}$$

where
i = rainfall intensity,
K, x, and n = constants for a given geographic location,
t = duration of storm in minutes, and
T = return period in years.

Equation 2.1 has not been widely adopted because of the difficulty in evaluating the constants. Statistical analysis of records from all stations in the United States have been made, but the relationship of the variables is not as simple as Eq. 2.1 would indicate.

The most complete analysis of rainfall frequency data has been prepared by Hershfield (1961). Maps of the United States with isohyet (lines of equal rainfall) include durations from 30 min to 24 hr and return periods from 1 to 100 years. The data were obtained primarily from 200 long-record first-order stations that have recording gages. Records were processed through 1958, and the average length of record was 48 years. In addition, more than 8400 other station records were evaluated through 1957.

Records of intense precipitation have been published by the Weather Bureau since 1895. All storms with a duration of 5 min or longer have been recorded if the amount of rainfall in millimeters exceeds $(5 + 0.25t)$, where t is the duration in minutes. Such a storm is often referred to as an excessive storm.

Rainfall frequency maps for 1- and 24-hr duration storms and for return periods of 2 and 100 years are given in Fig. 2.6. By linearizing the return period and the duration Weiss (1962) developed the following equation for the rainfall amount at a given location for partial duration series values:

$$I = 0.0256(C - A)x + 0.000256\big[(D - C) - (B - A)\big]xy$$
$$+ 0.01(B - A)y + A \tag{2.2}$$

where I = rainfall amount in inches,
   x = return period variate from Table 2.1,
   y = duration variate from Table 2.2,
   A = 2-year, 1-hour rainfall from Fig. 2.6a,
   B = 2-year, 24-hour rainfall from Fig. 2.6b,
   C = 100-year, 1-hour rainfall from Fig. 2.6c, and
   D = 100-year, 24-hour rainfall from Fig. 2.6d.

Equation 2.2 is valid if rainfall is in millimeters or other units, and is useful for programming on computers to calculate the entire array of frequency-duration values. The values of x and y should be evaluated for each region of the country (see Frederick et al., 1977).

Example 2.1.   Determine the rainfall intensity for a 20-min and for a 6-hr storm that will occur once in 50 years at Chicago, Ill.
Solution. Read from each of the four maps in Fig. 2.6 the following values for Chicago:

Duration (hr)	T (2-yr)	T (100-yr)
1	A = 1.43 in.	C = 2.75 in.
24	B = 2.80 in.	D = 5.70 in.

Read from Table 2.1, $x = 32.1$ for $T = 50$ years and from Table 2.2, $y = -24.0$ and 49.9 for 20-min and 6-hr storm duration, respectively. Substitute the above variables in Eq. 2.2 and obtain I to compute i.

Duration (hr)	T (yr)	I (in.)	I (mm)	i (mm/h)
0.33 (20 min)	50	1.87	47.5	143
6	50	3.84	97.5	16.3

From Eq. 2.2 curves similar to those in Fig. 2.7 for St. Louis, Mo., can be obtained. As will be explained in the next section, the curves in Fig. 2.7 were developed from the highest annual values for each year of record.

Table 2.1 Linearized Rainfall Frequency Variate for Eq. 2.2

Return Period in Years	1	2	5	10	25	50	100
Linearized Variate, x	−6.93	0	9.2	16.1	25.3	32.1	39.1

Source: From Weiss (1962).

Table 2.2 Linearized Rainfall Duration Variate for Eq. 2.2

Duration in Hours (min.)	0.17 (10)	0.33 (20)	0.5 (30)	0.67 (40)	1 (60)
Linearized Variate, y	−37.0	−24.0	−15.6	−9.4	0
Duration in Hours	2	3	6	12	24
Linearized Variate, y	17.6	28.8	49.9	73.4	100.0

Source: From Weiss (1962).

2.10. Hydrologic Frequency Analysis. The rainfall data shown in Fig. 2.6 were developed by treating the measured values as statistical variables. Statistical methods for determining frequency distributions have been adopted for many other applications varying from the size distribution of sand grains to the distribution of flood flows. Because of this possibility for wide application, a few of the more common methods will be presented.

The relationship between return period and probability of occurrence can be expressed by

$$T = 100/P \qquad (2.3)$$

where T = return period in years

          P = probability in percent that an observed event in a given year is equal to or greater than a given event.

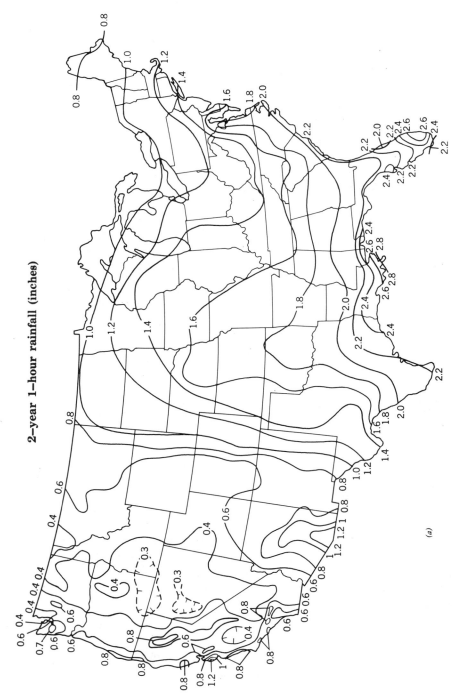

2–year 1–hour rainfall (inches)

(a)

Fig. 2.6. One- and 24-hour rainfall in inches to be expected at return periods of 2 and 100 years. (Note: 1 in. = 25.4 mm.) (Redrawn from Hershfield, 1961.)

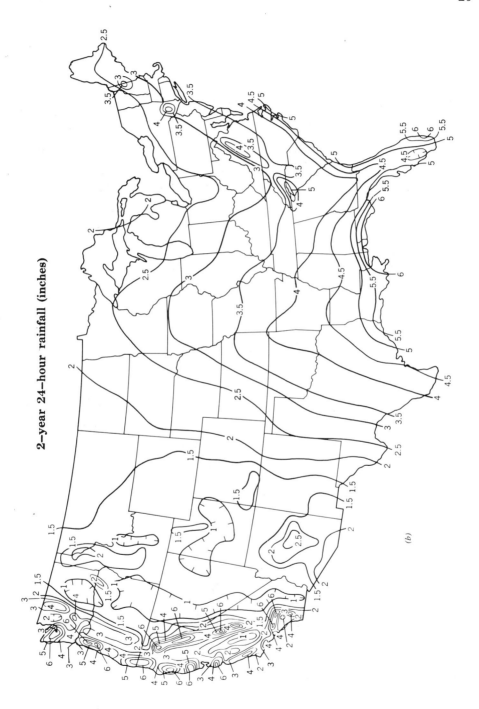

2–year 24–hour rainfall (inches)

(b)

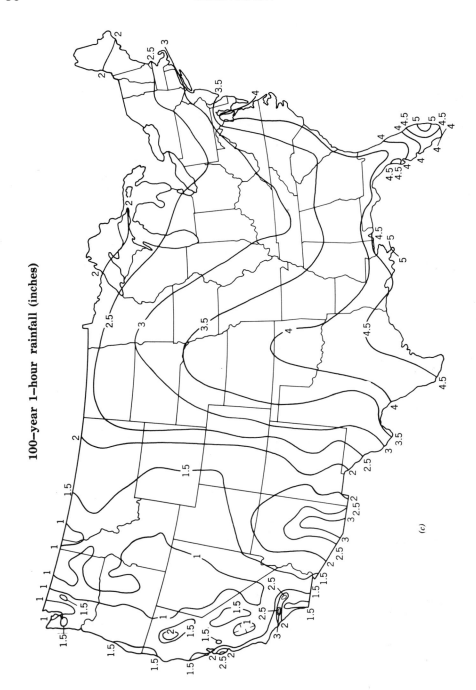

100–year 1–hour rainfall (inches)

(c)

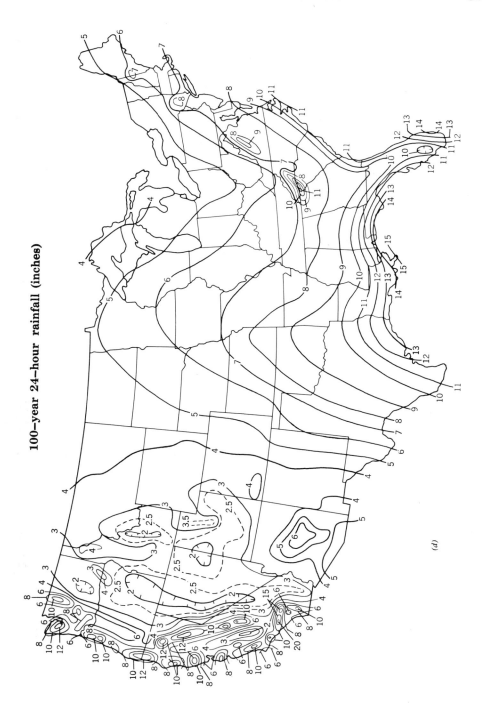

100–year 24–hour rainfall (inches)

(d)

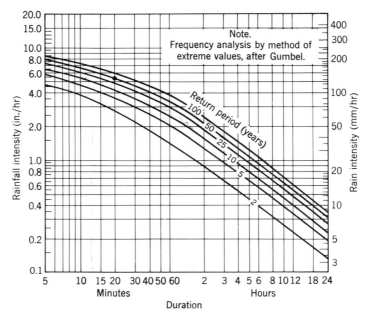

Fig. 2.7. Rainfall intensity-duration-frequency data for St. Louis Mo. (1903–1951). (*Note:* Frequency analysis by method of extreme values, after Gumbel.) (From U.S. Weather Bureau.)

Selection of Data. Experience has indicated that many hydrologic events have practically no significant value in the analysis because the hydrologic design of a structure is usually governed only by a few of the extreme conditions. Therefore, the portion of the data that is of insignificant value can be excluded.

One of two methods of selecting data is adopted, either the annual series or the partial-duration series. In the annual series, only the largest single event for each year is selected for analysis. Thus, for 20 years of record, only 20 values would be analyzed. With the partial-duration series, all values above a given base are chosen regardless of the number within a given time period. The partial-duration series is applicable if the second largest value (or lower) of the year would affect the design. An example is the design of drainage channels where damage may be due to flooding caused largely by flows lower than the annual peak flow. The partial-duration series was selected for the analysis of rainfall given in Fig. 2.6.

The annual and partial-duration series give essentially identical results for return periods greater than 10 years. Hershfield (1961), Langbein (1949), and others have shown this to be true for rainfall and floods. For example, if the 2-, 5-, and 10-year partial-duration series values selected from the maps in Fig. 2.6

are as given in the table below, the annual series values for corresponding return periods using conversion factors from Hershfield (1961) are as follows:

Return Period (yr)	Conversion Factor	Partial Series Values (in.)	Annual Series Values (in.)
2	0.88	3.00	2.65
5	0.96	3.75	3.60
10	0.99	4.21	4.17

Regardless of the method of selecting the data, the values must satisfy two important criteria, namely, (1) that the events be independent of a previous or subsequent event, and (2) that the data for the period of record for analysis must be representative of the long-time record. The first criteria is necessary from a statistical point of view, and the second implies that the predicted values will reflect only the pattern of occurrences from the period of record. Where the data are obtained from the highest annual values (extreme value law), the number of observations during the year should be large. Selection of a water year, such as October 1 to September 30, rather than the calendar year has proved beneficial for some types of data. Brakensiek (1959) found that starting the year on March 1 gave 15 percent better correlation than the calendar year for minimum water yields in Ohio.

Determination of Statistical Parameters. The next step is to calculate the mean or average value and to compute the standard deviation,

$$s = \left[\frac{\sum X^2 - (\sum X)^2/n}{n - 1} \right]^{1/2} \tag{2.4}$$

where X = measured value,
n = number of values,
$\bar{x}$ = mean value.

The coefficient of variation is $C_v = s/\bar{x}$ and $s^3 = C_v^3(\bar{x})^3$.
Then the unbiased estimate of

$$C_s = \frac{n}{(n - 1)(n - 2)} \frac{\sum(X^3)/n - 3(\bar{x})\sum(X^2)/n + 2(\bar{x})^3}{s^3} \tag{2.5}$$

where C_s = coefficient of skew.

Determination of Plotting Positions and Plotting of Data. Before proceeding, however, a brief understanding of the normal-probability curve is necessary. Figure 2.8a shows the normal curve. Deviation of the variable is plotted on the x-axis and probability of occurrence on the y-axis. Many phenomena in

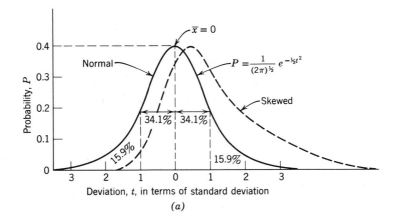

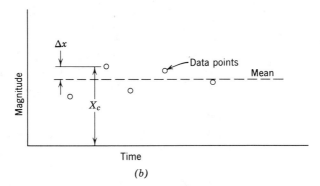

Fig. 2.8. (a) Normal and skewed probability distributions. (b) Occurrence of hydrologic events.

nature and some hydrologic events follow this distribution. Note that the mean occurs at zero deviation and that 34.1 percent of the values are within one standard deviation in each direction from the mean. Often the data are skewed as shown by the dashed curve in Fig. 2.8a.

Data are generally analyzed by one of two probability laws, the extreme value law, or the log-probability law. The extreme value law postulates that the annual maximum values approach a definite pattern of frequency distribution when the number of observations in each year becomes large, while the log-probability law states that the logarithms of the values are normally distributed.

A statistical approach for determining the plotting position as described by Chow (1951) will be followed. For an understanding of this method some hypothetical data shown in Fig. 2.8b will illustrate the method. Any point $X_c = \bar{x} + \Delta X$. The departure from the mean, ΔX, may be positive or negative and

may be very irregular and variable. From a statistical point of view, X_c posses-
ses the following two important properties: (1) the tendency to deviate from the
mean and (2) the frequency of occurrence. The first property is measured by
the standard deviation defined in Eq. 2.4. The second property is measured by
a term called the *frequency factor*, K, which depends on the law of occurrence
of a particular hydrologic event under consideration. This frequency factor is
defined by the equation,

$$\Delta X = sK \tag{2.6}$$

By substituting for ΔX and $s = C_v \bar{x}$,

$$X_c = \bar{x}(1 + C_v K) \tag{2.7}$$

The extreme value distribution postulates a coefficient of skewness C_s of
1.139 and $C_v = 0.363$. In this sense, it is a special case of the log-probability
law, which can be applied for any coefficient of skewness. For most hydrologic
data the log-probability law is suitable. The procedure for using Chow's method
is illustrated in Example 2.2.

For the log-probability law the frequency factor K is a function of the return
period (also P), the coefficient of variation, and the coefficient of skew.
Theoretically, Chow (1954) has shown that

$$C_s = 3C_v + C_v^3 \tag{2.8}$$

These corresponding values are shown in Table 2.3. In the log-probability law
the coefficient of skew is not a constant as in the extreme value law. By the
selection of the appropriate scale factor for the probability graph, Chow (1954)
has proposed a method whereby data with various coefficients of skew can be
plotted as a straight line. When data do not fit a log normal statistical distribution, McGuiness and Brakensiek (1964) have developed another procedure for
computing a rectifying constant to obtain a straight line plot.

A number of empirical equations have been developed for plotting the probability of observed events on probability paper. The following formula for the
return period has been adopted by the American Society of Civil Engineers and
is sometimes referred to as Gumbel's equation:

$$T = \frac{N + 1}{m} \tag{2.9}$$

where N = total number of statistical events,
         m = rank of events arranged in descending order of magnitude.

Table 2.3 Theoretical Log-Probability Frequency Factors

	Return Period (yr)					
	1.01	2	5	20	100	
C_s	Probability in Percentage Equal to or Greater Than Given Variate					Corresponding[a] C_v
	99	50	20	5	1	
0	−2.33	0	0.84	1.64	2.33	0
0.5	−1.98	−0.09	0.80	1.77	2.70	0.166
1.0	−1.68	−0.15	0.75	1.85	3.03	0.324
1.139[b]	−1.61	−0.16	0.73	1.86	3.11	0.363
1.4	−1.49	−0.19	0.69	1.88	3.26	0.436
1.5	−1.45	−0.20	0.68	1.89	3.31	0.462
2.0	−1.28	−0.24	0.61	1.89	3.52	0.596
3.0	−1.04	−0.28	0.51	1.85	3.78	0.818
4.0	−0.90	−0.29	0.42	1.78	3.91	1.000

[a] C_v applies for log-probability law only.
[b] For this value of C_s, the extreme value law also applies.
Source: From Chow (1954).

In Eq. 2.9, $m = 1$ for the largest value and $m = N$ for the smallest value. This equation has a statistical basis (Gumbel, 1954) for extreme values, but its greatest merit is its simplicity. For purposes of comparison with the theoretical curve in Example 2.2, the observed data are plotted according to Eq. 2.9 in Fig. 2.9.

Adequacy of Length of Record.   The adequacy of the length of record for a given level of significance is given by Mockus (1960) as

$$Y = (4.30t \, \log_{10}R)^2 + 6 \qquad (2.10)$$

where Y = minimum acceptable years of record,
$\quad\quad\quad t$ = Student's statistical value at the 90 percent level of significance with $(Y - 6)$ degrees of freedom,
$\quad\quad\quad R$ = ratio of magnitude of the 100-year event to the 2-year event.

Example 2.2.   Determine the annual precipitation for return periods of 1, 2, 5, 20, and 100 years at Los Angeles by the log-probability law based on 20 years of annual rainfall shown below. For a 90 percent probability of occurrence, is the length of record adequate?

Solution.   By calculation, using Eq. 2.4,

$$s = 176 \text{ mm (6.92 in.)}$$

Year	Millimeters (in.)/year	Year	Millimeters (in.)/year
1934	371 (14.6)	1944	488 (19.2)
1935	551 (21.7)	1945	295 (11.6)
1936	307 (12.1)	1946	295 (11.6)
1937	569 (22.4)	1947	323 (12.7)
1938	594 (23.4)	1948	183 (7.2)
1939	333 (13.1)	1949	203 (8.0)
1940	488 (19.2)	1950	269 (10.6)
1941	833 (32.8)	1951	208 (8.2)
1942	284 (11.2)	1952	665 (26.2)
1943	462 (18.2)	1953	241 (9.5)

mean annual rainfall,

$$\bar{x} = 7962/20 = 398 \ (15.68 \ \text{in.})$$

and $C_v = s/\bar{x} = 176/398 = 0.442$. From Eq. 2.7, $X_c = 398(1 + 0.442K)$.
By interpolation from Table 2.3 for $C_v = 0.442$, read values of K for each P

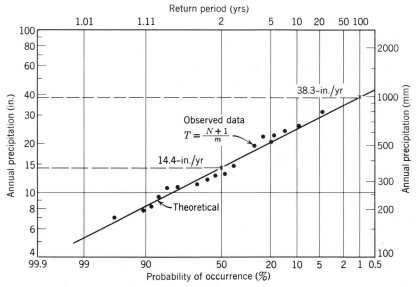

Fig. 2.9. Log-probability of annual precipitation at Los Angeles as computed in Example 2.2.

value and compute X_c as shown below. These X_c values are plotted in Fig. 2.9 on log-probability paper.

P (%)	T (yr)	K	X_c $\left[mm \ (in.) \right]$
99	1.01	-1.48	138 (5.43)
50	2	-0.19	365 (14.36)
20	5	0.69	520 (20.46)
5	20	1.88	729 (28.71)
1	100	3.27	973 (38.31)

The adequacy of length of record from Eq. 2.10 can be computed where the Student $t = 1.796$ for $(17 - 6) = 11$ degrees of freedom at the 90 percent level of significance. By trial and error, 17 for the minimum acceptable years of record was selected to agree with 16.8 as computed below. Student t values can be obtained from most statistics textbooks. From Fig. 2.9 or from X_c above

$$R = 973/365 = 2.67$$

and substituting in Eq. 2.10

$$Y = (4.30 \times 1.796 \times \log_{10} 2.67)^2 + 6 = 16.8 \text{ yr}$$

Since the actual length of record of 20 years is greater than the minimum acceptable, the estimate of 973 mm (38.31 in.) for $T = 100$ years can be expected to be reasonably reliable. Extreme caution should be taken when the computed Y value is greater than the length of record.

2.11. Point Rainfall Analysis.   A typical recording rain gage chart is given in Fig. 2.10. The line on the chart is a cumulative rainfall curve, the slope of the line being proportional to the intensity of the rainfall. The peak is the point of reversal of the recording gage. To analyze the chart, the time and amount of rain should be selected from representative points where the rainfall rate changes so that the data will represent the curve on the chart. These points may be tabulated as in Table 2.4, with cumulative rainfall and intensity for various periods of time also being recorded.

To determine the highest return period for a desired duration, the time period must be selected from the most intense portion of the storm. By referring to Fig. 2.7, the appropriate return period can be determined.

Example 2.3.   Determine the return period for the maximum rainfall intensity occurring for any 20-min period and for the first 140 min (2.33 hr) during the storm shown in Table 2.4 for St. Louis, Mo.

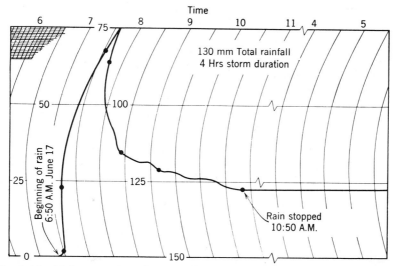

Fig. 2.10. Rain gage chart from a rain gage of the reversible, recording type.

Solution. From Table 2.4 the maximum rainfall intensity for 20 min is 138 mm/h (5.43 iph). Interpolating from Fig. 2.7, read a return period of 50 years. For the 140-min storm, the average intensity is 124/2.33 = 53 mm/h (2.1 iph), for which the return period is nearly 100 years.

Mass rainfall curves, required for some types of analyses, may be obtained by plotting the cumulative rainfall against time as in Fig. 2.11a. It is also often convenient to plot the rainfall intensity for increments of time as illustrated in Fig. 2.11b.

Table 2.4 Rain Gage Chart Analysis

Time (A.M.)	Time Interval (min)	Cumulative Time (min)	Rainfall during Interval[a] (mm)	Cumulative Rainfall (mm)	Rainfall Intensity for Interval (mm/h)
6.50					
7:00	10	10	1	1	6
7:10	10	20	10	11	60
7:15	5	25	11	22	132
7:35	20	45	46	68	138
7:45	10	55	19	87	114
8:25	40	95	31	118	47
9:10	45	140	6	124	8
10:50	100	240	6	130	4

[a] Corrected rainfall based on nonrecording gage depth.

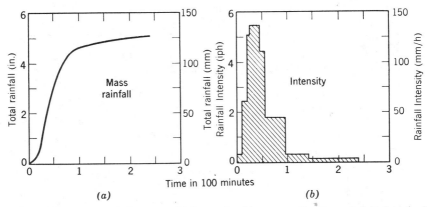

Fig. 2.11.   (a) Mass rainfall curve. (b) Intensity histogram for the rainfall data in Fig. 2.10.

2.12.   Classification of Storms.   Since no two rainstorms have exactly the same time-intensity relationships, it is often convenient to group storms with regard to their characteristics. The most common characteristics used in such groupings are the intensity of the storm and the pattern of the rainfall intensity histogram.

The pattern of a storm is determined by the arrangement of the rainfall intensity histogram. Storm patterns are important because they are one of the factors determining the shape of the runoff hydrograph. Arbitrarily selected storm patterns of rainfall intensities shown in Fig. 2.12 are uniform intensity, advanced pattern, intermediate pattern, and delayed pattern. The advanced pattern of rainfall brings higher intensities when the infiltration rate is the greatest (Chapter 3), thus causing some reduction in the runoff peaks. On the other hand, the delayed pattern causes higher runoff peaks, as the high intensities occur when the infiltration is at a minimum and depression storage has been largely satisfied. In general, the cold front produces a storm of an advanced type, and the warm front a uniform or intermediate pattern. In Ohio, in a study of 1-hour storms of all intensity classes, the advanced pattern was found to be the most common.

2.13.   Average Depth Over Area.   The rainfall depths given in Fig. 2.6 are point rainfall. Where the average depth over a watershed area must be determined, such amounts can be adjusted for different duration storms as indicated in Fig. 2.13. The design rainfall may be considered as the maximum for the storm, and thus the average over a watershed will be less than the maximum, as the curves show.

Where a gaging network has been established or records are available for a given watershed, the following procedures are applicable for determining the

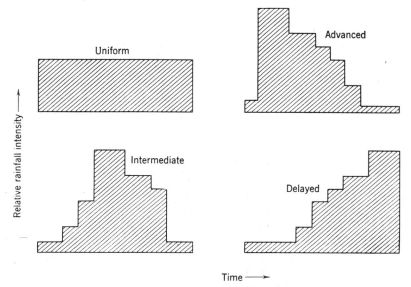

Fig. 2.12.   Rainfall intensity patterns. (From Horner and Jens, 1942).

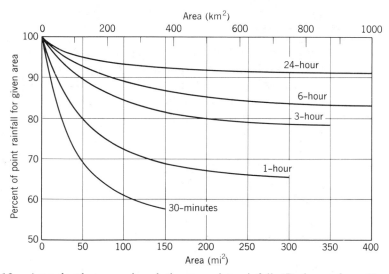

Fig. 2.13.   Area-depth curves in relation to point rainfall. (Redrawn from Hershfield, 1961.)

average depth of precipitation over an area. If only one rain gage is used, the rainfall is applied over the entire area. Where several gages are available, the simplest method is to take the arithmetic mean. Since each gage may not represent equal areas, other methods often give greater accuracy.

2.14. Thiessen Method. The use of the Thiessen method is illustrated in Fig. 2.14. The location of the rain gages is plotted on a map of the watershed. Straight lines are then drawn between the rain gages. Perpendicular bisectors are then constructed on these connecting lines in such a way that the bisectors enclose areas referred to as Thiessen polygons. All points within one polygon will be closer to its rain gage than to any of the others. The rain recorded is then considered to represent the precipitation within the appropriate polygon area.

Some difficulty may be encountered in determining which connecting lines to construct in forming the sides of the polygon. Though in general the shorter lines are used, the proper lines can best be determined by a trial-and-error procedure. Since only one set of Thiessen polygons generally needs to be drawn for a given watershed and set of rain gage locations, this procedure does not present a serious limitation. The average precipitation over a watershed can be determined by using the following equation:

$$P = \frac{A_1 P_1 + A_2 P_2 + \ldots + A_n P_n}{A} \tag{2.11}$$

where P represents the average depth of rainfall in a watershed of area A and $P_1, P_2, \ldots, P_n$ represent the rainfall depth in the polygon having areas $A_1, A_2, \ldots, A_n$ within the watershed.

Example 2.4. A storm on the watershed illustrated in Fig. 2.14 produces rainfall at the various gage locations as indicated. Compare the average precipitation as determined by the average depth and by the Thiessen methods.
Solution. By the average depth method the arithmetic mean is 50 mm (1.97 in.). By the Thiessen method, the areas represented by the various rain gages are determined with a planimeter and substituted in Eq. 2.12,

$$P = \frac{(65)(46) + (150)(55) + (269)(57) + (216)(55) + (56)(41) + (136)(46)}{892}$$

$$P = 53 \text{ mm (2.08 in.)}$$

2.15. Isohyetal Method. The isohyetal method consists of the depth of rainfall at the location of the various rain gages and plotting isohyets (lines of

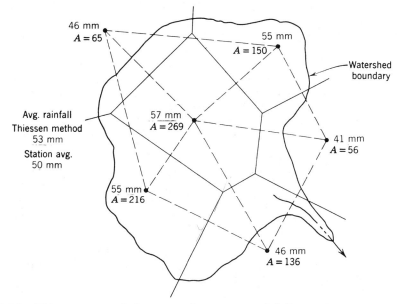

Fig. 2.14.   Thiessen network for computing average rainfall amount over a watershed.

equal rainfall) by the method used in drawing topographic maps. The area between isohyetals may then be planimetered and the average rainfall determined by the above equation.

The choice of the method of analysis will depend partly on the area of the watershed, the number of rain gages, the distribution of the rain gages, and in some situations, the character of the rainstorm. Depth-area curves, where needed, can be constructed from isohyetal maps.

DISTRIBUTION OF PRECIPITATION IN THE UNITED STATES

2.16. Time Distribution.   *Diurnal.*   The time of day in which precipitation may be expected to occur will depend on the type of precipitation. Frontal storms are not much influenced by diurnal effects. Storms of the convective type, since they are due to surface heating, are much more likely to occur in the afternoon.

Seasonal.   That rainfall be distributed throughout the growing season is important. A considerable difference in the seasonal distribution of precipitation throughout the United States is shown in Fig. 2.15. Even in the areas of the West Coast, where annual precipitation is high, summertime precipitation is generally very low, making irrigation necessary. In the Middle West and South the monthly summertime precipitation is generally somewhat higher than the

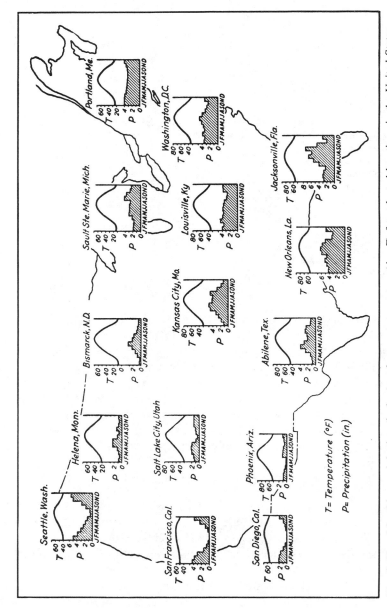

Fig. 2.15. Monthly precipitation (inches) and mean temperature (deg F) for selected locations in the United States. (Redrawn from Rouse, 1950.) (Note: 1 in. = 25.4 mm.)

monthly average, and in the eastern portion of the United States there is little difference between summer and winter precipitation.

Annual.   The annual rainfall over the United States is shown in Fig. 2.16. Annual rainfall amounts vary from less than 100 mm to over 2500 mm in some mountainous areas. Annual precipitation is not in itself a good index of the amount of moisture available for plant growth because evaporation, seasonal distribution, and water-holding capacity of the soil vary with geographical locations.

Cycles.   That precipitation occurs in cycles has often been suggested. However, as yet there has been no statistical proof that such cycles exist or that there is any relationship between such cycles and other natural phenomena. Some evidence exists that sunspot activity is related to summer temperature and severe droughts. Thompson (1973) showed that average July – August temperatures in the Corn Belt since 1900 follow roughly about a 20-year cycle of sunspot numbers. Similar observations have been made in other countries at the same latitudes. The most widely held view is that weather is a random variable.

2.17. Geographical Distribution.   The geographical distribution of rainfall over the United States is largely determined by the location of large bodies of water, by the movement of the major air masses, and by changes in elevation. Figure 2.17 illustrates the effect of elevation and of moist air-mass movement

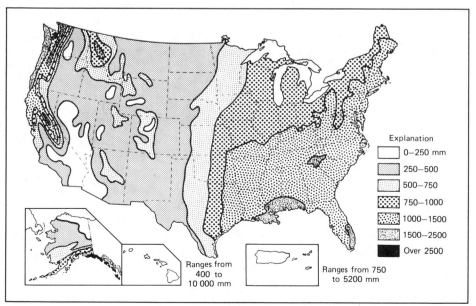

Explanation

☐	0–250 mm
	250–500
	500–750
	750–1000
	1000–1500
	1500–2500
■	Over 2500

Ranges from 400 to 10 000 mm

Ranges from 750 to 5200 mm

Fig. 2.16.   Average annual precipitation in the United States in millimeters. (Redrawn from U.S. Water Resources Council, 1978.)

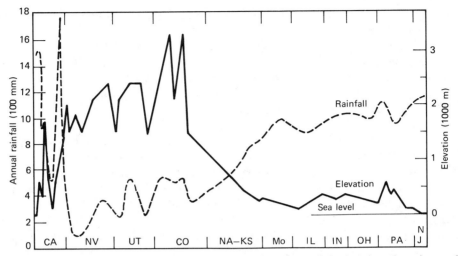

Fig. 2.17. Section of the United States along the 40th parallel, showing elevation and average annual rainfall.

on annual rainfall. It presents a section of the United States along the 40th parallel. Moving from west to east, one notes that the highest rainfall occurs as the air is first pushed up by the mountains, with lesser rises as the dryer air is pushed to higher elevations. As the air moves down the mountain slopes, lower annual rainfall is generally observed. The rainfall does not increase until the effects of the maritime tropical air moving up from the Gulf of Mexico become apparent. Then the rainfall gradually increases as the eastern boundary of the United States is approached, with the effect of the Appalachian mountains again apparent.

REFERENCES

Brakensiek, D. L. (1959). "Selecting the Water Year for Small Agricultural Watersheds." *Am. Soc. Agr. Eng. Trans.* **2** (1), 5–8, 10.

Brawand, H., and H. Kohnke (1958). "Microclimate and Water Vapor Exchange at the Soil Surface." *Soil Sci. Soc. Am. Proc.* **16**, 195–198.

Chow, V. T. (1951). "General Formula for Hydrologic Frequency Analysis." *Am. Geophys. Union Trans.* **32**, 231–237, April.

Chow, V. T. (1954). "The Log-Probability Law and Its Engineering Applications." *Am. Soc. Civil Eng.* (Separate No. 536) **80**, Nov.

Gumbel, E. J. (1954). *Statistical Theory of Extreme Values and Some Practical Applications.* U.S. Bureau Standards, Applied Mathematics Series 33.

Frederick, R. H., V. A. Myers, and E. P. Auciello (1977). Five- to 60-Minute Precipitation Frequency for the Eastern and Central U.S. NOAA Tech. Memo. HYDRO-35, Silver Spring Md.

Haan, C. T. (1977). *Statistical Methods in Hydrology.* Iowa State Univ. Press, Ames, Iowa.

Hershfield, D. N. (1961). "Rainfall Frequency Atlas of the United States." U.S. Weather Bureau Tech. Paper 40, May.

Horner, W. W., and S. W. Jens (1942). "Surface Runoff Determination from Rainfall without Using Coefficients." *Trans. Am. Soc. Civil Engrs.,* **107,** 1039–1117.

Langbein, W. B. (1949). "Annual Floods and the Partial Duration Series." *Am Geophys. Union Trans.* **30,** 879–881.

Laws, J. O. (1941). "Measurements of the Fall-Velocity of Water-Drops and Raindrops." *Trans. Am. Geophys. Union* 22, 709–721.

Laws, J. O., and D. A. Parsons (1943). "The Relation of Raindrop-Size to Intensity. *Am Geophys. Union Hyd. Rpts.,* Pt. 2, 452–460.

Linsley, R. K., M. A. Kohler, and J. L. H. Paulhus (1975). *Hydrology for Engineers,* 2nd ed. McGraw-Hill, New York.

McGuinness, J. L., and D. L. Brakensiek (1964). "Simplified Techniques for Fitting Frequency Distributions to Hydrologic Data." U.S. Agr. Res. Service, Agr. Hbk. 259.

Mockus, V. (1960). "Selecting a Flood-Frequency Method." *Am. Soc. Agr. Eng. Trans.* **3,** 48–51, 54.

Thiessen, A. H. (1911). "Precipitation Averages for Large Areas." *Monthly Weather Rev.* 39, 1082–1084.

Thompson, L. M. (1973). "Cyclical Weather Patterns in the Middle Latitudes." *J. Soil and Water Conservation* 28, 87–89.

U.S. Department Agriculture (1941). "Climate and Man." Yearbook of Agriculture.

Weiss, L. L. (1962). "A General Relation between Frequency and Duration of Precipitation." *Monthly Weather Rev.,* 87–88.

Wischmeier, W. W., and D. D. Smith (1958). "Rainfall Energy and Its Relation to Soil Loss." *Trans. Am. Geophys. Union* 39, 285–291.

PROBLEMS

2.1. Determine the total rainfall to be expected once in 5, 25, and 100 years for a 60-min storm at your present location.

2.2. Determine the maximum rainfall intensity to be expected once in 10 years for storms of durations of 10, 30, 120, and 360 min, respectively, at your present location.

2.3. From rainfall data given in Table 2.4 determine the maximum rainfall intensity for any 5-, 30-, and 240-min period. Determine the return period for these intensities if the storm occurred at St. Louis, Mo.

2.4. Compute the average rainfall for a given watershed by the Thiessen method from the following data. How do the weighted average and the station average compare?

Rain Gage	Area [ha (ac)]	Rainfall [mm (in.)]
A	14.0 (34.6)	58 (2.30)
B	4.5 (11.2)	41 (1.60)
C	5.3 (13.2)	51 (2.02)
D	4.9 (12.1)	43 (1.71)

2.5.   During a 60-min storm the following amounts of rain fell during successive 15-min intervals: 33 mm (1.3 in.), 23 (0.9), 15 (0.6), and 5 (0.2). What is the maximum intensity for 15 min and the average intensity? If the storm had occurred at St. Louis, Mo., how often would you expect such a 60-min storm to occur? What type of storm pattern was it? From what type of rain gage was the data obtained?

2.6.   From the local rainfall data supplied by your instructor, calculate and plot the best theoretical probability curve for return periods of 2, 5, 20, and 100 years. For a 90 percent chance of occurrence is the period of record adequate?

2.7.   Plot the data in Problem 2.6 using the ASCE empirical equation for plotting positions.

2.8.   If the average rainfall intensity for 50 years of record is 51 mm/h (2.0 iph) for a storm duration of 60 min and the coefficient of variation of the data is 0.324, compute the theoretical rainfall intensities for return periods of 2, 5, 20, and 100 years. Assume that the log-probability law is applicable.

2.9.   Compute the mean, standard deviation, and coefficient of variation for the following maximum rainfall intensity for each year of a 5-year period: 178, 127, 126, 102, 102 mm/h (7, 5, 5, 4, 4 iph).

CHAPTER 3

Infiltration, Evaporation, and Transpiration

Three phases of the hydrologic cycle of particular interest in agriculture are infiltration, evaporation, and transpiration. Infiltration is the passage of water into the soil surface and is distinguished from percolation, which is the movement of water through the soil profile. Evaporation is the process by which moisture is returned to the air from a liquid to a gaseous state. Transpiration is evaporation from plants. About three fourths of the total precipitation on the land areas of the world returns directly to the atmosphere by evaporation or transpiration. Most of the balance returns to the ocean as surface or subsurface flow. Evaporation and transpiration are difficult to separate and are often considered together and called evapotranspiration.

Infiltration is of particular interest, for if water is to be conserved in the soil and made available to plants, it must first pass through the soil surface. If the infiltration rate is high, less water will pass over the soil surface and erosion will be reduced. In this way runoff quantities and peaks are lowered.

Evaporation, which may occur either from the water surface or from moisture on soil particles, is important for water conservation. Evapotranspiration is required for determining irrigation requirements for crops as well as water storage in ponds and reservoirs. High evapotranspiration from such crops as grass is beneficial for drainage because of the increased capacity of the soil for storing moisture.

INFILTRATION $\left(depth/time \right)$

The term infiltration refers specifically to entry of water into the soil surface. Infiltration rate has the dimensions of volume per unit of time per unit of area. These units reduce to depth per unit time. Infiltration should not be confused with hydraulic conductivity nor with soil capillary conductivity. Infiltration is the sole source of soil moisture to sustain the growth of vegetation and of the ground water supply of wells, springs, and streams.

The movement of water into the soil by infiltration may be limited by any restriction to the flow of water through the soil profile. Although such restriction often occurs at the soil surface, it may occur at some point in the lower ranges of the profile. The most important items influencing the rate of infiltration have to do with the physical characteristics of the soil and the cover on the soil surface, but such other factors as soil moisture, temperature, and rainfall intensity are also involved.

Infiltration into unsaturated soil is defined by the differential equation (Klute, 1952)

$$\frac{\partial \theta}{\partial t} = \frac{\partial}{\partial z}\left(K \frac{\partial \phi}{\partial z}\right) + \frac{\partial}{\partial z}(Kg) \tag{3.1}$$

where θ = the moisture content in volume of water per unit volume of soil,

K = the unsaturated hydraulic conductivity (LT^{-1}),

ϕ = the capillary potential (L),

g = gravitational constant (LT^{-2}),

z = the coordinate in the vertical direction (L).

A general analytical solution to Eq. 3.1 cannot be obtained because both K and ϕ are functions of θ. Graphical solutions have been made and numerical solutions are practical with computers.

3.1. Soil Factors.   Soil functions essentially as a pervious medium that provides a large number of passageways for water to move into the surface. The effectiveness of the soil as an agent for transporting water depends largely upon the size and permanency of these channels. In general the size of the passageways and the infiltration into the soil is dependent upon: (1) the size of the particles that make up the soil, (2) the degree of aggregation between the individual particles, and (3) the arrangement of the particles and aggregates. The larger the pore size that can be maintained, the greater is the resulting infiltration rate.

The importance of maintaining permanent channels, particularly at the soil surface, is critical. The usual rapid reduction in the rate of intake of water through the surface is accompanied by the formation of a thin compact layer on the surface. This layer is a result of severe breakdown of soil structure due in part to the beating action of raindrops and in part to an assorting action of the water flowing over the surface, fitting the fine particles around the larger ones to form a relatively impervious seal giving the surface of the soil a slick appearance.

This surface-sealing effect can largely by eliminated when the soil surface is protected by mulch or by some other permeable mechanical protection. The effectiveness of such protection is illustrated in Fig. 3.1, which first shows the constant infiltration rate of soil covered by straw. After 40 min of infiltration at

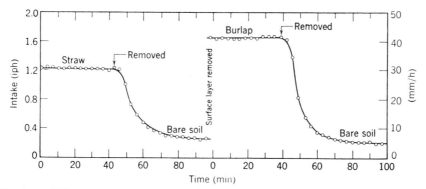

Fig. 3.1. Effect of protective cover on infiltration. (Redrawn from Duley, 1939.)

a constant rate, the straw was removed and the infiltration rate dropped to about one sixth of its original value. The straw had protected against the formation of the impervious surface layer, and when the straw was removed the impermeable layer developed quickly through the beating action of the raindrops. By removing the puddled surface layer of soil and protecting the newly exposed soil surface with a layer of burlap, the infiltration rate increased to a new high value. When after 40 min the burlap was again removed, the soil surface puddled, and the infiltration rate fell to a new low.

3.2. Vegetation. Surface sealing can be greatly reduced by vegetation. In general vegetative cover and surface condition have more influence on infiltration rates than do the soil type and texture. The protective cover may be grasses or other close-growing vegetation as well as mulches. It has been shown that when infiltration rates are determined for soil protected by vegetation and the vegetation is removed, surface sealing occurs and infiltration drops much as illustrated in Fig. 3.1. Figure 3.2 gives a number of infiltration curves for unprotected soil and for several surface cover conditions as determined for three South Carolina soils.

3.3. Other Factors. Other factors affecting infiltration include land slope, antecedent soil moisture, and water temperature (a special case being frozen soil). The effect of slope on rate of infiltration has generally been shown to be small, and to be more important on slopes less than 2 percent than on steeper gradients. The effect of slopes steeper than 2 percent on infiltration is not significant.

Soil moisture generally reduces or limits the infiltration rate. The reduction is due in large part to the fact that moisture causes some of the colloids in the soil to swell, and thereby reduce both the pore space and the rate of water movement. Consequently, in making infiltration tests in the field, it is customary to make both a dry soil run and a wet run, often 24 hours later. Design is usually

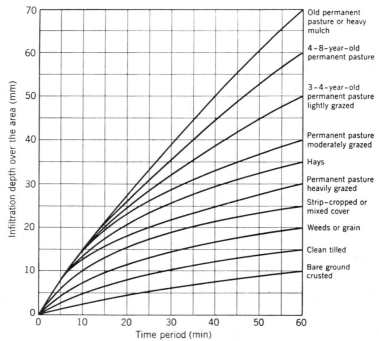

Fig. 3.2. Typical mass infiltration curves. (Redrawn from Holtan and Kirkpatrick, 1950.)

based on the minimum values obtained. In a completely saturated soil underlain with an impervious layer or layers, infiltration will be zero.

The effect of water temperature on infiltration is not significant, perhaps because the soil changes the temperature of the entering water and the size of the pore spaces may change with temperature changes. Although freezing of the soil surface greatly reduces its infiltration rate, freezing does not necessarily render the soil impervious.

3.4. Soil Additives. The physical characteristics of the soil, including the infiltration capacity, can be changed by adding chemicals. In general these additives are one of two types. The first type consists of materials that add to the permanency of the soil aggregate formations, and thereby generally improve the soil structure. This improved structure causes considerable increases in both the infiltration and percolation rates. The second type of additive is essentially a wetting agent that does not change the soil but instead changes the angle of contact of the soil water with the soil surface and thereby the rate at which water can move through the soil. It therefore affects water movement at depths greater than the zone of application. In general, it may be necessary to reapply these wetting agents periodically as they leach out with continued water application.

Additives are also applied which decrease infiltration rates. One group of chemical additives reduces infiltration capacity by causing soil particles to swell and to become hydrophilic. Fine clays are sometimes added to soils. These swell and seal soil pores to reduce infiltration rates. Partial or complete sealants, such as petroleum or plastic films are applied to soil surfaces to decrease or prevent infiltration.

Decreased infiltration is desired to decrease losses of water from reservoirs or irrigation canals or to increase runoff to surface water supplies for direct use or for ground water recharge.

3.5. Predicting Infiltration. Infiltration data are commonly expressed graphically with the rate as the ordinate and time as the abscissa. Figure 3.3 presents a typical infiltration curve. Here, as usual, the potential infiltration capacity initially exceeds the rate of water application. However, as the soil pores fill with water, and as surface sealing takes place, the rate of water intake gradually decreases. It then normally approaches a constant value which may be taken as the infiltration rate of the soil.

The infiltration curve in Fig. 3.3 can be expressed by (Horton, 1939)

$$f = f_c + (f_0 - f_c)e^{-kt} \tag{3.2}$$

where f = infiltration capacity or the maximum rate at which soil under
          a given condition can take water through its surface (LT^{-1}),
      f_c = the constant infiltration capacity as t approaches infinity,
      f_0 = infiltration capacity at the onset of infiltration,
      k = a positive constant for a given soil and initial condition,
      t = time.

In Fig. 3.3 the infiltration capacity did not drop to equal the rainfall rate until several minutes after the onset of infiltration. During this initial time period the infiltration rate was equal to the rainfall rate and less than the infiltration capacity. Up to this time the runoff rate was thus zero. As the infiltration capacity fell, below the rainfall rate, runoff occurred.

Because of the difficulty in evaluating infiltration, the U.S. Soil Conservation Service (1972) divided all soils into four hydrologic groups A, B, C, and D on the basis of infiltration rates. The procedure for applying infiltration data to obtain runoff is discussed in Chapter 4.

EVAPORATION AND TRANSPIRATION

Evaporation is the transfer of liquid water into the atmosphere. The water molecules, both in the air and in the water, are in rapid motion. Evaporation occurs when a larger number of the moving molecules break from the water

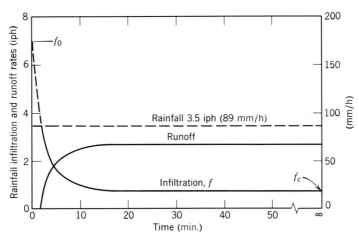

Fig. 3.3. Typical infiltration and runoff curves developed from infiltrometer data.

surface and escape into the air as vapor than the number that reenter the water surface from the air and become entrapped in the liquid.

Transpiration is the process through which water vapor passes into the atmosphere through the tissues of living plants. In areas of growing plants, moisture passes into the atmosphere by evaporation from soil surfaces and by transpiration from plants. For convenience in analyzing moisture transfer in this common situation, the two are combined and referred to as evapotranspiration.

The amount of water that passes through plants by the transpiration process is often a substantial portion of the total moisture available during the growing season. It can vary from practically nothing to as much as 635 mm (25 in.) in depth, depending largely upon the moisture available, the kind of plant, density of plant growth, the amount of sunshine, and the soil fertility and structure. Less than 1 percent is actually retained by the growing organisms. That the rate of evaporation or transpiration increases with the rise in temperature of the surface is to be expected as vapor pressure increases with increases in temperature. It has been shown that mean monthly air temperatures do not alone provide a satisfactory means of predicting mean monthly evaporation.

Wind increases the rate of evaporation, particularly as it disperses the moist layer found directly over the evaporating water surface under stagnant conditions. Because of this mixing, the characteristics of the atmosphere above the surface are of interest. As might be expected, from the decreased concentration of water molecules, evaporation increases with decreased barometric pressure. Likewise, if other conditions are unchanged, there is greater evaporation at higher elevations. Also the rate of evaporation has been found to decrease with increases in the salt content of the water.

Basic methods of predicting evaporation and evapotranspiration can be grouped into three categories:

1. Mass Transfer. This approach recognizes that moisture moves away from evaporating and transpiring surfaces in response to the combined phenomena of turbulent mixing of the air and the vapor pressure gradient. Thornthwaite and Holzman (1942) proposed such a method. Application of methods based on mass transfer principles requires measurements of wind velocity and humidity at two or more elevations and is seldom practical.

2. Energy Balance. Heat is required for evaporation of water, so if there is no change in water temperature the net radiation or heat supplied is a measure of evaporation. Energy balance methods are proving to be practical in application. The Penman (1956) equation is an example of this approach.

3. Empirical Methods. Several such methods have been developed from experience and field research, which are based primarily on the assumption that the energy available for evaporation is proportional to the temperature. Blaney and Criddle (1950) and Thornthwaite (1948) have proposed equations of this type.

3.6. Evaporation from Water Surfaces. Many evaporation formulas for free-water surfaces are based upon Dalton's Law:

$$E = C(e_s - e_d) \tag{3.3}$$

where e_s = the saturated vapor pressure at the temperature of the water surface in mm of Hg,

e_d = the actual vapor pressure of the air (e_s times relative humidity) in mm of Hg,

E = the rate of evaporation,

C = a constant.

Rohwer (1931) evaluated the constant in Eq. 3.3 as

$$C = (0.44 + 0.073W)(1.465 - 0.00073p) \tag{3.4}$$

where W = average wind velocity in km/h at a height of 0.15 m (6 in.),

p = atmospheric pressure in mm Hg at 0°C.

With these units E is in mm/d. To find the evaporation from reservoirs, the calculated E should be multiplied by 0.77.

Meyer (1942) evaluated the constant for pans and shallow ponds (E as mm/mo.),

$$C = 15 + 0.93W \tag{3.5}$$

and for small lakes and reservoirs

$$C = 11 + 0.68W \tag{3.6}$$

where W = average wind velocity for the period in km/h at a height of 7.6 m (25 ft). The vapor pressure, e_d, should be measured at 7.6 m height and the air temperature is the average of the daily minimum and maximum.

Example 3.1.   Compute the evaporation for the month of June from a shallow pond if the surface water temperature is 15.6°C (60°F), the average wind speed is 4.8 km/h (3 mph), and the average temperature and relative humidity at 7.6 m (25 ft) height are 21.1°C (70°F) and 40 percent, respectively.
Solution.   Substituting in Eqs. 3.3 and 3.5 (Meyer equation) and the vapor pressures from Fig. 3.4,

$$E = (15 + 0.93 \times 4.8)(13.0 - 18.5 \times 0.40)$$
$$= 109 \text{ mm/mo. } (4.3 \text{ in./mo.})$$

Evaporation measurements from free water surfaces are commonly made using evaporation tanks or pans. The Class A pan, accepted as standard by the

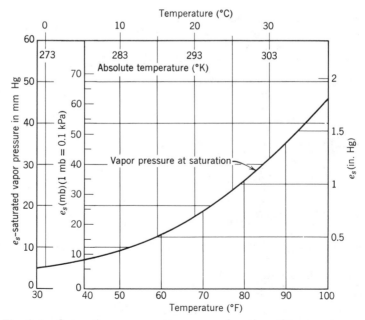

Fig. 3.4.   Saturation vapor pressure as a function of temperature.

U.S. Weather Bureau, is 1.22 m (4 ft.) in diameter 254 mm (10 in.) deep, and requires a water depth of about 190 mm (25 in.). The pan is supported about 150 mm (6 in.) above the ground so that the air may circulate under it, and the materials and color of the pan are specified. This pan is widely used in the United States. Descriptions of other styles of pans and correction coefficients for converting evaporation data from a pan of one type to that of another are given by Meinzer (1949). These pans have higher rates of evaporation than do larger free water surfaces, a factor of about 0.7 being recommended for converting observed evaporation rates to those for large surface areas. Because of the "oasis" effect and likely higher water temperatures, evaporation depth is generally higher on small ponds than on large lakes. Although the vapor pressure of highly saline water is slightly less than pure water at the same temperature, the effect can be neglected in estimating reservoir evaporation. The geographical distribution of average annual evaporation from shallow lakes is shown in Fig. 3.5.

3.7. Evaporation from Land Surfaces. Because of differences in soil texture and in expected soil moisture movement, it is difficult to generalize on the amounts of evaporation from soil surfaces. For saturated soils, the evaporation may be expected to be essentially the same as from open free-water surfaces.

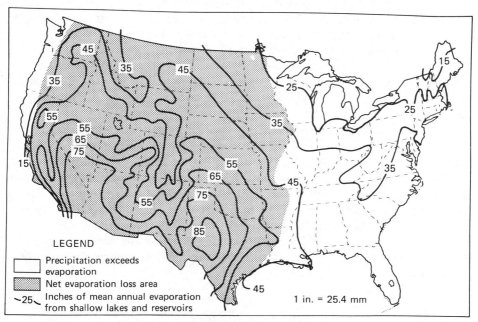

Fig. 3.5. Average annual evaporation in inches from shallow lakes and net evaporative loss area. (Redrawn from USDA, RCA Review Draft, 1980.)

As the water table drops, however, the evaporation rate will decrease greatly. Evaporation from the soil surface is generally unimportant at moisture levels below field capacity, as soil moisture movement is very slow when the soil surface is relatively dry. Mulches are effective for several days after a rain. A mulch restricts air movement, maintains a high air vapor pressure near the soil surface, and shields the soil from solar energy, all of which reduces evaporation. Freezing of a bare soil surface causes the surface to become wet, and greatly increases the evaporation rate after thawing.

3.8. Transpiration Ratio. The effectiveness of the plant's use of water in producing dry matter is often given in terms of its transpiration ratio. This is the ratio of the weight of water transpired to the weight of dry matter in the plant. It therefore varies with the same factors as does transpiration. Approximate transpiration ratios for several common plants are: 250 for sorghum, 350 for corn, 450 for red clover, 500 for wheat, 640 for potatoes, and 900 for alfalfa.

3.9. Evapotranspiration. For convenience, evaporation and transpiration are combined into evapotranspiration (*ET*), often referred to as consumptive use. Various methods for determining evapotranspiration include: (1) tank and lysimeter experiments; (2) field experimental plots where the quantity of water applied is kept small to avoid deep percolation losses and surface runoff is measured; (3) soil moisture studies, a large number of moisture samples being taken at various depths in the root zone; (4) analysis of climatological data; (5) integration methods where the water used by plants and evaporation from the water and soil surfaces are combined for the entire area involved; and (6) inflow-outflow method for large areas where yearly inflow into the area, annual precipitation, yearly outflow from the area, and the change in ground water level are evaluated.

Many practical applications can be made of evapotranspiration estimates, but the principal use is to predict soil moisture deficits for irrigation. Analyzing weather records and estimating evapotranspiration rates, drought frequencies, and excess moisture periods can show potential needs for irrigation and drainage. Similar studies to determine available tillage and harvesting days are of value in selecting optimum size for farm machines. The average daily evapotranspiration during the year, obtained for corn from lysimeters at Coshocton, Ohio, is shown in Fig. 3.6. The greatly decreased *ET* at the end of the summer may delay maturation of corn and harvesting and tillage operations, should heavy rainfall occur.

EVAPOTRANSPIRATION

The basic approaches to prediction of evaporation have been developed into several usable methods for estimating evapotranspiration.

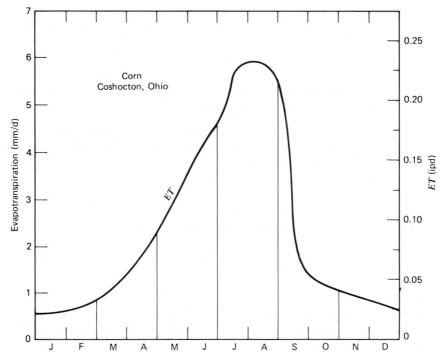

Fig. 3.6. Average evapotranspiration from corn at Coshocton, Ohio, from lysimeter data.

3.10. Blaney–Criddle Method. Blaney and Criddle (1950) developed a method that is widely used for determining evapotranspiration from climatological and irrigation data. The procedure is to correlate existing evapotranspiration data for different crops with the monthly temperature, percent of daytime hours, and length of growing season. The correlation coefficients are then applied to determine the evapotranspiration for other areas where only climatological data are available. The monthly evapotranspiration can be computed by the empirical formula

$$u = 25.4ktp/100$$
$$= 25.4kf \tag{3.7}$$

where u = monthly evapotranspiration in mm,
        k = monthly evapotranspiration (consumptive use) coefficient (function of crop type and temperature),
        t = mean monthly temperature in degrees Fahrenheit,
        p = monthly percent of total daytime hours of year,
        $f = (tp)/100$ = monthly evapotranspiration (consumptive use) factor.

Total evapotranspiration for the growing season or other period is the sum of the monthly totals. For temperatures in °C the above equation becomes

$$u = kp(0.46t_c + 8.13) \qquad (3.8)$$

Mean monthly temperatures and percent of daytime hours for each month can be determined from local weather records.

Example 3.2. Compute the evapotranspiration for corn at Sandusky, Ohio, latitude 41½° N, for the month of July using the Blaney–Criddle equation. Assume a crop coefficient, $k = 0.6$.
Solution. From U.S. Weather Bureau Monthly Climatological Data the long-time average monthly temperature (average of daily minimum and maximum values) is 23.9°C (75.0°F) and the average percentage of the daytime hours of the year for July is 10.32. Substituting in Eq. 3.8,

$$u = 0.6 \times 10.32 \,(0.46 \times 23.9 + 8.13) = 118 \text{ mm/mo. (4.64 in.)}$$

3.11. Penman Method. Penman (1948, 1956), approached the problem of estimating the evaporation from a free-water surface by examining the energy balance at the water surface expressed by

$$R_n = E + A + S + C \qquad (3.9a)$$

where R_n = net radiant energy available at the earth's surface,
  E = energy used in evaporating water,
  A = energy used in heating air,
  S = energy used in heating the water,
  C = energy used in heating the surroundings of the water.

He reasoned that energy used in heating the water and its container could be neglected and that the evaporation of water could be predicted from the equation

$$E = R_n - A. \qquad (3.9b)$$

Combination of this equation with Dalton's law (Eq. 3.3) (energy balance and mass transfer) results in an equation for evapotranspiration in which the needed data are available from meteorological records. The equation for well-watered short grass (called potential ET) presented by Penman (1963) is

$$ET = \frac{\Delta}{\Delta + \gamma}(R_n - G) + \frac{\gamma}{\Delta + \gamma}15.36\,(1 + 0.0062v_2)\,(e_s - e_d) \qquad (3.10)$$

where ET = potential evapotranspiration in cal cm^{-2}day^{-1},
   Δ = slope of saturation vapor pressure curve in mb/°C at mean air temperature,
   γ = psychrometric constant in mb/°C,
   R_n = net radiant energy available at the surface in cal cm^{-2}day^{-1},
   G = energy into the soil in cal cm^{-2}day^{-1},
   v_2 = average wind velocity in km/day at a height of 2 m,
   e_s = mean saturated vapor pressure in mb (average of pressure at maximum temperature and at minimum daily air temperature),
   e_d = saturated vapor pressure at mean dew point temperature in mb (also e_s × relative humidity).

Net radiant energy may be calculated from

$$R_n = R_a(1 - r)\left(0.18 + \frac{0.55\, n}{N}\right)$$

$$- \sigma T_a{}^4 (0.56 - 0.08 \sqrt{e_d})\left(0.10 + \frac{0.9\, n}{N}\right) \tag{3.11}$$

where R_a = mean extraterrestial radiation in cal cm^{-2}day^{-1},
   r = radiation reflection coefficient (0.05 for water surface and 0.2 for green crops),
   n/N = ratio of actual to possible hours of sunshine,
   σ = Stefan–Boltzmann constant (11.71×10^{-8} cal cm^{-2}day^{-1}),
   T_a = air temperature °K (°C + 273),
   e_d = actual vapor pressure of air, mb (also saturated vapor pressure at dew point temperature).

Useful values for calculating ET from the Penman equation are given in Tables 3.1, 3.2, and 3.3. Since the equation estimates potential evapotranspiration for a well-watered crop, actual ET from other crops must be modified because soil moisture may not be adequate or available and the crop may be mature or undeveloped. Many complex factors are involved, but conversion constants suggested by Penman (1948) will give a rough estimate of ET. The equation below applies only if $r = 0.05$ for a water surface in Eq. 3.11,

$$u = \sum_{j=1}^{n} c(ET)_j \tag{3.12}$$

where c = seasonal coefficient ranging from 0.6 in winter to 0.8 in summer at latitudes of 50° north and converging on 0.7 toward the equator in humid regions and on a somewhat smaller value than 0.7 in semiarid regions,
   n = days of the month or period.

Table 3.1 Mean Air Temperature Weighing Factors in the Penman Equation

Air Temperature		$\dfrac{\Delta}{\Delta + \gamma}$	$\dfrac{\gamma}{\Delta + \gamma}$
°C	°F		
1	33.8	0.417	0.583
5	41	0.478	0.522
10	50	0.552	0.448
15	59	0.621	0.379
20	68	0.682	0.318
25	77	0.735	0.265
30	86	0.781	0.219
35	95	0.819	0.181
40	104	0.851	0.149

Source: From Jensen (1966) as computed from *Smithsonian Met. Tables,* 6th ed., 1958, Eq. (2), p. 365 and Table 103, p. 372.

Plant growth models have been developed that provide more precise solutions of actual *ET*. This subject is beyond the scope of this text.

Example 3.3.   Compute the evapotranspiration for the conditions in Example 3.2 using the Penman equation with $G = 0$ and the radiation coefficient of 0.2 for green crops.

Solution.   Compute the net radiant energy from Eq. 3.11. From Table 3.2 read $R_a = 937$ cal cm^{-2}day^{-1}. From U.S. Weather Bureau Climatological Data or

Table 3.2 Extraterrestrial Radiation in cal cm^{-2} day^{-1a}

Month	Degrees North Latitude		
	60°	40°	20°
Jan.	86.0	358.5	630.9
Feb.	235.0	539.8	796.9
Mar.	424.4	662.4	820.2
Apr.	687.5	847.5	915.7
May	866.0	929.1	911.9
June	983.8	1001.6	948.3
July	891.9	940.6	911.9
Aug.	714.1	843.1	886.1
Sept.	494.9	720.1	856.4
Oct.	258.1	527.7	739.9
Nov.	112.6	397.1	666.7
Dec.	54.5	318.3	599.3

[a] Values in this table are computed from the expression:

$$889 \left(\frac{\text{value from Brunt (p. 112, 1944)}}{\text{days per month}} \right).$$

Table 3.3 Maximum Possible Hours of Sunshine, Average Monthly Values[a]

	Degrees North Latitude				
Month	25°	30°	35°	40°	45°
Jan.	10.8	10.4	10.1	9.7	9.2
Feb.	11.3	11.1	10.9	10.7	10.4
Mar.	12.0	12.0	12.0	11.9	11.9
Apr.	12.7	12.9	13.1	13.3	13.5
May	13.4	13.7	14.0	14.4	14.9
June	13.7	14.1	14.5	15.0	15.6
July	13.6	13.9	14.3	14.7	15.3
Aug.	13.0	13.2	13.5	13.8	14.1
Sept.	12.3	12.4	12.4	12.5	12.6
Oct.	11.6	11.5	11.3	11.2	11.0
Nov.	10.9	10.6	10.3	10.0	9.6
Dec.	10.6	10.2	9.8	9.3	8.8

[a] Average monthly values taken at 15th of month. From Marvin (1905).
Note: See U.S. Weather Bureau Climatological Data for actual hours of sunshine at each geographical location.

NOAA records, the 8-year average minimum temperature is 18.9°C (e_s = 21.9 mb), average maximum temperature is 28.9°C (e_s = 39.9 mb), from which the mean temperature is 23.9°C (297°K) and the mean vapor pressure is 30.9 mb. Vapor pressure can be read from Fig. 3.4. The average monthly relative humidity is 65.5 percent and n/N is 0.745.

$$R_n = 937(1 - 0.2) (0.18 + 0.55 \times 0.745)$$
$$- (11.71 \times 10^{-8}) (297^4) (0.56 - 0.08 \sqrt{30.9 \times 0.655}) (0.1 + 0.9 \times 0.745)$$
$$= 302 \text{ cal cm}^{-2}\text{day}^{-1}$$

From Weather Bureau monthly climatological data or NOAA, the wind velocity at 21.3 m (70 ft) height is 9.75 km/h (6.1 mph), or 234 km/d. Assuming wind variation with height is logarithmic, at 2 m height assuming a roughness height for corn in July of 1 m,

$$v_2 = 234 \times \ln 2/ \ln 21.3 = 53 \text{ km/d}$$

Substituting in Eq. 3.10 using air temperature weighing factors from Table 3.1 (by interpolation),

$$ET = (0.723 \times 302) + 0.277 \times 15.36 (1 + 0.0062 \times 53) (30.9 - 0.655 \times 30.9)$$
$$= 279 \text{ cal cm}^{-2}\text{day}^{-1}$$

or

$$279 \times 10/590 = 4.7 \text{ mm/d}$$
$$= (4.7 \times 31)$$
$$= 146 \text{ mm/mo.}$$
$$= (5.74 \text{ in./mo.})$$

3.12. Empirical Solar Radiation Method. Jensen and Haise (1963, 1966) have presented an energy-balance approach to estimating evapotranspiration that is simpler in application than Penman's equation. The Jensen–Haise method for potential ET for well-watered alfalfa (30- to 50-cm height) developed in the western U.S. is

$$ET = C_t(t_c - t_x)R_s = 0.025(t_c + 3)R_s \qquad (3.13)$$

where ET = potential evapotranspiration in mm/d,
$\quad\ C_t$ = temperature coefficient,
$\quad\ t_c$ = mean air temperature in °C,
$\quad\ t_x$ = intercept of ET/R_s vs t regression line with the temperature axis,
$\quad\ R_s$ = solar radiation as equivalent depth of evaporation in mm/d.

The coefficients C_t and t_x can be obtained from accurate ET and climatic data for a local area. Actual ET for a crop is obtained by multiplying the potential ET from Eq. 3.13 by a crop coefficient. This coefficient varies with the stage of growth of the plant as shown in Fig. 3.7. Jensen and Haise (1963), Stegman et al. (1977), and others have developed coefficients for many crops and locations. Such data are being used by commercial irrigation scheduling services in western U.S. Solar radiation can be obtained from Table 3.4 and mean temperature is available from weather records.

Most of the ET methods will give reliable estimates for 10-day periods or longer, but daily ET predictions are less reliable. For the entire growing season estimates by the different methods will be close.

Example 3.4. Using the solar radiation method estimate total evapotranspiration for the month of August for irrigated grain sorghum near Dodge City, Kansas. The following crop data are typical for the area.

Planting date	June 10	0 days
Heading date	August 15	66 days

Solution.   Read R_s = 10.2 mm/day from Table 3.4. August 1 is (51/66) 77 percent of time planting to heading and August 31 is 16 days after heading. For

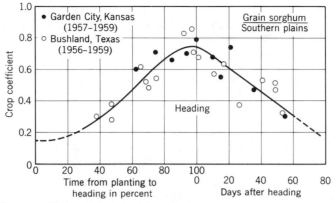

Fig. 3.7. Crop coefficient for the solar radiation method in relation to stage of growth of sorghum. (From Jensen and Haise, 1963.)

Table 3.4 Mean Measured Daily Total Solar and Sky Radiation (short wave), $R_s{}^a$

North Latitude Deg.–Min.	*Location*	*Equivalent R_s (mm/day)*						
		March	*April*	*May*	*June*	*July*	*Aug.*	*Sept.*
25-54	Brownsville, TX	6.9	7.5	9.4	10.2	10.6	9.6	8.0
31-56	Midland, TX	8.3	9.5	10.2	10.9	10.6	10.0	8.7
33-26	Phoenix, AZ	9.0	11.2	12.5	12.6	11.3	10.5	9.7
36-08	Stillwater, OK	6.6	8.0	8.4	10.3	9.6	9.3	8.1
37-46	Dodge City, KS	7.1	9.0	9.7	11.2	11.0	10.2	8.6
39-07	Grand Jct., CO	7.5	9.1	10.3	12.0	11.5	10.3	8.6
42-48	Lander, WY	7.7	9.3	10.1	11.5	11.2	9.9	8.4
44-02	Rapid City, SD	6.8	8.3	9.2	10.2	10.2	9.3	7.7
46-02	Astoria, OR	4.6	6.4	8.4	8.0	9.1	8.0	6.1
48-13	Glasgow, MT	6.6	8.0	9.7	10.2	10.4	9.4	6.9

a Computed from weekly mean values from U.S. Weather Bureau records and adapted from Jensen and Haise (1963). (1 gram water = 590 calories, heat of vaporization)

August read an average crop coefficient of about 0.7 from Fig. 3.7. Assuming mean August temperature of 25°C and substituting in Eqs. 3.13 and 3.12,

$$ET = 0.025 (25 + 3) 10.2 = 7.1 \text{ mm/day}$$
$$u = 0.7 \times 31 \times 7.1 = 154 \text{ mm/mo.}$$

REFERENCES

Blaney, H. F., and W. D. Criddle (1950). "Determining Water Requirements in Irrigated Areas from Climatological and Irrigation Data." *U.S. Dept. Agr. SCS-TP-96.*

Brunt, D. (1944). *Physical and Dynamical Meteorology* (2nd ed.). Cambridge University Press, Cambridge.

Duley, F. L. (1939). "Surface Factors Affecting the Rate of Intake of Water by Soils." *Soil Sci. Soc. Am. Proc.* **4**, 60–64.

Holtan, H. N., and M. H. Kirkpatrick, Jr. (1950). "Rainfall, Infiltration and Hydraulics of Flow in Runoff Computation." *Trans. Am. Geophys. Union* **31**, 771–779.

Horton, R. E. (1939). "Analysis of Runoff-Plot Experiments with Varying Infiltration Capacity." *Trans. Am. Geophys. Union* **20**, 693–711.

Jensen, M. E. (1966). "Empirical Methods of Estimating or Predicting Evapotranspiration Using Radiation." In ASAE Conference Proceedings, Evapotranspiration and Its Role in Water Resources Management, Dec., 49–53, 64.

Jensen, M. E., and H. R. Haise (1963). "Estimating Evapotranspiration from Solar Radiation." *Proc. ASCE J. Irrigation Drainage Div.* **89** (IR4), 15–41.

Klute, A. (1952). "A Numerical Method for Solving the Flow Equation for Water in Unsaturated Materials." *Soil Sci.* **73**, 105–116.

Marvin, C. F. (1905). *Sunshine Tables.* U.S. Weather Bureau, W. B. no. 805, Pts. I, II, and III; Reprinted 1944.

Meinzer, O. E. (1949). *Hydrology: Physics of the Earth—IX.* Dover Publications, New York.

Meyer, A. F. (1942). *Evaporation from Lakes and Reservoirs.* Minnesota Resources Commission, St. Paul.

Penman, H. L. (1948). "Natural Evapotranspiration from Open Water, Bare Soil and Grass." *Proc. Roy. Soc. of London* **193**, 120–145.

Penman, H. L. (1956). "Estimating Evapotranspiration." *Trans. Am. Geophys. Union* **37**, 43–46.

Penman, H. L. (1963). "Vegetation and Hydrology." Tech. Comm. No. 53, Commonwealth Bureau Soils. Harpenden, England.

Rohwer, C. (1931). "Evaporation from Free Water Surfaces." U.S. Dept. Agr. Tech. Bull., 271.

Stegman, E. C. et al. (1977). "Crop Curves for Water Balance Irrigation Scheduling in S.E. North Dakota." N. Dak. Agr. Expt. Sta. Res. Report 66.

Thornthwaite, C. W., and B. Holzman (1942). "Measurement of Evaporation from Land and Water Surfaces." *U.S. Dept. Agr. Tech. Bull.,* **817**.

Thornthwaite, C. W. (1948). "An Approach toward a Rational Classification of Climate." *Geograph. Rev.* **38**, 55–94.

U. S. Soil Conservation Service (1972). Hydrology, National Engineering Handbook, Section 4, Washington, D. C.

PROBLEMS

3.1. Compute the daily evaporation from a free-water surface if the wind speed is 32 km/h (20 mph) at 0.15 m height, water and air temperature are both 29°C (80°F) and the relative humidity of air is 50 percent. Atmospheric pressure is 757 mm Hg. (29.8 in. Hg.). How would your computed value compare to the evaporation from a dry-soil surface? From a Class A Weather Bureau pan?

3.2. Assuming that the Horton infiltration curve is valid, determine the constant infiltration rate if $f_0 = 51$ mm/h (2.0 iph), f at 10 min is 13 mm/h (0.5 iph), and $k = 12.9$. What is the infiltration rate at 20 min?

3.3. Using the Blaney–Criddle method, determine the growing season (May through August) evapotranspiration for corn at your present location, assuming a crop coefficient, k of 0.7.

3.4. Using Penman's equation, determine the growing season (May through August) evapotranspiration for corn at your present location, assuming a radiation reflection coefficient of 0.05 for water. The correction coefficient for season is 0.8.

3.5. Determine the *ET* for sorghum at Bushland, TX, for the months of June and August. Planting date is June 1 and heading date is August 1. Assume mean temperatures for June and August are 27 and 30°C, respectively. See Fig. 3.7.

3.6. Using Penman's equation determine the evaporation from a free water surface for the month of June of the last year and at the nearest station where climatological data and pan evaporation are available. How does your calculated value compare with pan evaporation for this month?

CHAPTER 4

Runoff

Conservation structures and channels must be designed to handle natural flows of water from rainfall or melting snow. Runoff constitutes the hydraulic "load" that the structure or channel must withstand.

4.1. Definition. Runoff is that portion of the precipitation that makes its way toward stream channels, lakes, or oceans as surface or subsurface flow. The term "runoff" usually means surface flow. The engineer designing channels and structures to handle natural surface flows is concerned with peak rates of runoff, with runoff volumes, and with temporal distribution of runoff rates and volumes.

4.2. The Runoff Process. Before runoff can occur, precipitation must satisfy the demands of evaporation, interception, infiltration, surface storage, surface detention, and channel detention.

Interception may be so great as to prevent a light rain from wetting the soil. Interception by dense covers of forest or shrubs commonly amounts to 25 percent of the annual precipitation. A good stand of mature corn may have a net interception storage capacity of 0.5 mm (0.02 in.). Trees, such as willows, may intercept nearly 13 mm (0.5 in.) from a long, gentle storm. Interception also has a detention storage effect, delaying the progress of precipitation that reaches the ground only after running down the plant or dropping from the leaves.

Runoff will occur only when the rate of precipitation exceeds the rate at which water may infiltrate into the soil (see Chapter 3). After the infiltration rate is satisfied, water begins to fill the depressions, small and large, on the soil surface. As the depressions are filled overland flow begins. The depth of water builds up on the surface until it is sufficient to result in runoff in equilibrium with the rate of precipitation less infiltration and interception. The volume of water involved in the depth build-up is surface detention. As the flow moves into defined channels there is a similar build-up of water in channel detention. The volume of water in surface and channel detention is returned to runoff as the runoff rate subsides. The water in surface storage eventually goes into infiltration or is evaporated.

FACTORS AFFECTING RUNOFF

The factors affecting runoff may be divided into those factors associated with the precipitation and those associated with the watershed.

4.3.   Rainfall.   Rainfall duration, intensity, and areal distribution (see Chapter 2) influence the rate and volume of runoff. Total runoff for a storm is clearly related to the duration for a given intensity. Infiltration will decrease with time in the initial stages of a storm. Thus a storm of short duration may produce no runoff, whereas a storm of the same intensity but of long duration will result in runoff.

Rainfall intensity influences both the rate and the volume of runoff. An intense storm exceeds the infiltration capacity by a greater margin than does a gentle rain; thus the total volume of runoff is greater for the intense storm even though total precipitation for the two rains is the same. The intense storm actually may decrease the infiltration rate because of its destructive action on the soil structure at the surface.

Figure 4.1 shows relationships of runoff to rainfall intensity and to total rainfall per storm. The data summarize 8 years of records on bare plots at

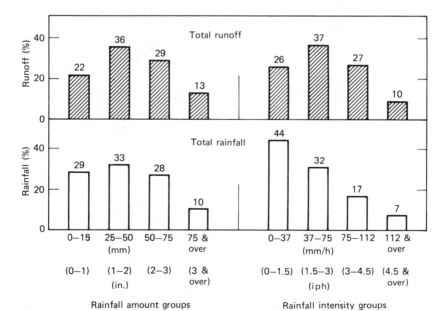

Fig. 4.1.   Total storm runoff related to rainfall amounts and intensities. (Redrawn from North Carolina Agr. Expt. Sta. Bull. 347, 1944.)

Statesville, North Carolina. In this study rainfall amounts greater than 25 mm and rainfall intensities greater than 37 mm/h gave higher percentages of runoff than the corresponding percentages for the rainfall groups.

Rate and volume of runoff from a given watershed are influenced by the distribution of rainfall and of rainfall intensity over the watershed. Generally the maximum rate and volume of runoff occurs when the entire watershed contributes. However, an intense storm on one portion of the watershed may result in greater runoff than a moderate storm over the entire watershed.

That runoff does not necessarily correspond to monthly precipitation is shown in Fig. 4.2, especially for the 7000-ha (27-square-mile) watershed. Runoff from this area is high during the late winter and early spring months because of soil moisture that seeps below the root zone and results in interflow (ground water) into the larger streams. The runoff from the 0.8-ha (2-acre) watershed more nearly reflects precipitation particularly for amounts that fall at rates greater than 25 mm/h (1.0 iph).

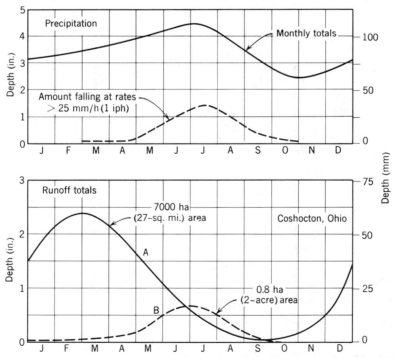

Fig. 4.2. Monthly precipitation and runoff from small and large drainage areas. (Redrawn from Harrold, 1961.)

4.4. Watershed. Watershed factors affecting runoff are size, shape, orientation, topography, geology, and surface culture. Both runoff volumes and rates increase as watershed size increases. However, both rate and volume per unit of watershed area decrease as the runoff area increases. Watershed size may determine the season at which high runoff may be expected to occur. Harrold (1950) has observed that on watersheds in the Ohio River basin 99 percent of the floods from drainage areas of 259 ha (1 square mile) occur in May through September; 95 percent of the floods on drainage areas of 26 million ha (100,000 square miles) occur in October through April.

Long, narrow watersheds are likely to have lower runoff rates than more compact watersheds of the same size. The runoff from the former does not concentrate as quickly as it does from the compact areas, and long watersheds are less likely to be covered uniformly by intense storms. When the long axis of a watershed is parallel to the storm path, storms moving upstream cause a lower peak runoff rate than storms moving downstream. For storms moving upstream, runoff from the lower end of the watershed is diminished before the peak contribution from the headwaters arrives at the outlet. However, a storm moving downstream causes a high runoff from the lower portions coincident with high runoff arriving from the headwaters.

Topographic features, such as slope of upland areas, the degree of development and gradients of channels, and the extent and number of depressed areas affect rates and volumes of runoff. Watersheds having extensive flat areas or depressed areas without surface outlets have lower runoff than areas with steep, well-defined drainage patterns. The geologic or soil materials determine to a large degree the infiltration rate, and thus affect runoff. Vegetation and the practices incident to agriculture and forestry also influence infiltration (see Chapter 3). Vegetation retards overland flow and increases surface detention to reduce peak runoff rates. Works of man, such as dams, levees, bridges, and culverts, all influence runoff rates.

PREDICTING RUNOFF

Methods of runoff estimation necessarily neglect some factors and make simplifying assumptions regarding the influence of others. Methods presented here are applicable to small agricultural watersheds less than a few hundred hectares.

DESIGN RUNOFF RATES

The capacity to be provided in a structure that must carry runoff may be termed the design runoff rate. Structures and channels are planned to carry runoff that occurs within a specified return period. Vegetated controls and temporary structures are usually designed for a runoff that may be expected to occur once in 10 years; expensive, permanent structures will be designed for runoffs expected only once in 50 or 100 years. Selection of the design return period, also called recurrence interval, depends on the economic balance between the cost of periodic repair or replacement of the facility, and the cost of providing additional capacity to reduce the frequency of repair or replacement. In some instances the downstream damage potentially resulting from failure of the structure may dictate the choice of the design frequency.

4.5. Rational Method. The rational method of predicting a design peak runoff rate is expressed by the equation

$q = CiA$
< 2000 acre

C — ratio, peak runoff rate to rainfall intensity

$$q = 0.0028\, CiA \tag{4.1}$$

where q = the design peak runoff rate in m³/s,

i — inches/hr
A — acres

C = the runoff coefficient,

i = rainfall intensity in mm/h for the design return period and for a duration equal to the "time of concentration" of the watershed,

A = the watershed area in hectares.

t_c

The time of concentration of a watershed is the time required for water to flow from the most remote (in time of flow) point of the area to the outlet once the soil has become saturated and minor depressions filled. It is assumed that, when the duration of a storm equals the time of concentration, all parts of the watershed are contributing simultaneously to the discharge at the outlet. One of the most widely accepted methods of computing the time of concentration was developed by Kirpich (1940),

$T_c = 0.0078\, L^{0.77} S^{-0.385}$

$$T_c = 0.0195\, L^{0.77} S^{-0.385} \tag{4.2}$$

where T_c = time of concentration in min (see Appendix A),

L — ft
S — ft/ft

L = maximum length of flow in m,

T_c — MINUTE

S = the watershed gradient in m per m or the difference in elevation between the outlet and the most remote point divided by the length, L.

Hydrologists are not in agreement as to the best procedure for computing the time of concentration. Mockus (1961) prepared a nomograph (see Appendix A) for computing the time of concentration which considers length of the main channel, topography, vegetal cover, and infiltration rate. Horn and Schwab (1963) found that Mockus' values of watershed lag gave slightly better estimates of the actual runoff than several other methods when taken equal to the time of concentration.

U.S. SCS (1972) developed the "Upland Method" for estimating time of concentration. With this method the length of flow is divided by an estimated velocity of flow to obtain the travel time. The sum of the travel times for overland flow (sheet runoff) and for all channel flow equals the time of concentration. For such estimates the flow path is taken from the most remote point in the watershed to the outlet. In small watersheds of a few hectares, where a well-defined channel does not exist, runoff occurs mostly as overland flow. The upland method should be limited to small watersheds less than 800 hectares (2000 acres). Manning's equation (see Chapter 7) is suitable for estimating flow velocities. Overland flow velocities are difficult to estimate as the rainfall intensity will influence the flow depth. U.S. SCS (1972) has developed velocity-slope curves for several watershed conditions. Better methods for determining time of concentration are needed.

The runoff coefficient C is defined as the ratio of the peak runoff rate to the rainfall intensity and is dimensionless. Estimates of the runoff coefficient from small single-crop watersheds at Coshocton, Ohio, showed that the primary effects were attributed to the infiltration rate, surface cover, and rainfall intensity. These estimates are presented in Table 4.1 for hydrologic soil group B. The runoff coefficient can be converted to other hydrologic soil groups by referring to Table 4.2. These soil groups are defined in Table 4.3.

Equation 4.1 may not appear to be dimensionally correct. In English units i is specified in inches per hour and numerically 1 inch per hour is nearly equal (1.008) to cubic feet per second per acre. The constant 0.0028 converts the equation to SI units.

The rational method assumes that the frequency of rainfall and runoff are similar, which has been confirmed by Larson and Reich (1973). The method is a great oversimplification of a complicated process. However, the method is considered sufficiently accurate for runoff estimation in the design of relatively inexpensive structures where the consequences of failure are limited. Application of the rational method as presented here is normally limited to watersheds of less than 800 ha (2000 ac).

Table 4.1 Runoff Coefficient "C" for Agricultural Watersheds (Soil Group B)

Cover and hydrologic Condition	Coefficient C for rainfall rates of		
	25 mm/h (1 iph)	100 mm/h (4 iph)	200 mm/h (8 iph)
Row crop, poor practice	0.63	0.65	0.66
Row crop, good practice	0.47	0.56	0.62
Small grain, poor practice	0.38	0.38	0.38
Small grain, good practice	0.18	0.21	0.22
Meadow, rotation, good	0.29	0.36	0.39
Pasture, permanent, good	0.02	0.17	0.23
Woodland, mature, good	0.02	0.10	0.15

Source: Horn and Schwab (1963).

Table 4.2 Hydrologic Soil Group Conversion Factors

Cover and hydrologic condition	Factors for converting the runoff coefficient C from group B soils to[a]		
	Group A	Group C	Group D
Row crop, poor practice	0.89	1.09	1.12
Row crop, Good practice	0.86	1.09	1.14
Small grain, poor practice	0.86	1.11	1.16
Small grain, good practice	0.84	1.11	1.16
Meadow, rotation, good	0.81	1.13	1.18
Pasture, permanent, good	0.64	1.21	1.31
Woodland, mature, good	0.45	1.27	1.40

[a] Factors were computed from Table 4.3 by dividing the curve number for the desired soil group by the curve number for group B.

Table 4.3 Runoff Curve Numbers for Hydrologic Soil-Cover Complexes for Antecedent Rainfall Condition II, and $I_a = 0.2S$

Land Use or Cover	Treatment or Practice	Hydrologic Condition	*Hydrologic Soil Group			
			A	B	C	D
Fallow	Straight row	—	77	86	91	94
Row crops	Straight row	Poor	72	81	88	91
	Straight row	Good	67	78	85	89
	Contoured	Poor	70	79	84	88
	Contoured	Good	65	75	82	86
	Terraced	Poor	66	74	80	82
	Terraced	Good	62	71	78	81
Small grain	Straight row	Poor	65	76	84	88
	Straight row	Good	63	75	83	87
	Contoured	Poor	63	74	82	85
	Contoured	Good	61	73	81	84
	Terraced	Poor	61	72	79	82
	Terraced	Good	59	70	78	81
Close-seeded	Straight row	Poor	66	77	85	89
legumes or	Straight row	Good	58	72	81	85
rotation	Contoured	Poor	64	75	83	85
meadow	Contoured	Good	55	69	78	83
	Terraced	Poor	63	73	80	83
	Terraced	Good	51	67	76	80
Pasture or		Poor	68	79	86	89
range		Fair	49	69	79	84
		Good	39	61	74	80
	Contoured	Poor	47	67	81	88
	Contoured	Fair	25	59	75	83
	Contoured	Good	6	35	70	79

Table 4.3 (Continued)

Land Use or Cover	Treatment or Practice	Hydrologic Condition	*Hydrologic Soil Group			
			A	B	C	D
Meadow (permanent)		Good	30	58	71	78
Woods		Poor	45	66	77	83
(farm wood-		Fair	36	60	73	79
lots)		Good	25	55	70	77
Farmsteads		—	59	74	82	86
Roads and right-of-way (hard surface)		—	74	84	90	92

*Soil Group	Description	Final infiltration Rate (mm/h)
A	*Lowest Runoff Potential.* Includes deep sands with very little silt and clay, also deep, rapidly permeable loess.	8–12
B	*Moderately Low Runoff Potential.* Mostly sandy soils less deep than *A*, and loess less deep or less aggregated than *A*, but the group as a whole has above-average infiltration after thorough wetting.	4–8
C	*Moderately High Runoff Potential.* Comprises shallow soils and soils containing considerable clay and colloids, though less than those of group *D*. The group has below-average infiltration after pre-saturation.	1–4
D	*Highest Runoff Potential.* Includes mostly clays of high swelling percent, but the group also includes some shallow soils with nearly impermeable sub-horizons near the surface.	0–1

Source: U. S. Soil Conservation Service, *National Engineering Handbook, Hydrology*, Section 4 (1972) and U. S. Dept. Agr. ARS 41-172 (1970).

The rational method is developed from the assumptions that: (1) rainfall occurs at uniform intensity for a duration at least equal to the time of concentration of the watershed, and (2) rainfall occurs at a uniform intensity over the entire area of the watershed. If these assumptions were fulfilled, the rainfall and runoff for the watershed would be represented graphically by Fig. 4.3a. The figure shows a rain of uniform intensity for a duration equal to the time of concentration, T_c. If a storm of duration greater than T_c occurred, the runoff rate would be less than q because the rainfall intensity would be less than i (see Chapter 2 for relationships between rainfall intensity and duration). A rain of

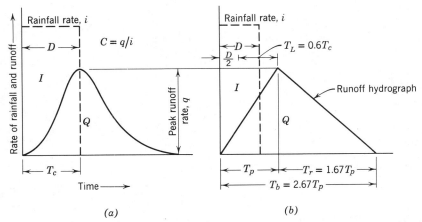

Fig. 4.3. Rainfall and runoff with assumptions (a) for the rational equation and (b) for the Soil Conservation Service triangular hydrograph method of runoff estimation.

duration less than T_c would result in a runoff rate less than q because the entire watershed would not contribute simultaneously to the discharge at the outlet.

The rational method is illustrated by the following problem.

Example 4.1.   Determine the design peak runoff rate for a 50-year return period storm from a 40.5-ha (100-ac) watershed near Chicago, Ill., with the following characteristics:

ha	Subarea (ac)	Topography percent slope	Soil Group (See Table 4.3)	Land Use, Treatment and Hydrologic Condition
24.3	(60)	flat	C	row crop, contoured, good
16.2	(40)	10 to 30	B	woodland, good

The maximum length of flow is 610 m (2000 ft) and the difference in elevation along this path is 3 m (10 ft).

Solution.   The watershed gradient is $(10/2000)100 = 0.5$ percent. From Appendix A or Eq. 4.2, $T_c = 20$ min. From Chapter 2 for a 50-year return period near Chicago, the 20-min rainfall is 97 mm/h (3.8 iph).

The runoff coefficient C from Table 4.1 for row crop, good practice, and woodland is 0.56 and 0.10, respectively, and the factor correcting hydrologic soil group C to group B for the 24.3-ha subarea from Table 4.2 is 1.09.

$$C = (24.3/40.5) \times 0.56 \times 1.09 + (16.2/40.5) \times 0.10 = 0.41$$

Substituting in Eq. 4.1,

$$q = 0.0028 \times 0.41 \times 97 \times 40.5 = 4.51 \text{ m}^3/\text{s (159 cfs)}$$

4.6. Soil Conservation Service Method. This method described by U.S. SCS (1973) was originally developed for uniform rainfall using the assumptions for a triangular hydrograph shown in Fig. 4.3b. The time to peak flow,

$$T_p = D/2 + T_L = D/2 + 0.6 \, T_c \qquad (4.3)$$

where T_p = time to peak,
  D = duration of excess rainfall,
  T_L = time of lag,
  T_c = time of concentration.

Time of concentration is the longest travel time and is not the time of peak as in the rational equation. Time of lag is an approximation of the mean travel time. It may be obtained from the nomograph in Appendix A. The time of peak is necessary to develop a design hydrograph for routing runoff through a storage reservoir or for combining hydrographs from several subwatersheds. For some small watersheds the time of peak may exceed the time of concentration. The time of recession for the triangular hydrograph is taken as $1.67 T_p$, thus the total time of flow is $2.67 T_p$. The peak runoff rate derived from the triangular hydrograph is

$$q = 0.0021 QA/T_p \qquad (4.4)$$

where Q = runoff volume in mm depth (area under the hydrograph),
  q = runoff rate in m³/s,
  A = watershed area in ha,
  T_p = time of peak in hours.

Example 4.2. Using the SCS method, determine the peak runoff rate from a 100-ha (247-ac) watershed from a uniform 36-min (0.6-h) storm that produced 10 mm (0.39 in.) of runoff volume. Assume T_c for the watershed is 0.5 h.
Solution. Substituting in Eq. 4.3,

$$T_p = 0.6/2 + 0.6 \times 0.5 = 0.6 \text{ h}$$

Substituting in Eq. 4.4,

$$q = 0.0021 \times 10 \times 100/0.6 = 3.50 \text{ m}^3/\text{s (124 cfs)}$$

For nonuniform rainfall the storm is divided into increments of duration ΔD and increments of runoff volume ΔQ for the same period. The ΔD is defined as $T_p/3$ for this purpose. The peak runoff for each increment is computed by Eq. 4.4 and added together in such a way as to give the peak runoff for the storm. Type I 24-hour rainfall distribution storms apply to Hawaii, Alaska, and the coastal side of the mountains along the west coast. Type II distributions represent the remainder of the U.S., Puerto Rico, and the Virgin Islands. The method is applicable for watersheds 800 ha (2000 acre) or less in area and for slopes less than 30 percent. The basic input is 24-hour rainfall and watershed characteristics by curve numbers as given in Table 4.3. Twenty-one peak runoff rate vs watershed area charts have been prepared, which simplify their use. The procedure is too lengthy and involved to be included, but it is described in more detail in U.S. SCS (1973).

4.7. Flood Frequency Analysis Method.   One method of runoff estimation, called flood frequency analysis, depends upon the existence of a number of years of record from the basin under study. These records then constitute a statistical array that defines the probable frequency of recurrence of floods of given magnitudes. Extrapolation of the frequency curves enables the hydrologist to predict flood peaks for a range of return periods. The procedure for this method is the same as that for rainfall described in Chapter 2.

4.8. Other Methods.   Many other methods have been proposed for estimating flood runoff. Chow (1962) made a thorough review of 66 such methods of runoff computation. He also developed a new procedure for Illinois based on rainfall, soil, and cover conditions.

A number of empirical formulas have been developed to describe the magnitude of extreme floods. These formulas take the form

$$q = KA^x \tag{4.5}$$

where q = the magnitude of the peak runoff,
K = a coefficient dependent upon various characteristics of the watershed,
A = the watershed area,
x = a constant for a given location.

Potter (1961) proposed a topographic index and delineated major geological zones for predicting peak runoff rates. This method was adopted by the U.S. Bureau of Public Roads for the central and eastern states.

RUNOFF VOLUME

It is often desirable to predict the total volume of runoff that may come from a watershed during a design flood. Total volume is of primary interest in the design of flood control reservoirs.

[handwritten top annotations:]
GRASS FILTER
w/
AMOUNT OF WATER TO INFILTRATE = RUNON + RAINFALL
= ACRE-IN + IN (ACRE)
↑ UNKNOWN
(Q)(WATERSHED SIZE) RUNOFF VOLUME — slopes not critical

4.9. Soil Conservation Service Method.

This method was developed from many years of storm flow records for agricultural watersheds in many parts of the United States. The equation, which applies to the curves in Fig. 4.4, is

[handwritten: USE FOR GRASS FILTER]

$$Q = \frac{(I - 0.2S)^2}{I + 0.8S} \qquad (4.6)$$

[handwritten: $S = \frac{1000}{CN} - 10$]

where Q = direct surface runoff in depth in mm,
I = storm rainfall in mm (see Chapter 2),
S = maximum potential difference between rainfall and runoff in mm, starting at the time the storm begins,

[handwritten: VOL = (acre)(inches) = acre-inch]

On gaged watersheds I can be plotted against Q and the value of S obtained directly. As pointed out in Chapter 3 and shown by Eq. 4.6, runoff decreases as S or infiltration increases. The initial abstraction, $I_a = 0.2S$, consists of interception losses, surface storage, and water that infiltrates into the soil prior to runoff.

[handwritten right margin:]
USING
— Ge + I
— Correct Source
— Got Weighted
N
— Ge + Q

[handwritten left margin:]
CN
- barnyard
- high 90's
etc.

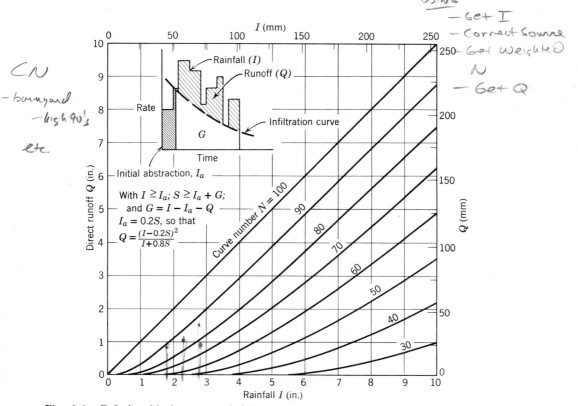

Fig. 4.4. Relationship between rainfall and runoff. (Redrawn from U.S. Soil Conservation Service, 1972.)

For convenience in evaluating antecedent moisture, soil conditions, land use, and conservation practices, the U.S. Soil Conservation Service (1972) defines

$$S = \frac{25\,400}{N} - 254 \tag{4.7}$$

where N = an arbitrary curve number varying from 0 to 100.

Thus, if

$$N = 100, \text{ then } S = 0 \text{ and } I = Q.$$

Curve numbers can be obtained from Table 4.3. These values apply to antecedent rainfall condition II, which is an average value for annual floods. Correction factors for other antecedent rainfall conditions are listed in Table 4.4.

Table 4.4   Antecedent Rainfall Conditions and Curve Numbers (for $I_a = 0.2S$)

Curve Number for Condition II	Factor to Convert Curve Number for Condition II to	
	Condition I	Condition III
10	0.40	2.22
20	0.45	1.85
30	0.50	1.67
40	0.55	1.50
50	0.62	1.40
60	0.67	1.30
70	0.73	1.21
80	0.79	1.14
90	0.87	1.07
100	1.00	1.00

Condition	General Description	5-Day Antecedent Rainfall (mm)	
		Dormant Season	Growing Season
I	Optimum soil condition from about lower plastic limit to wilting point	<13	<36
II	Average value for annual floods	13–28	36–53
III	Heavy rainfall or light rainfall and low temperatures within 5 days prior to the given storm	>28	>53

Source: U. S. Soil Conservation Service, *National Engineering Handbook, Hydrology,* Section 4 (1972).

Condition I is for low runoff potential with soil having low antecedent moisture suitable for cultivation. Condition III is for wet conditions prior to the storm. As indicated in Table 4.4 no upper limit for antecedent rainfall is intended. The limits for the "dormant season" apply when the soils are not frozen and when no snow is on the ground.

Since duration of a storm affects the amount of rainfall, runoff volume must be evaluated for each design application. The time of concentration is not a good criterion for the determination of storm volume since a short-duration, high-intensity storm may produce the peak flow, but not necessarily the maximum runoff volume. The Soil Conservation Service has established 6 hours duration as the minimum storm period for flood-water-retarding structures, but this time is modified for certain conditions where a greater runoff would result.

Example 4.3. Determine the estimated maximum volume of runoff during the growing season for a 50-year return period that may be expected from the watershed of Example 4.1. Assume that antecedent moisture 5 days prior to the storm was 40 mm (1.6 in.) of rainfall; and the critical duration of the storm is 6 hours.

Solution. From Chapter 2 the rainfall for a 6-hour storm is 107 mm (4.2 in.) at Chicago. Since the percent reduction for converting point rainfall to areal rainfall is less than 1 percent (see Chapter 2) in this case, no correction need be made. From Table 4.3 for Antecedent Rainfall Condition II, read the appropriate curve number and calculate the weighted value as follows:

Subarea, A ha (ac)	Soil Group	Land Use, Treatment, and Condition	Curve No., N	NA
24.3 (60)	C	row crop, contoured, good	82	1993
16.2 (40)	B	woodland, good	55	891
40.5 (100 ac)			Total	2884

Weighted curve number = 2884/40.5 = 71.2
Substituting in Eq. 4.7,

$$S = \frac{25\ 400}{71.2} - 254 = 103 \text{ mm}$$

$$Q = \frac{(107 - 0.2 \times 103)^2}{107 + (0.8 \times 103)} = 39 \text{ mm (1.54 in.)}$$

$$Q = \frac{39 \times 40.5}{1000} = 1.58 \text{ ha-m (12.8 ac-ft)}$$

Example 4.4.   If 51 mm (2.0 in.) of rainfall occurred the day after the 50-year storm in Example 4.3, what is the expected runoff?
Solution.   From Table 4.4 Antecedent Rainfall Condition III applies as (107 + 40) > 53 and the new curve number is (71.2)(1.20) = 86, and S = 41 mm. Substituting in Eq. 4.6, Q = 22 mm (0.87 in.). Although the rainfall was less than half that in Example 4.3, the runoff was 56 percent of the previous amount, illustrating the importance of antecedent moisture.

WATER YIELD

When surface runoff is to be stored in ponds or reservoirs, the total runoff volume for a period of several months, usually the annual volume, is of more interest than the runoff from a design storm. The annual runoff is often referred to as the water yield. For gaged watersheds, frequency analysis of flow records as described in Chapter 2 usually provides the best design data. Brakensiek (1959) has shown that the selection of runoff data for a water year rather than for a calendar year can greatly improve the reliability of results. The date for beginning the water year varies with geographical location, but in general it coincides with the season of maximum runoff. For example, at Coshocton, Ohio, March 1 was the best date. Since water yield records are limited, the results from one area must often be extended to other areas of similar hydrologic conditions.

4.10.   Estimation of Minimum Water Yield.   Minimum water yields obtained by frequency analysis of 18 years of record at Coshocton, Ohio, are shown in Table 4.5. Since the minimum yield is the least flow (dry year), it decreases with the return period, unlike rainfall and storm runoff, which increase with the return period. Minimum yield increases with watershed size because of the greater contribution of ground water flow, especially in humid areas.

Example 4.5.   Determine the minimum annual water yield volume from a 86-ha watershed at Coshocton, Ohio, for the driest year in 25 years.
Solution.   Read from Table 4.5 by interpolation half way between 130 and 178 mm (86 ha is half way between 31 and 141 ha) is 159 mm.

Minimum volume = 86 × 159 × 10⁻³ = 13.7 ha-m (111 ac-ft)

From water yield records in southern Iowa, Nixon and Schwab (1961) proposed a watershed rating as a means of estimating water yields from other similar watersheds. This rating included factors dependent on climate, land use, steepness of slope, soil series and type, and management practices. In this study the size of the watershed within the range of 121 to 4,048 ha (300 to 10,000 ac) was of less importance than the above factors. Flow records showed that in the dry

Table 4.5   Minimum Annual Water Yield at Coshocton, Ohio (Mixed vegetation and for water year beginning October 1.)

Watershed Area		Minimum Annual Water Yield for Return Periods of			
ha	*(ac)*	*2 yr*	*10 yr*	*25 yr*	*50 yr*
12	(29)	221 mm[a]	132 mm	107 mm	86 mm
31	(76)	244	157	130	109
141	(349)	330	221	178	152
1 041	(2,570)	356	234	185	160
7 045	(17,400)	394	267	216	183

[a] 25.4 mm = 1 in.
Source: Harrold (1957).

years, which are those of greatest interest, ground water contribution was small. In some geographic regions ground water flow, which is normally of long duration, may be a very important part of water yield. Springs or tile flow can also make a major contribution.

The minimum yield for a period greater than one year is of practical interest in reservoir design. If water needs are just equal to reservoir capacity, shortage of water will likely occur if the annual water yield is not sufficient to refill the reservoir. However, if storage is greater than the need, some water can be carried over to the dry year. The water yield for southern Iowa (Fig. 4.5) illustrates this point. For a return period of 10 years the water yield for 1-, 2-, 3- and 4-year periods was 46 (1.8), 117 (4.6), 198 (7.8), and 318 mm (12.5 in.), respectively. If storage was available, the annual average yield is increased from 46 to 58, 66, or 79 mm per year for the 2-, 3-, and 4-year periods, respectively. These values from above are 46/1, 117/2, and so on. For return periods greater than 10 years the average multiple-year yield becomes increasingly greater than the 1-year yield.

4.11.   Other Methods.   For small ponds and reservoirs the information in Fig. 10.11 in Chapter 10 will provide a conservative estimate of water yield.

Although rather lengthy calculations would be required, the Soil Conservation Service method of estimating storm runoff given in Section 4.9 could be applied to daily rainfall. For the winter season, snowmelt may be estimated by Eq. 4.8. The annual water yield would then be the sum of the daily runoffs and snowmelt.

4.12.   Snowmelt.   For ungaged watersheds, snowmelt may be determined from the equation

$$M = 45.72KD \qquad (4.8)$$

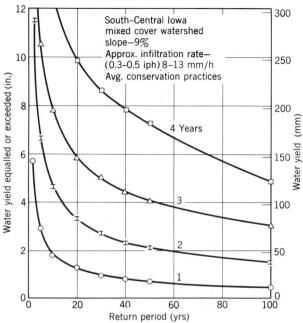

Fig. 4.5. Frequency of minimum water yield in south-central Iowa for flow periods from 1 to 4 years. (From Nixon and Schwab, 1961.)

where M = depth of snowmelt water from watershed in mm/d,
         K = a constant for watershed and climatic conditions (0.02 for low runoff potential, 0.06 for average, and 0.30 for high runoff potential),
         D = degree-days for a given day, in °C.

The average temperature is the average of maximum and minimum for the day. A day with an average temperature of 15°C thus has 15 degree-days. Adjustment of temperature to the average elevation of the watershed can be made by a decrease of 2.2°C (4°F) for every 305 m (1000 ft.) increase in elevation. Only temperatures above freezing are considered. Snowmelt is of most importance in the mountainous regions of western U.S.

RUNOFF HYDROGRAPHS

A hydrograph is a graphical or tabular representation of runoff rate against time. Figure 4.6 gives the hydrograph of the discharge from a 125-ha (309-ac) agricultural watershed. This discharge resulted from a storm of 43 mm (1.71 in.) in 30 min. For small agricultural watersheds, streamflow records from which

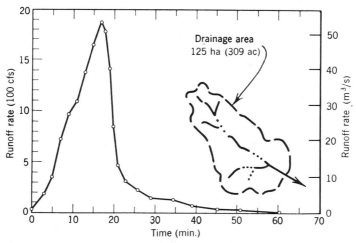

Fig. 4.6. Natural runoff hydrograph for the upper Theobold Watershed, Woodbury County, Iowa, June 1950.

hydrographs can be developed are generally not available. One of the characteristics of hydrographs for a given watershed is that the duration of flow is nearly constant for individual storms regardless of the peak flow. Sherman (1932) developed a hydrograph representing one inch of runoff, which is referred to as a unit hydrograph. Commons (1942) and later others developed dimensionless hydrographs.

4.13. Triangular Hydrographs.   The simplest runoff hydrograph is the triangular hydrograph described in Figs. 4.3*b* and 4.7 and Eq. 4.4. They are applied (a) to subareas of a watershed or (b) to time increments of a rainstorm. In either case the flow rates at a given time are added from the several hydrographs, thus producing a curvilinear hydrograph. Except for the tail end, it is a good approximation of an actual hydrograph. As shown in Fig. 4.7, the time of peak and the peak flow rate are the same as for the dimensionless hydrograph. The base of the triangular hydrograph was so chosen to give the same area (volume) as the dimensionless hydrograph. Equation 4.4 applies to both hydrographs; thus if any two of the three variables are known the third can be computed.

4.14. Dimensionless Hydrograph.   The dimensionless hydrograph shown by the smooth curve in Fig. 4.7 has a shape that approximates the flow from an intense storm from a small watershed. It has an idealized shape and is sometimes called a synthetic hydrograph. The SCS triangular hydrograph shown by the dashed lines closely approximates the dimensionless hydrograph and greatly simplifies calculation of the variables. The dimensionless hydrograph arbitrarily has 100 units of flow for the peak and 100 units of time for the

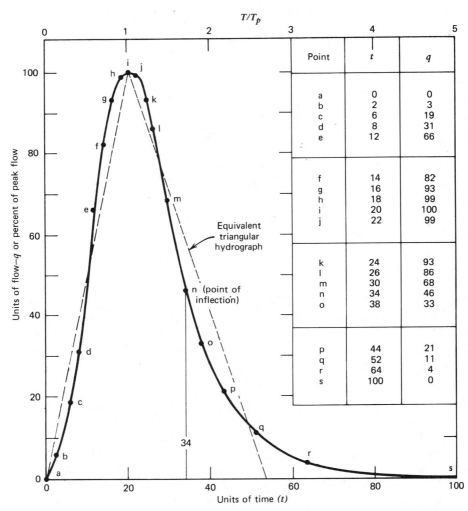

Fig. 4.7. Dimensionless and triangular flood hydrographs. (Adapted from U.S. Soil Conservation Service, 1972.)

duration of flow. As described in this text, it is the hydrograph to be developed for flood routing in Chapter 11.

To develop the design hydrograph for a watershed, the peak flow and the runoff volume must be known for the desired return period storm. The design hydrograph is developed from the dimensionless hydrograph by using appropriate conversion factors. The factor u is the ratio of the total runoff volume to the area under the dimensionless hydrograph. The area under the hydrograph is

2620 square units. Thus, each square unit under the basic hydrograph has a value of

$$u = Q/2620 \tag{4.9}$$

for the design storm having a total runoff volume Q. The factor w is the ratio of peak runoff for the design storm to the peak flow of 100 on the dimensionless hydrograph. Each unit of flow on the dimensionless hydrograph has a value of

$$w = q/100 \tag{4.10}$$

in the hydrograph of the design storm. The factor k is the value that each unit of time on the dimensionless hydrograph represents in the design hydrograph. On the design hydrograph 1/100 of the peak flow times 1/100 of the duration of runoff must equal 1/2620 of the flood volume just as it does on the dimensionless hydrograph. Since w is equal to 1/100 of the design peak flow, k must be equal to 1/100 of the design duration, and u is 1/2620 of the design flood volume; therefore

$$wk = u$$

and

$$k = u/w \tag{4.11}$$

When runoff rate is measured in cubic meters per second, runoff volume is measured in hectare-meters, and time is measured in minutes,

$$k = \frac{u(\text{ha-m}) \times 10\ 000\ (\text{m}^2/\text{ha})}{w(\text{m}^3/\text{s}) \times 60\ (\text{sec/min})} = 167\,\frac{u}{w}$$

The coordinates of the design hydrograph are obtained by multiplying the ordinates and abscissas of the dimensionless hydrograph by w and k, respectively.

Example 4.6. Develop a runoff hydrograph for a design return period of 50 years for the watershed of Examples 4.1 and 4.3.
Solution. The peak runoff is 4.51 m³/s and the flood volume is 1.58 ha-m. From Eqs. 4.9, 4.10, and 4.11,

$$u = 1.58/2620 = 0.0006 \text{ ha-m/unit}$$
$$w = 4.51/100 = 0.0451 \text{ m}^3/\text{s per unit}$$
$$k = 167 \times 0.0006/0.0451 = 2.22 \text{ min/unit}$$

Ordinates and abscissas of the design hydrograph are obtained by multiplying the values of q and t from Fig. 4.7 by w and k, respectively. The calculated coordinates are as follows:

Point	kt	wq	Point	kt	wq
a	0	0	j	48.8	4.46
b	4.4	0.14	k	53.3	4.19
c	13.3	.86	l	57.7	3.88
d	17.8	1.40	m	66.6	3.07
e	26.6	2.98	n	75.5	2.07
f	31.1	3.70	o	84.4	1.49
g	35.5	4.19	p	97.7	.95
h	40.0	4.46	q	115.4	.50
i	44.4	4.51	r	142.1	.18
Point i is at T_p			s	222.0	0

4.13. Other Hydrographs. A large number of synthetic hydrographs have been developed, but none of these has received general acceptance. Because of limited flow records, these hydrographs as well as peak flow estimates leave much to be desired.

Several theoretical hydrographs have been proposed based on different statistical frequency distributions. Dodge (1959) developed a unit hydrograph from the Poisson probability function, Gray (1970) employed a two-parameter gamma distribution, and Reich (1962) investigated a three-parameter Pearson type-III function. Linsley et al. (1975) described several methods of developing unit hydrographs and a hydrograph for overland flow. They defined the unit hydrograph as the hydrograph of one inch (or 1 cm) of direct runoff from a storm of specified duration.

REFERENCES

Brakensiek, D. L. (1959). "Selecting the Water Year for Small Agricultural Watersheds." *Am. Soc. Agr. Eng. Trans.* 2(no. 1), 5–8, 10.

Chow, Ven Te (1962). "Hydrologic Determination of Waterway Areas for the Design of Drainage Structures in Small Drainage Basins." *Ill. Eng. Expt. Sta. Bull.,* no. 462, March.

Commons, G. G. (1942). "Flood Hydrographs." *Civil Eng.* 12, 571–572.

Dodge, J. C. I. (1959). "A General Theory of the Unit Hydrograph." *J. Geophys. Res.,* 64, 241–256.

Gray, D. M. (ed.) (1970). *Handbook on the Principles of Hydrology.* Water information Center, Inc. Port Washington, N.Y.

Gray, D. M. (1961). "Synthetic Unit Hydrographs for Small Watersheds." *Am. Soc. Civil Eng. Proc.* **87**, HY4, Paper 2854, 33–54, July.

Harrold, L. L., G. O. Schwab, and B. L. Bondurant (1976). *Agricultural and Forest Hydrology.* Ohio State University Bookstore, Columbus, Ohio.

Harrold, L. L. (1957). "Minimum Water Yield from Small Agricultural Watersheds." *Am. Geophys. Union Trans.* **38**, 201–208.

Harrold, L. L. (1961). "Hydrologic Relationships on Watersheds in Ohio." *Soil Conserv.* **26**, 208–210.

Horn, D. L., and G. O. Schwab (1963). "Evaluation of Rational Runoff Coefficients for Small Agricultural Watersheds." *Am. Soc. Agr. Eng. Trans.* **6**(no. 3), 195–198, 201.

Kirpich, P. Z. (1940). "Time of Concentration of Small Agricultural Watersheds." *Civil Eng.* **10**, 362.

Larson, C. L. and B. M. Reich (1973). "Relationship of Observed Rainfall and Runoff Recurrence Intervals." In Schulz, E. F. et al. (eds.) *Flood and Droughts. Proc. 2nd Internat. Sym. in Hydrology, Sept. 1972.* Water Res. Publ. Ft. Collins, Colo.

Linsley, R. K., M. A. Kohler, and J. L. H. Paulhus (1975). *Hydrology for Engineers,* 2nd ed. McGraw–Hill, New York.

Nixon, P. R. and G. O. Schwab (1961). "Water Yield Prediction in Southern Iowa Based on Watershed Characteristics." *Iowa State J. Sci.,* **35**, 331–342.

Potter, W. D. (1961). *Peak Rates of Runoff from Small Watersheds.* U. S. Bureau Public Roads, Hydraulic Design Series no. 2, April.

Reich, B. M. (1962). *Design Hydrographs for Very Small Watersheds from Rainfall, Civil Engineering Section.* Colorado State University, July.

Sherman, L. K. (1932). "Stream Flow from Rainfall by Unit Graph Method." *Eng. News-Record* **108**, 501–505.

U. S. Soil Conservation Service (1973). A Method for Estimating Volume and Rate of Runoff in Small Watersheds. SCS-TP-149, Washington, D.C.

——(1972). "Hydrology." *National Engineering Handbook,* Section 4. Washington, D.C.

PROBLEMS

4.1. Calculate the 10-year return period peak runoff for an 81-ha (200-ac) watershed having a runoff coefficient of 0.4. The maximum length of flow of water is 610 m (2000 ft) and the fall along this path is 6.1 m (20 ft). Watershed is at your present location.

4.2. Determine the peak rate of runoff from a 8.1-ha (20-ac) watershed during a 30-min storm that gave the following amounts of rain during successive 5-min periods: 3, 8, 15, 15, 8, and 3 mm (0.1, 0.3, 0.6, 0.6, 0.3, and 0.1 in.). The time of concentration for the watershed is 10 min. The runoff coefficient for the

2 ha (5 ac) is 0.2 and 0.6 for the remaining 6.1 ha (15 ac). What is the approximate return period for this runoff if the watershed is at your present location?

4.3. By the rational method, determine the design peak runoff for a 50-year return period. The watershed consists of 48.6 ha (120 ac), one-third in rotation meadow and the remainder is in row crops on the contour. The hydrologic soil group is B, and the time of concentration is 30 min. The watershed is at your present location.

4.4. If the watershed in Problem 4.3 is in southern Alabama, what would be the peak runoff rate? If the return period was decreased to 2 years, what would be the rate of runoff?

4.5. Determine the depth (water) of snowmelt in one day for average conditions if a snow field is at an average elevation of 1830 m (6000 ft.). The nearest weather station at an elevation of 305 m (1000 ft) had a minimum temperature of 12°C and a maximum of 16°C.

4.6. Assuming that 15 degree-days (°C) were accumulated in 4 days, what depth of snow would be melted under average conditions if the snow had a water equivalent of 10 percent?

4.7. Determine the runoff volume in mm depth (in.) from a 50-year return period storm at your present location assuming Antecedent Rainfall Condition III, Hydrologic Soil-Cover Complex Curve Number 70, and duration of the critical storm 3 hr.

4.8. Determine the flood volume in ha-m (ac-ft) for a 50-year return period storm for the dormant season at your present location. The 364-ha (900-ac) watershed is in row crops, terraced on Hydrologic Soil Group C in good condition for one-third of the area, and the remaining two-thirds is in woodland on Hydrologic Soil Group D in poor condition. Assume 25 mm (1 in.) of rainfall occurred 3 days prior to the design storm and the critical storm duration is 24 hr.

4.9. Determine the minimum annual water yield from a 40.5-ha (100-ac) watershed near Coshocton, Ohio, which will occur once in 25 years. What would be the equivalent surface area of a reservoir to store this quantity of water, if the average depth was 2.4 m (8 ft)?

4.10. From Fig. 10.11 determine the minimum size drainage area for a farm pond at your present location if the storage capacity is 1.23 ha-m (10 ac-ft) and if seepage and evaporation are neglected. What is the equivalent water yield in depth over the drainage area?

4.11. Assuming the dimensionless hydrograph is applicable, determine the duration of flow if the peak runoff is 8.5 m³/s (300 cfs) and the volume of flow is 102 mm (4 in.) for a 50-year storm from an 85 ha (210-ac) watershed. What are the coordinates for point n on the design hydrograph?

4.12. Determine q, T_p, and T_b for a triangular unit hydrograph 25 mm (1 in.) of runoff for a 40.5-ha (100-ac) watershed at your present location if the storm duration is 2 hr and T_c is 2 hr (see Fig. 4.3b).

4.13. By frequency analysis methods discussed in Chapter 2 determine the estimated maximum annual discharge for return periods of 2 and 100 years from the following 18 years of maximum annual floods (1939–1956) from a gaged watershed: 53, 91, 109, 0.8, 0.8, 0.3, 20, 36, 0.5, 0.3, 56, 38, 0.3, 0.8, 0.3, 5, 0.3, and 3 mm (2.1, 3.6, 4.3, 0.03, 0.03, 0.01, 0.8, 1.4, 0.02, 0.01, 2.2, 1.5, 0.01, 0.03, 0.01, 0.2, 0.01 and 0.1 in.). Is the length of record adequate for these estimates to be reliable 90 percent of the time?

CHAPTER 5

Water Erosion and Control Practices

The two major types of erosion are geological erosion and man-made erosion. Geological erosion includes soil-forming as well as soil-eroding processes that maintain the soil in a favorable balance, suitable for the growth of most plants. Man-made erosion includes breakdown of soil aggregates and accelerated removal of organic and mineral particles resulting from improper tillage and removal of natural vegetation. Geological erosion has contributed to the formation of our soils and their distribution on the surface of the earth. This long-time eroding process caused most of our present topographic features, such as canyons, stream channels, and valleys.

Man-made erosion is caused primarily by water and wind. The forces involved are (1) attacking forces, which remove and transport the soil particles and (2) resisting forces, which retard erosion. Water erosion is the removal of soil from the land by running water, including runoff from melted snow and ice. Types of water erosion include raindrop, sheet, rill, gully, and stream channel erosion. Wind erosion is discussed in Chapter 6.

Erosion is one of the most important agricultural problems in the world. It is a primary source of sediments that pollutes streams and fills reservoirs. Erosion also adds to the removal of valuable plant nutrients lost with the runoff. Some estimates in the 1970s were as high as 4 billion Mg (metric tons) annually in the United States. This amount is about a 30 percent increase over that in the 1930s even though government subsidies, educational programs, and new practices have been instigated. In the early 1970s greater emphasis was given to erosion as a contributor to nonpoint pollution. Nonpoint refers to erosion from the land surface rather than from channels and gullies.

5.1. Factors Affecting Erosion by Water. The major variables affecting soil erosion are climate, soil, vegetation, and topography. Of these the vegetation and to some extent the soil may be controlled. The climatic factors and the topographic factors, except slope length, are beyond the power of man to control.

Climate. Climatic factors affecting erosion are precipitation, temperature, wind, humidity, and solar radiation. Temperature and wind are most evident through their effects on evaporation and transpiration. However, wind also changes raindrop velocities and the angle of impact. Humidity and solar radiation are somewhat less directly involved in that they are associated with temperature.

The relationship between precipitation characteristics, and runoff and soil loss is complex. In a comprehensive study of 19 independent variables, all measuring rainfall characteristics or interactions of combined characteristics, the most important single measure of the erosion-producing power of a rainstorm was the product, rainfall energy times maximum 30-min intensity; for runoff, it was rainfall energy for 24-hr antecedent precipitation. The results of these tests for three soils are reported in Table 5.1.

Soil.   Physical properties of soil affect the infiltration capacity and the extent to which it can be dispersed and transported. These properties that influence erosion include soil structure, texture, organic matter, moisture content, and density or compactness, as well as chemical and biological characteristics of the soil. As yet no one soil characteristic or index provides a satisfactory means of predicting erodibility.

Vegetation.   The major effects of vegetation in reducing erosion are (1) interception of rainfall by absorbing the energy of the raindrops and thus reducing runoff, (2) retardation of erosion by decreased surface velocity, (3) physical restraint of soil movement, (4) improvement of aggregation and porosity of the soil by roots and plant residue, (5) increased biological activity in the soil, and (6) transpiration, which decreases soil moisture, resulting in increased storage capacity. These vegetative influences vary with the season, crops, degree of maturity, soil, and climate, as well as with the kind of vegetative material, namely, roots, plant tops, and plant residues.

Table 5.1   Percent of Total Soil Loss Variation for Fallow Plots Explained by Various Rainfall Characteristics and Combinations of Characteristics

Variable	Shelby Soil 8% Slope (%)	Marshall Soil 9% Slope (%)	Fayette Soil 16% Slope (%)
(1) Amount of rain	73	39	42
(2) 15-min intensity	43	50	55
(3) 30-min intensity	56	56	80
(4) (1), (2), and (3)	79	66	83
(5) Rainfall energy	82	55	62
(6) Interaction product of (3) × (5)	89	71	88
(7) Four variables[a]	92	79	88

[a] Rainfall energy, energy times 30-min intensity, total energy since last cultivation, and antecedent precipitation index.

Source: Wischmeier and Smith (1958).

Topography.   Topographic features that influence erosion are degree of slope, length of slope, and size and shape of the watershed. On steep slopes high velocities cause serious erosion by scour and by sediment transportation.

5.2.   Raindrop Erosion.   Raindrop erosion is soil splash resulting from the impact of water drops directly on soil particles or on thin water surfaces. Although the impact on water in shallow streams may not splash soil, it does cause turbulence, providing a greater sediment-carrying capacity.

Tremendous quantities of soil are splashed into the air, most of it more than once. The amount of soil splashed into the air as indicated by the splash losses from small elevated pans was found to be 50 to 90 times greater than the washoff losses. On bare soil it is estimated that as much as 224 Mg/ha (100 tons/ac) are splashed into the air by heavy rains. The relationship between erosion and rainfall momentum and energy is determined by raindrop mass, size, size distribution, shape, velocity, and direction.

The energy equation that has been developed by Wischmeier and Smith (1958) is

$$E = 12.1 + 8.9 \log i \qquad (5.1)$$

where E = kinetic energy in m-Mg/ha-mm,
          i = intensity in mm/h.

The process of soil erosion involves soil detachment and soil transportation. The corresponding soil characteristics that describe the ease with which soil particles may be detached and transported are soil detachability and soil transportability. In general, soil detachability increases as the size of the soil particles increase, and soil transportability increases with a decrease in particle size. That is, clay particles are more difficult to detach than sand, but clay is more easily transported.

On level land raindrop splash is not serious, but on sloping fields considerably more soil is splashed downhill than uphill. This may account to a large extent for serious erosion on short, steep slopes. Rill erosion increases with the length of slope and is more serious at the lower end of the field, whereas raindrop erosion occurs over the entire area. The effects of slope or inclined rain are depicted in Fig. 5.1. The effect of a single drop as it strikes is shown in the high speed photograph in Fig. 5.2a; the successive steps that take place in drop-crater formation are illustrated in Fig. 5.2b. Splashed particles may move more than 0.6 m (2 ft) in height and more than 1.5 m (5 ft) laterally on level surfaces.

Factors affecting the direction and distance of soil splash are slope, wind, surface condition, and such impediments to splash as vegetative cover and mulches. On sloping land the splash moves farther downhill than uphill not only because the soil particles travel further, but also because the angle of impact

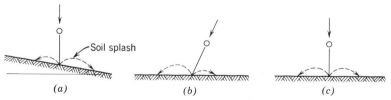

Fig. 5.1. Differential soil movement caused by raindrop splash. (a) Sloping land. (b) Inclined rainfall. (c) Vertical rainfall. (From Kohnke and Bertrand, 1959.)

causes the splash reaction to be in a downhill direction. Components of wind velocity up or down the slope have an important effect on soil movement by splash. Surface roughness and impediments to splash tend to counteract the effects of slope and wind. Contour furrows and ridges break up the slope and cause more of the soil to be splashed uphill. If raindrops fall on crop residue or growing plants, the energy is absorbed and thus soil splash is reduced. Raindrop impact on bare soil not only causes splash but also decreases aggregation and causes deterioration of soil structure. The washing-out of fine materials results in so-called erosion pavement, which is the result of an accumulation of coarse particles or rock fragments at the surface. This effect may take place with either of the principal erosive agents, water or wind.

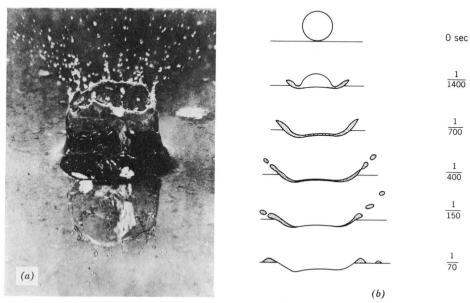

Fig. 5.2. Raindrop characteristics when striking moist soil. (a) The drop bursts upward and outward. (Courtesy ARS.) (b) Steps in drop crater formation. (After Mihara, 1952.)

5.3. Sheet Erosion. The idealized concept of sheet erosion has been that it was the uniform removal of soil in thin layers from sloping land, resulting from sheet or overland flow occurring in thin layers. Current fundamental studies of the mechanism of erosion, in which both time lapse and high speed photographic techniques have been used, indicate that this idealized form of erosion rarely occurs. Minute rilling takes place almost simultaneously with the first detachment and movement of soil particles. The constant meander and change of position of these microscopic rills obscure their presence from normal observation, hence establishing the false concept of sheet erosion.

The beating action of raindrops combined with surface flow causes this initial microscopic rilling. From an energy standpoint raindrop erosion is far more important because raindrops have velocities of about 6 to 9 m/s (20 to 30 fps), whereas overland flow velocities are about 0.3 to 0.6 m/s (1 to 2 fps). Raindrops cause the soil particles to be detached and the increased sediment reduces the infiltration rate by sealing the soil pores. Areas where loose shallow topsoil overlies a tight subsoil are most susceptible to erosion. The eroding and transporting power of sheet flow are functions of the depth and velocity of runoff for a given size, shape, and density of soil particle or aggregate.

5.4. Rill Erosion. Rill erosion is the removal of soil by water from small but well-defined channels or streamlets when there is a concentration of overland flow. Conventionally rill erosion occurs when these channels have become sufficiently large and stable to be readily seen. Rills are small enough to be easily removed by normal tillage operations. Although rill erosion is often overlooked, it is the form of erosion in which most soil erosion occurs. Detachability and transportability in rill erosion are more serious because of the higher runoff velocities that are involved. Rill erosion is most serious where intense storms occur on soils having high runoff-producing characteristics and loose shallow topsoil.

5.5. Gully Erosion. Gully erosion produces channels larger than rills. These channels carry water during and immediately after rains, and, as distinguished from rills, gullies cannot be obliterated by tillage. Thus, gully erosion is an advanced stage of rill erosion much as rill erosion is an advanced stage of sheet erosion.

The rate of gully erosion depends primarily on the runoff-producing characteristics of the watershed, the drainage area, soil characteristics, the alignment, size, and shape of the gully, and the slope in the channel.

A gully develops by processes that may take place either simultaneously or during different periods of its growth. These processes are (1) waterfall erosion at the gully head, (2) channel erosion caused by water flowing through the gully or by raindrop splash on unprotected soil, (3) alternate freezing and thawing of the exposed soil banks, and (4) slides or mass movement of soil in the gully. Four stages of gully development are generally recognized:

Stage 1.   Channel erosion by downward scour of the topsoil. This stage normally proceeds slowly where the topsoil is fairly resistant to erosion.

Stage 2.   Upstream movement of the gully head and enlargement of the gully in width and depth. The gully cuts to the *C* horizon, and the weak parent material is rapidly removed. A waterfall often develops where the flow plunges from the upstream segment to the eroded channel below.

Stage 3.   Healing stage with vegetation beginning to grow in the channel.

Stage 4.   Stabilization of the gully. The channel reaches a stable gradient, gully walls reach a stable slope, and vegetation begins to grow in sufficient abundance to anchor the soil and permit development of new topsoil. The healing stage is a necessary prelude to stabilization and one stage grades into the other.

During the two latter stages the gully head has progressed toward the upper end of the watershed, and the rate of runoff into the gully head decreases because the drainage area is reduced. The remainder of the runoff enters at many points along the length of the gully.

Evaluation and prediction of gully development is difficult because the factors are not well defined and field records of gullying are inadequate. From aerial photographs and field topographic surveys Beer and Johnson (1963) developed a prediction equation for the deep loess region in western Iowa.

Of the several systems of gully classification, the one given in Table 5.2 is based on an arbitrary classification of gully sizes and drainage areas. Another system classifies gullies with respect to their cross sections. Gully cross sections may be V- or U-shaped, depending upon soil and climatic conditions, age of the gully, and type of erosion. U-shaped gullies may be found in loessial regions and alluvial valleys where both the surface soil and the subsoil are easily eroded. Under such conditions gullies tend to develop vertical walls that result from undermining and collapse of the banks. The scouring of soil by concentrated runoff in unprotected depressions results in V-shaped gullies

Table 5.2   Description of Gullies and Drainage Areas

Relative size	Gully Depth		Drainage Area	
	(m)	(ft)	(ha)	(ac)
Small	1 or less	(3 or less)	2 or less	(5 or less)
Medium	1 to 5	(3 to 15)	2 to 20	(5 to 50)
Large	5+	(15+)	20+	(50+)

having sloping heads. Such gullies may develop where the subsoil is resistant to erosion. Both V and U shapes are commonly found in the same channel. More precise classifications based on gully shape have been proposed by Ireland et al. (1939).

As gullies work upstream, the most active portion is near the upper end or at the gully head, whereas the most stable section of the gully is generally at the lower end. Active gullies are gullies that continue to enlarge. They may be identified by the presence of bare soil exposed on the side slopes.

5.6. Stream Channel Erosion.   Stream channel erosion consists of soil removal from stream banks or soil movement in the channel. Stream channel erosion and gully erosion are distinguished primarily in that channel erosion applies to the lower end of headwater tributaries and to streams that have nearly continuous flow and relatively flat gradients whereas gully erosion generally occurs in intermittent streams near the upper ends of headwater tributaries.

Stream banks erode either by runoff flowing over the side of the stream bank or by scouring and undercutting below the water surface. Stream bank erosion, less serious than scour erosion, is often increased by the removal of vegetation, by overgrazing, or by tilling too near the banks. Scour erosion is influenced by the velocity and direction of flow, depth and width of the channel, and soil texture. Poor alignment and the presence of obstructions such as sandbars increase meandering, the major cause of erosion along the bank.

Sediment in streams is transported by suspension, by saltation, and by bed load movement. Sediments loads may be estimated by resurveying reservoir bottoms and by sampling the flow of streams. Variables affecting sediment movement include velocity of flow; turbulence; size distribution, diameter, cohesiveness, and specific gravity of transported materials; channel roughness; obstructions to flow; and the availability of materials for movement.

Suspension.   Suspended sediment is that which remains in suspension in flowing water for a considerable period of time without contact with the stream bed. Formulas have been derived to express the concentration of sediment at any point in the vertical direction when the concentration at some reference level is known.

A typical relationship showing velocity, sediment concentration, and sediment discharge for different depths at the center of a straight natural channel is shown in Fig. 5.3. The velocity and concentration of sediment vary with depth as shown in (a) and (b); the quantity of sediment discharge is a product of velocity and concentration, as illustrated in (c). Although the velocity is a minimum at the channel floor, the high concentration of sediment near the bed accounts for the maximum sediment discharge at a point just above the bottom of the channel. The distribution of fine sediment is more nearly uniform with depth than is that of coarse material.

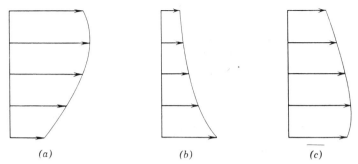

(a)   (b)   (c)

Fig. 5.3. (a) Velocity, (b) sediment concentration, and (c) suspended sediment discharge distribution in a straight natural stream. (Redrawn from Subcommittee on Sedimentation, 1948.)

Saltation. Sediment movement by saltation occurs where the particles skip or bounce along the stream bed. The height of bounce, expressed in mathematical form, is directly proportional to the ratio of particle density to fluid density. Particles in water rise only about 1/1000 of that in air (Fig. 6.5), or, for most practical conditions, a few particle diameters. In comparison to the total sediment transported, saltation is considered relatively unimportant.

Bed Load. Bed load is sediment that moves in almost continuous contact with the stream bed, being rolled or pushed along the bottom by the force of the water. Laboratory studies have shown that the critical threshold velocity (competent velocity) required to initiate movement of particles in the bottom of a stream is expressed by the empirical equation

$$v_t = 0.152 \, d^{4/9} \, (G - 1)^{1/2} \tag{5.2}$$

where v_t = threshold velocity in m/s,
    d = particle diameter in mm,
    G = specific gravity of the particles.

This equation was developed by Mavis (1935) for unigranular materials ranging in diameter from 0.35 to 5.7 mm and in specific gravity from 1.83 to 2.64.

 Though saltation and bed load are distinct types of movement, saltation is usually included in bed load when sediment transportation is described. Although considerable data have been gathered on sediment transported by streams, most measurements are limited to the collection of suspended sediment since bed load is more difficult to determine. Bed loads in the Colorado River varied from 12 to 50 percent of the total load and the percentage by weight of suspended material in the water ranged from 0.13 to 2.81 percent based on monthly averages over a period of 15 years.

SOIL LOSSES

Soil losses vary considerably with the type of erosion. Often both wind and water erosion are present in the same area. For example, in the Texas Panhandle region Finnell (1951) reported that on slopes of 2 percent or less, 89 percent of erosion was caused by wind and 11 percent by water. In more humid areas losses are nearly all water related.

The importance of soil losses is indicated by the effect of topsoil depth on crop yield, as shown in Fig. 5.4. On some soils these crop yield decreases can be largely overcome by high fertilization.

From watershed studies in Ohio the annual and monthly soil losses for the corn years are shown in Fig. 5.5. These areas were approximately 0.6 ha (1.5 ac) in size and were located on slopes from 6 to 15 percent. The soil losses from watersheds with recommended conservation practices are compared to the soil losses from the corresponding check watersheds representing typical Ohio practices. In the example in Fig. 5.5 the losses from the check watershed are more than double those from the corresponding conservation watershed. With a few exceptions, erosion from the conservation watersheds is considerable less than that from the check areas.

5.7. Universal Soil Loss Equation (USLE).   This equation was developed from more than 40 years of data measured from small plots located in many states. It is useful to determine the adequacy of conservation measures in farm planning

USLE — for single slope areas
— assesses relative effects of cons. practices

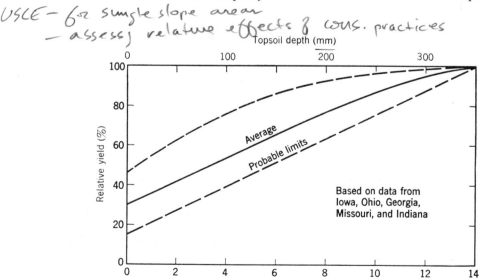

Fig. 5.4.   Effect of topsoil depth on relative crop yield. (Based on data by Stallings, 1950.)

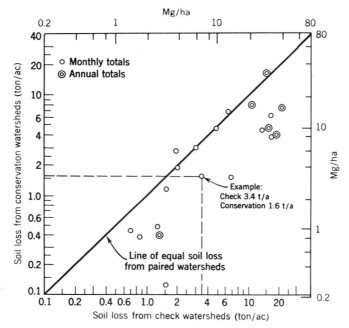

Fig. 5.5. Annual and monthly soil loss from small paired watersheds (corn years only). (Redrawn from Harrold, 1949.)

and to predict nonpoint sediment losses in pollution control programs. Despite its simplification of the many variables involved, the USLE is the most widely accepted method of estimating sediment loss.

The average annual soil loss, as determined by Wischmeier (1976), can be estimated from the equation

$$A = 2.24RKLSCP \tag{5.3}$$

where A = average annual soil loss in Mg/ha (metric tons/ha),

R = rainfall and runoff erosivity index by geographic location as given in Fig. 5.6 or Table 5.3,

K = soil-erodibility factor (see Table 5.4), which is the average soil loss in t/a per unit of erosion index for a particular soil in cultivated continuous fallow with an arbitrarily selected slope length l of 22 m (73 ft) and slope steepness S, of 9 percent (if K is Mg/ha, change constant 2.24 to 1.0),

LS = topographic factor evaluated in Fig. 5.7 and Eqs. 5.4 and 5.5,

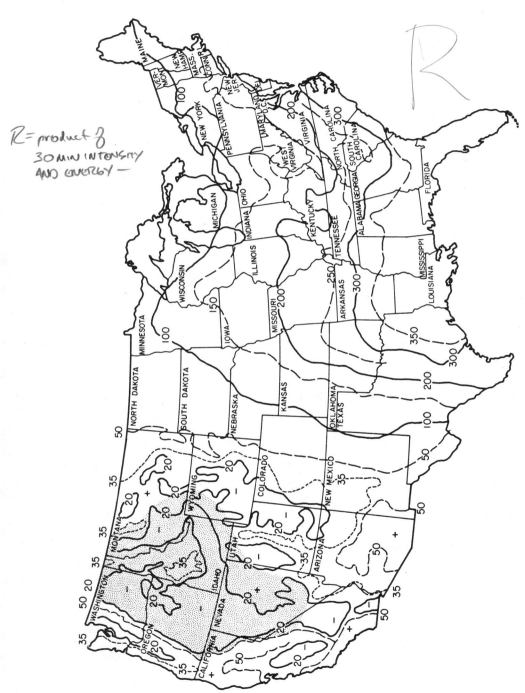

Fig. 5.6. Rainfall and runoff erosivity index R by geographic location. (From USDA and EPA, 1975.)

R = product of
30 MIN INTENSITY
AND ENERGY —

Table 5.3   Frequency of Annual and Single-Storm Erosion Index *R*

	Return Period in Years			
Location	2	5	10	20
ANNUAL EROSION INDEX, *R*				
Little Rock, Ark.	308	422	510[a]	569
Indianapolis, Ind.	166	225	275[a]	302
Devils Lake, N.D.	56	90	120[a]	142
SINGLE-STORM EROSION INDEX, *R*				
Little Rock, Ark.	69	115	158	211
Indianapolis, Ind.	41	60	75	90
Devils Lake, N.D.	27	39	49	59

[a] Interpolated values.
Source: Wischmeier and Smith (1965).

C = cropping-management factor, which is the ratio of soil loss for given conditions to soil loss from cultivated continuous fallow as given in Table 5.5,

P = conservation practice factor, which is the ratio of soil loss for a given practice to that for up and down the slope farming as given in Table 5.6.

The index factor R found by Wischmeier (1959) to be most highly correlated with soil loss for a fallow condition (see (6) in Table 5.1), was a product of the kinetic energy of the storm and the maximum 30-min intensity. This product

Table 5.4 K, Soil-Erodibility Factor by Soil Texture in t/a[a]

	Organic Matter Content (%)		
Textural Class	0.5	2	4
Fine sand	0.16	0.14	0.10
Very fine sand	0.42	0.36	0.28
Loamy sand	0.12	0.10	0.08
Loamy very fine sand	0.44	0.38	0.30
Sandy loam	0.27	0.24	0.19
Very fine sandy loam	0.47	0.41	0.33
Silt loam	0.48	0.42	0.33
Clay loam	0.28	0.25	0.21
Silty clay loam	0.37	0.32	0.26
Silty clay	0.25	0.23	0.19

[a] Selected from USDA-EPA, Vol. I (1975) and are estimated averages of specific soil values. For more accurate values by soil types use local recommendations of Soil Conservation Service or state agencies. (1 t/a = 2.24 Mg/ha)

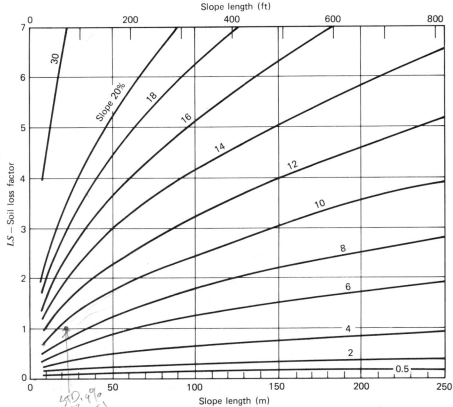

LS

Fig. 5.7. LS soil loss factor by length and steepness of slope. (From Wischmeier and Smith, 1965.) *length is from ridge to point where deposition begins*

was called the rainfall and runoff erosivity index given in Fig. 5.6. For estimation purposes these values are satisfactory for average annual losses. For annual soil losses for return periods from 2 to 20 years as well as for single-storm losses use R values given in Table 5.3. For the single-storm losses other factors in the soil loss equation should be determined for the specific conditions existing in the field at the time of the storm. USDA Hbk. 282 gives R values for 181 locations in the United States of which only three are shown in Table 5.3.

The soil-erodibility factor K for a series of bench-mark soils was obtained by direct soil loss measurements. Correlation of soil loss with physical and chemical properties is not yet fully developed. Average K values for soi's of different texture are given in Table 5.4. This factor is the only one in Eq. 5.3 that has dimensions.

The topographic factor LS as given by Smith and Wischmeier (1962) adjusts

Table 5.5 Ratio of Soil Loss from Crops to Corresponding Loss from Continuous Fallow[a]

Cover, Sequence, and Management	Crop Yields		Crop-Stage Period[b]				
	Meadow (tons)	Corn (bu)	0 (%)	1 (%)	2 (%)	3 (%)	4 (%)
1st-yr corn after meadow, RdL[c]	2	60	15	30	27	15	22
2nd-yr corn after meadow, RdL	3	70	32	51	41	22	26
2nd-yr corn after meadow, RdR[d]	3	70	60	65	51	24	65
3rd- or more yr corn, RdL	—	70	36	63	50	26	30
Small grain w/meadow seeding:							
(1) In disked corn residues							
After 1st-corn after meadow	2	60	—	30	18	3	2
After 2nd corn after meadow	2	60	—	40	24	5	3
(2) On disked corn stubble, RdR							
After 1st corn after meadow	2	—	—	50	40	5	3
After 2nd corn after meadow	2	—	—	80	50	7	3
Established grass and legume meadow	3	—	—	—	0.4	—	—

[a] Portion of 100-line published table (Wischmeier, 1960).

[b] Crop-stage periods are defined below:

    0 Turnplowing to seedbed preparation.

    1 Seedbed—first month after seeding.

    2 Establishment—second month after seeding.

    3 Growing cover—from 2 months after seeding to harvest.

    4 Stubble or residue—harvest to plowing or new seedbed.

[c] RdL, crop residues left and incorporated by plowing.

[d] RdR, crop residues removed.

Source: Smith and Wischmeier (1962).

Table 5.6   Recommended Conservation Practice Factors P^a

Percent Slope	P_c Contouring (maximum slope length in m)	P_{sc} Strip Cropping[b]	P_{tc} Terracing and Contouring[c]
Parallel to Field Boundary	0.8[d]	—	—
1.1–2	0.6 (150)	0.30	—
2.1–7	0.5 (100)	0.25	0.10
7.1–12	0.6 (60)	0.30	0.12
12.1–18	0.8 (20)	0.40	0.16
18.1–24	0.9 (18)	0.45	—

[a] Factor for up and down slope is 1.0.
[b] A system using 4-year rotation of corn, small grain, meadow, meadow. Use with terraces for farm planning.
[c] Recommended only for computing soil loss from the field or loss to the terrace channel with upslope plowing.
[d] For slopes up to 12% only.
Source: Wischmeier and Smith (1965).

the soil loss from the standard length of 22 m (73 ft) and 9 percent slope. These factors can be calculated from the equations

$$L = (l/22)^x \tag{5.4}$$

and

$$S = \frac{(0.43 + 0.30s + 0.043s^2)}{6.574} \tag{5.5}$$

where x = a constant, 0.5 for slopes >4 percent, 0.4 for 4 percent, and 0.3 for <3 percent
   l = slope length in m, (Top of ridge to place where deposition begins)
   s = field slope in percent

The product *LS* can be read directly from Fig. 5.7 for $x = 0.5$. The slope length is measured from the point where surface flow originates (usually the top of the ridge) to the outlet channel or a point down slope where deposition begins.

The cropping-management factor *C* includes the effects of cover, crop sequence, productivity level, length of growing season, tillage practices, residue management, and the expected time distribution of erosive rainstorms. The values for the cropping and residue variables as a ratio of the soil loss for crops to that for continuous fallow are given in Table 5.5. The time distribution of the rainfall-erosion index will vary with geographic location. Examples are given in Fig. 5.8.

The conservation practice factor P is given in Table 5.6 for contouring, strip cropping, and terracing. The reduction in soil loss at a given slope is about 50 percent for the next more intensive practice. In humid regions strip cropping and terracing by definition include contouring. These practices involve engineering design, while the other erosion factors are more agronomic in nature. As will be discussed in detail in Chapter 8, terracing affects the slope length so that the factor L must also be adjusted to the terrace interval when computing soil loss for terracing. The terracing factor applies to the soil loss from the outlet of the terrace. Compared to up- and down-slope farming, terraces are extremely effective. If the soil loss to the terrace channel is desired, the contouring factor applies rather than that for terracing. If part of the soil deposited in the channel is not considered lost or if upslope plowing is practiced, use the strip cropping factor. In either case the slope length factor must also be reduced to that for the horizontal terrace interval. Terraces by themselves do not control erosion from the slope; this control must be obtained by contouring or good crop and soil management. For soil losses with terraces for farm planning

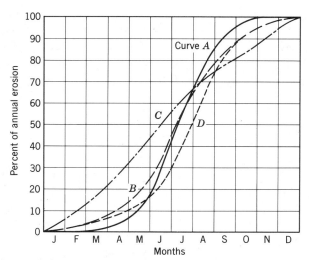

Fig. 5.8. Monthly distribution of the rainfall and runoff erosivity index: Curve A, northwestern Iowa, northern Nebraska, southeastern South Dakota. Curve B, northern Missouri, and central Illinois, Indiana, and Ohio. Curve C, Louisiana, Mississippi, western Tennessee, and eastern Arkansas. Curve D, Atlantic Coastal Plains of Georgia and the Carolinas. (Redrawn from Smith and Wischmeier, 1962.)

purposes, the strip cropping factor should be selected. The maximum slope lengths for which the contouring factors apply are shown in Table 5.6.

By evaluating the factors in Eq. 5.3 the predicted soil loss can be determined for a given set of conditions. If the soil loss is higher than the minimum required to maintain productivity, it may be reduced by changing the crop management practices or the conservation practices. Both physical and economic factors as well as social problems need to be considered in establishing soil-loss tolerances, called T values. These values vary by topsoil depth from 4.5 Mg/ha (2 t/a) to 11 Mg/ha (5 t/a). They are the maximum rates of soil erosion that will permit a high level of crop productivity to be sustained economically and indefinitely. Criteria for control of sediment pollution may dictate even lower T values.

Example 5.1.   Determine the soil loss for the following conditions: Location Memphis, Tennessee, $K = 0.1$ t/a, $l = 122$ m (400 ft), $s = 10$ percent, $C = 0.18$ (approximate for corn-corn-oats-meadow rotation with good management), and the field is to be contoured.
Solution.   From Fig. 5.6 read $R = 310$ as the rainfall erosion index, from Fig. 5.7 read $LS = 2.7$, and from Table 5.6 read $P_c = 0.6$. Substituting in Eq. 5.3,

$$A = 2.24 \times 310 \times 0.1 \times 2.7 \times 0.18 \times 0.6 = 20.2 \text{ Mg/ha (9.0 t/a)}$$

Example 5.2.   If the soil-loss tolerance for the conditions in Example 5.1 is 3.0 t/a, what practices could be adopted to accomplish this reduction?
Solution.   Since C and P are the only practices that can be changed, the following combinations are possible by substituting the appropriate factors in Eq. 5.3:

Conservation Practice	P factor	C factor	Soil Loss (t/a)	Remarks
(1) Strip crop	0.3	0.18	4.5	Soil loss too high
(2) Strip crop	0.3	0.12 max.	3.0	Satisfactory
(3) Terracing	0.12	0.18	0.7	Soil loss from field or same as that to the terrace channel with upslope plowing
(4) Terracing	0.12	0.81 max.	3.0	Soil loss from field
(5) Terracing	0.6	0.16 max.	3.0	P_c for soil loss to terrace channel
(6) Terracing	0.3	0.32 max.	3.0	P_{sc} for farm planning

Note that strip cropping (1) is not adequate; however, in line (2) the 3.0 t/a soil loss with strip cropping is possible by reducing the cropping-management factor C to 0.12 or less. In line (3), terracing with $C = 0.18$ would give a field soil loss lower than required. Thus, with terracing, the cropping-management factor may be increased to 0.81 without exceeding the permissible soil loss of 3.0 t/a from the field. In lines (3), (4), (5), and (6) the terrace slope length is assumed to be 60 feet (see Chapter 8), which reduces the LS factor to 1.0. If the soil loss to the terrace channel is to be 3.0 t/a or less, C factor in line (5) should not be exceeded.

Example 5.3. If strip cropping is selected as the most desirable conservation practice in Example 5.2, what recommendations could be made for the cropping-management practice?
Solution. From line (2) in Example 5.2 the maximum C factor to keep soil loss to 3.0 t/a is 0.12. Assuming a corn-oats-meadow rotation, fall plowing, and corn residue plowed under, the following calculations were made using values for Columns 3 and 4 from Table 5.5 and Fig. 5.8 (Curve C), respectively:

Crop	Months	Soil Loss in Percent of Continuous Fallow (Table 5.5)	Percent of Annual Erosion (Fig. 5.8)	C Factor
Corn, 1st year	Jan–May	15	43	0.065
	June	30	13	0.039
	July	27	10	0.027
	Aug–Sept	15	22	0.033
	Oct–Dec	22	12	0.026
Small grain,	Jan–Feb	22	13	0.029
with	Mar	30	9	0.027
meadow	Apr	18	10	0.018
seeding	May–Oct	3	63	0.019
	Nov–Dec	0.4	5	0.000
Meadow	Jan–Oct	0.4	95	0.004
	Nov–Dec	22	5	0.011
			Total	0.298

Average C for three years = 0.100

Since $0.100 < 0.12$, solution is satisfactory.

The C factor in Column 5 is the product of the values in Columns 3 and 4. Other combinations of crops and management practices could be selected. If terraces were installed as shown in Example 5.2, continuous corn could be grown without exceeding a C factor of 0.81. In addition to soil losses other considerations, such as cost of terracing, would enter into the decision.

Example 5.4.   Determine the soil loss from a single 10-year return period storm for the conditions in Example 5.1 except for the crop management factor. The 0.18 factor is an average value for the 4-year rotation and does not apply at the time the storm occurred. The storm came during the second corn year in the rotation and during the first month (Period 1) after seeding. All residue from the previous crop was left and incorporated by plowing.
Solution.   From Table 5.3 read $R = 158$ for nearest station at Little Rock, Ark., and from Table 5.5 read $C = 0.51$. Substitute in Eq. 5.3,

$$A = 2.24 \times 158 \times 0.1 \times 2.7 \times 0.51 \times 0.6 = 29.2 \text{ Mg/ha (13.1 t/a)}$$

Sediment production downstream in a watershed may be estimated from the USLE and the sediment delivery ratio. The soil loss equation estimates gross sheet and rill erosion, but does not account for sediment deposited enroute to the place of measurement nor for gully or channel erosion downstream. The sediment delivery ratio is defined as the ratio of sediment delivered at a location in the stream system to the gross erosion from the drainage area above that point. This ratio varies widely with size of area, steepness, density of drainage network, and many other factors. The sediment delivery ratio varies roughly as the inverse of the 0.2 power of the drainage area. For watersheds of 2.6, 130, 1300, and 26 000 ha (0.01, 0.5, 5, and 100 square miles), ratios of 0.65, 0.33, 0.22, and 0.10, respectively, were suggested as average values by Roehl (1962).

EROSION CONTROL PRACTICES

In this chapter the primary emphasis is on controlling erosion by water from cultivated farm land and other nonpoint sources. Vegetated waterways and terraces are important control measures, but they will be discussed in later chapters, as specialized design procedures are required. Control measures are often necessary on range and pasture lands and in forests even though erosion may not be a serious problem. Construction sites, roadways, mine spoil banks, or other exposed soil surfaces produce much sediment. Growing vegetation, crop residues, and mulches that protect the soil surface are the most effective control measures.

5.8. Contouring.   This practice is that of performing field operations, such as plowing, planting, cultivating, and harvesting approximately on the contour. It reduces surface runoff by impounding water in small depressions and decreases the development of rills in which the high water velocity results in destructive erosion. The distribution of contouring in the United States is shown in Fig. 5.9. The greatest concentration is in the midcentral states where the major benefit is the conservation of moisture.

The relative effectiveness of contouring for controlling erosion on various

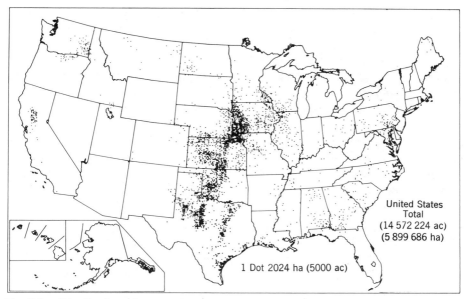

United States
Total
(14 572 224 ac)
(5 899 686 ha)

1 Dot 2024 ha (5000 ac)

Fig. 5.9.   Distribution of contouring in the United States (1969). (Courtesy U.S. Bureau of the Census.)

slopes is shown by the conservation practice factor P given in Table 5.6. These values are applicable to fields relatively free from gullies and depressions other than grassed waterways and for slope lengths shown in Table 5.6. If lister or ridge cultivation is practiced, the storage capacity of the furrows is materially increased and the conservation practice factor is reduced to about half that shown for contouring. Contouring alone on steep slopes or under conditions of high rainfall intensity and soil erodibility will increase gullying because row breaks may release the stored water. Break-overs cause cumulative damage as the volume of water increases with each succeeding row.

The effectiveness of contouring is also impaired by changes in infiltration capacity of the soil owing to surface sealing. Depression storage is reduced after tillage operations cease and settlement takes place. Studies by Harrold (1947) showed that contour cultivation together with good sod waterways reduced watershed runoff 75 to 80 percent at the beginning of the season. This reduction dropped to as low as 20 percent at the end of the year, leaving an annual average reduction in runoff, due to contouring, of 66 percent.

Because of varying slopes in most fields, all crop rows cannot be on the true contour. To establish row directions a guide line (true contour) is laid out at one or more elevations in the field. On small fields of uniform slope, one guide line may be sufficient. Another guide line should be established if the slope along the row direction exceeds 1 to 2 percent in any row laid out parallel to the guide line. A small slope along the row is desirable to prevent runoff from a large storm breaking over the small ridges. Where practical, field boundaries for contour farming should be relocated on the contour or moved so as to eliminate odd-shaped fields that would result in short, variable-length rows, called point rows. Hand levels or Abney levels are satisfactory for establishing contour guide lines and strip cropping.

In plowing a field start backfurrows on the guide lines and work out from them until about half the area is plowed. Next plow the land between, leaving deadfurrows as shown in Fig. 5.10, Method 1. The advantage of this method is that it practically eliminates turning on plowed ground. If the field has been in an intertilled crop and the point rows are visible, Method 2 in Fig. 5.10 may be more satisfactory. This procedure will establish the deadfurrows on the guide line where the backfurrows were located on the previous plowing.

5.9. Strip cropping.   Strip cropping is the practice of growing alternate strips of different crops in the same field. For controlling water erosion, the strips are always on the contour, but in dry regions strips are placed cross-wise to the prevailing wind direction for wind erosion control. The distribution of strip

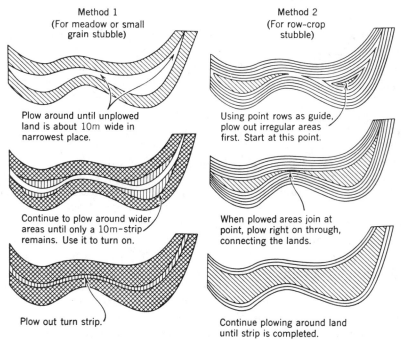

Fig. 5.10.   Methods of plowing out point row areas. (Redrawn from Hay, 1948).

cropping in the United States is shown in Fig. 5.11. The greatest concentration is in Montana and North Dakota where wind erosion is prevalent.

The three general types of strip cropping shown in Fig. 5.12 are (a) contour strip cropping with layout and tillage held closely to the exact contour and with the crops following a definite rotational sequence, (b) field strip cropping with strips of a uniform width placed across the general slope; when used with adequate grassed waterways the strips may be placed where the topography is too irregular to make contour strip cropping practical (Field strip cropping for wind erosion control consists of parallel strips placed crosswise to the direction of the prevailing wind.), and (c) buffer strip cropping with strips of some grass or legume crop laid out between contour strips of crops in the regular rotations; they may be even or irregular in width; they may be placed on critical slope areas of the field. Their main purpose is to give protection from erosion. The type used depends on cropping systems, topography, and type of erosion hazards.

Rotations that provide strips of close-growing perennial grasses and legumes alternating with grain and intertilled crops are the most effective. Their effectiveness in reducing runoff and soil loss is illustrated in Table 5.7 in which a 4-year rotation is compared with continuous cotton.

The three general methods of laying out contour strip cropping are (1) both

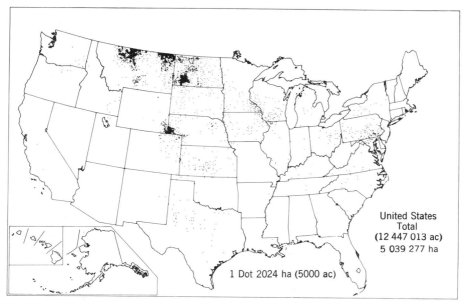

United States
Total
(12 447 013 ac)
5 039 277 ha

1 Dot 2024 ha (5000 ac)

Fig. 5.11. Distribution of strip cropping in the United States for both water and wind erosion control (1969). (Courtesy U.S. Bureau of the Census.)

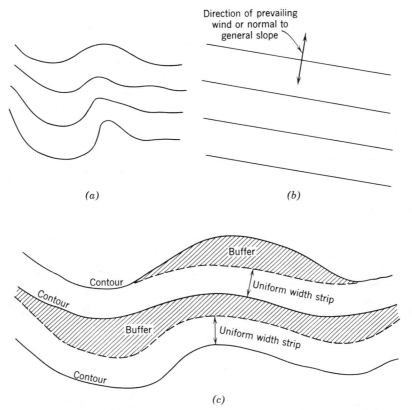

Fig. 5.12. Three types of strip cropping: (a) contour, (b) field, and (c) buffer.

Table 5.7 Runoff and Soil Losses on Class III Land with 7 Percent Slope at Watkinsville, Georgia

Rotation	Runoff (mm) (in.)	Soil Loss (Mg/ha) (t/a)
First year fescue	104 (4.1)	2.2 (1.0)
Second year fescue	15 (0.6)	0 (0)
Corn	48 (1.9)	2.0 (0.9)
Cotton	99 (3.9)	7.4 (3.3)
Rotation average	66 (2.6)	2.9 (1.3)
Continuous cotton	254 (10.0)	30.9 (13.8)

Source: USDA-ARS, Agr. Inf. Bull. 269 (1963).

edges of the strips on the contour; (2) one or more strips of uniform width laid out from a key or base contour line; and (3) alternate uniform width and variable width correction or buffer strips. Methods of layout vary with topography and individual's preference. The initial guide line may be laid out from top, bottom, or middle of the slope. A maximum deviation of 1 to 2 percent slope from the true contour is permitted.

The recommended widths for contour strip cropping shown in Table 5.8 will reduce the computed soil loss as indicated by the corresponding conservation practice factor P_{sc}. As strips should be of a width that is convenient to farm with multiple-row equipment, they should correspond to some multiple of implement width.

In some areas insect damage, due to the extended exposed crop borders, has proved to be a serious disadvantage of strip cropping. Establishment of rotations that give a minimum of protective harbor to insects, spray programs, and other approved insect-control measures reduce this problem. Crop damage from chinch bugs can often be avoided by growing corn or small grain with meadow in alternating strips in separate fields, thus eliminating the bordering corn and small grains.

Grass strips and large meadow outlets can be grazed through use of portable electric fencing. Grazing may also be facilitated by establishing interfield rotations that provide that each year one field will be entirely in meadow owing to the first and second years' meadows being adjoining.

There are a number of secondary factors that should be taken into account when considering contour systems of farming. Barger (1938) found that while power requirements are about equal for contour versus up- and downgrade operations, savings in time and fuel consumption averaged 13 and 9 percent, respectively, in favor of contour operations.

Up- and downgrade operations contribute to other inefficiencies such as wheel slippage, stops for gear changes, and tractor and implement wear due to

— MAKE WIDTH FIT EQUIP
— SHOOT FOR EVEN NUMBER OF ROWS

Table 5.8 Recommended Widths for Contour Strips

Slope (%)	P_{sc}	Width of Strips[a]		
		(m)	(ft)	FIELD STRIP
2	0.30	30	100	
6	0.25	27	90	SAME
10	0.30	24	80	
14	0.40	21	70	NOT
18	0.45	18	60	RCMMND

[a] Note: Width of four 30-in. crop rows is 3.0 m (10 ft) and of four 42-in. rows is 4.3 m (14 ft). Make strip widths multiples of these for 4-row equipment.

Source: Wischmeier and Smith (1965).

WI SCS

SLOPE	CONTOUR	FIELD STRIP
2-6	110'	110'
6-12	88'	88'
12-20	74'	NR
>20	60	

FIELD STRIP IS ⊥ TO WIND,
NO RESPECT TO CONTOUR

R WIDTH = (168-75)
         / %

starts and stops. Operation of powered implements such as combines and balers is simplified when implement speed is uniform.

5.8. Tillage Practices. Tillage is the mechanical manipulation of the soil to provide soil conditions suited to the growth of crops, the control of weeds, and for the maintenance of infiltration capacity and aeration. Traditionally tillage has consisted of cutting loose, granulating, and inverting the plow furrow slice, thus turning under the residues. While the essential basis for tillage is the preparation of a seedbed, the role of tillage has become more important as a conservation measure.

Reduced losses of water, soil, and plant nutrients as the result of contouring or strip cropping reflect only those that do not actually leave the field. Excess soil and water movement is still found within the cultivated strips. The soil from this year's cultivated strip is deposited at the top of the close-growing strip below it. Next year, when this second strip is cultivated, this and other soil will move on down the hill another 18 to 30 m. Similarly, there is movement of soil from the terrace intervals to the terrace channels. On some topography, contouring, strip cropping, and terracing may be impractical, thus necessitating alternative conservation measures.

Tillage is a primary tool in applying effective conservation to the land. Its primary purpose is to provide an adequate soil and water environment for the plant. Its role as a means of weed control has diminished with increased use of herbicides and improved timing of operations.

The effect of tillage upon erosion is a function of its effect upon such factors as aggregation, surface sealing, infiltration, and resistance to wind movement. Where intensive or excessive tillage has destroyed structure, increase in the erosion hazard results.

Tillage may also contribute heavily to erosion through its mechanical movement of the soil during the tillage process. A slope of about 10 percent, when plowed on the contour throwing the furrow uphill, gave 19 percent more soil movement uphill than soil movement downhill compared to throwing the furrows downhill; however, the width of cut was the same (Mech, 1942). Soil movement downhill during all harrowing and cultivating operations increased as the slope increased. It was concluded that turning furrows uphill during contour plowing was the most satisfactory means of compensating for the downhill movement of the soil during cultivation, harrowing, and the normal seasonal erosion processes.

Studies in Indiana utilizing artificial rainfall, showed that minimum tillage significantly influenced infiltration rates and erosion losses. Tests two to three weeks after corn planting and with three different antecedent moisture levels showed that of the 132 mm of water applied, infiltration totaled 56 mm for the conventional treatment and 84 mm for the minimum tillage treatment. Thus, runoff was 76 mm and 48 mm, respectively. According to Meyer (1961), soil loss totaled 37.4 Mg/ha from conventional tillage and 19.5 Mg/ha for minimum tillage. One of the major benefits of minimum tillage is the increased residue left on the surface. Such residue is extremely effective in reducing erosion.

The effectiveness of listing as a conservation measure has been shown in many studies. Iowa reports that on an erosive loess soil, over a 5-year period, contour listing of corn cut soil loss to about one-ninth that of uphill and downhill planting; water losses were reduced 61 mm.

As a general practice, subsoiling has not resulted in large yield increases or vastly improved soil conditions. Where subsoiling has been applied to problems of pans, soles, and other specialized profile conditions, more significant results have been obtained. For example, in western irrigated lands, it has been found that on soils having a compacted plow pan, on stratified soils, and on soils having relatively thin compacted or cemented layers, subsoiling or deep plowing does loosen the profile to improve leaching with irrigation water, at least in the initial years after treatment. The length of the period of effectiveness varies widely with soil conditions.

The effectiveness of all subsoiling and deep tillage is highly dependent upon the specific soil characteristics of treated areas, moisture conditions, crop management practices, secondary tillage operations, etc. Time of treatment is also important, all studies having indicated that the most effective results are obtained when the soil conditions are dry, thus contributing to the shattering action of subsoiling.

5.10. Soil and Water Conservation Districts. The purpose of the soil and water conservation district in the U.S. is to provide a local group organization for the conservation of soil, moisture, and related resources and to promote better land use. As a condition to receiving benefits under the Soil Conservation and Domestic Allotment Act, passed by Congress in 1935, the states were required to enact suitable laws providing for the establishment of soil conservation districts. The first district was organized in 1937, and all states have enacted district laws.

This legislation has been patterned largely after the standard state soil conservation districts law prepared by the U.S. Soil Conservation Service (1936). Most districts are called soil and water conservation districts. After such districts are established in local communities, they may request technical assistance from such agencies as the Soil Conservation Service, State Cooperative Extension Service, and county officials for carrying out erosion control and other land-use management activities.

Provided they are not in contradiction to other state laws a soil and water conservation district may have the power, (1) to conduct surveys, investigations, and research relating to soil erosion control programs; (2) to conduct demonstrational projects; (3) to carry out preventative and control measures on the land; (4) to cooperate and make agreements with farmers and to furnish technical and financial aid; (5) to make available to land occupiers, through sale or rent, machinery, equipment, fertilizers, and so on; (6) to develop conservation plans for farms; (7) to take over erosion control projects either state or federal; (8) to lease, purchase, or acquire property in order to carry out objectives of the program; and (9) to sue or be sued in the name of the district. In many states the district has the power of eminent domain. In other states the powers of these districts have been broadened. In practice the development of conservation farm plans and technical assistance in adopting conservation practices have been the principal results of the district setup.

The supervisors also have authority in some states to formulate land-use regulations, provided they are approved by a majority of the land occupiers by a referendum vote. In three states the vote must carry by 90 percent. These regulations may include provisions for carrying out terracing, building ponds, installing conservation structures, and adopting various types of tillage practices and cropping programs. They may also specify that certain lands should be retired from cultivation. Few land-use regulations have been adopted by soil and water conservation districts.

REFERENCES

ASAE (1977). "Soil Erosion and Sedimentation." In Proc. National Symposium on Soil Erosion and Sedimentation by Water. ASAE Publication 4-77 (Dec.).

Beasley, R. P. (1972). *Erosion and Sediment Pollution Control.* Iowa State University Press, Ames, Iowa.

Beer, C. E., and H. P. Johnson (1963). "Factors in Gully Growth in the Deep Loess Area of Western Iowa." *Am. Soc. Agr. Eng. Trans.* **6**, (no. 3), 237–240.

Finnell, H. H. (1951). "Depletion of High Plains Wheatlands." U.S. Dept. Agr. Circ. 871.

Harrold, L. L. (1947). "Land-Use Practices on Runoff and Erosion from Agricultural Watersheds." *Agr. Eng.* **28**, 563–566.

Harrold, L. L. (1949). Soil Loss as Determined by Watershed Measurements." *Agr. Eng.* **30**, 137–140.

Hay, R. C. (1948). "How to Farm on the Contour." Univ. Illinois Agr. Ext. Circ. 575 (revised).

Hudson, N. (1971). *Soil Conservation.* Cornell University Press, Ithaca, New York.

Ireland, H. A., et al. (1939). "Principles of Gully Erosion in the Piedmont of South Carolina." U.S. Dept. Agr. Tech. Bull. 633.

Jacobson, P. (1961). "Mechanics of Water Erosion" (Chapter 37), in C. B. Richey, et al., *Agricultural Engineer's Handbook.* McGraw-Hill, New York.

Kohnke, H., and A. R. Bertrand (1959). *Soil Conservation*. McGraw-Hill, New York.

Mavis, F. T., et al. (1935). "The Transportation of Detritus by Flowing Water." State University, Iowa, Studies in Eng. Bull. 5.

Mech, S. J., and G. R. Free (1942). "Movement of Soil during Tillage Operations." *Agr. Eng.* **23**, 379–382.

Meyer, L. D., and J. V. Mannering (1961). "Minimum Tillage for Corn: Its Effect on Infiltration and Erosion," *Agr. Eng.* **42**, 72–75, 86, 87.

Mihara, Y. (1952). "Effect of Raindrops and Grass on Soil Erosion." In Proc. 6th Intern. Grassland Congr., 987–990.

Roehl, J. N. (1962). "Sediment Source Areas, Delivery Ratios and Influencing Morphological Factors." Intern. Assoc. Scientific Hydrology, Commission of Land Erosion. Publ. No 59.

Smith, D. D., and W. H. Wischmeier (1962). "Rainfall Erosion." *Advan. Agron.* **14**, 109–148.

Soil Conservation Society of America (1978). "Soil Erosion: Prediction and Control." Special Publication No. 21., The Society, Ankeny, IA.

Stallings, J. H. (1950). "Erosion of Topsoil Reduces Productivity." U.S. Dept. Agr. SCS-TP-98.

Subcommittee on Sedimentation (1948). Federal Inter-Agency River Basin Committee, A Study of Methods Used in Measurement and Analysis of Sediment Loads in Streams: Report No. 8. Measurement of the Sediment Discharge of Streams, Univ. Iowa, St. Paul District Sub-Office, Corps of Engineers.

U.S. Dept. Agr. Res. Serv. and Environmental Protection Agency (1975). "Control of Water Pollution from Cropland," Vol. I, U.S. GPO, Washington, D.C.

——— (1976). "Control of Water Pollution from Cropland," Vol. II. U.S. GPO, Washington, D.C.

U.S. Soil Conservation Service (1936). "A Standard State Soil Conservation Districts Law." U.S. GPO, Washington, D.C.

Wischmeier, W. H. (1976). "Cropland Erosion and Sedimentation." Chapter 3 in USDA-EPA Control of Water Pollution, Vol. II. U.S. GPO, Washington, D.C.

——— (1959). "A Rainfall Erosion Index for a Universal Soil-Loss Equation." *Soil Sci. Soc. Am. Proc.* **23**, 246–249.

——— (1960). "Cropping-Management Factor Evaluations for a Universal Soil-Loss Equation." *Soil Sci. Soc. Am. Proc.* **24**, 322–326.

Wischmeier, W. H. and D. D. Smith. (1978). "Predicting Rainfall Erosion Losses—A Guide to Conservation Planning." U.S. Dept. Agr., Agr. Hdbk. 537.

——— (1965). "Predicting Rainfall-Erosion Losses from Cropland East of the Rocky Mountains." USDA-ARS Agr. Handbook 282.

——— (1958). "Rainfall Energy and Its Relation to Soil Loss." *Am. Geophys. Union Trans.* **39**, 285–291.

Wischmeier, W. H., D. D. Smith, and R. E. Uhland (1958). "Evaluation of Factors in the Soil-Loss Equation." *Agr. Eng.* **39**, 458–462, 474.

PROBLEMS

5.1. If the specific gravity of moving sediment is 2.5 and the velocity of the stream along its bed is 0.15 m (0.5 fps), what is the maximum size of soil particle that can be moved?

5.2. If the soil loss at Memphis, Tenn., for a given set of conditions is 11.2 Mg/ha (5 t/a), what is the expected soil loss at your present location if all factors are the same except the rainfall factor?

5.3. If the degree of slope is increased from 2 to 10 percent, what is the relative increase in erosion caused by water? Assume other factors are constant.

5.4. If the soil loss for a given set of conditions is 4.5 Mg/ha (2 t/a) for a 61-m (200-ft) length of slope, what soil loss could be expected for a 244-m (800-ft) slope length?

5.5. Determine the soil loss for a field at your present location if $K = 0.15$ t/a, $l = 91$ m (300 ft), $s = 10$ percent, $C = 0.2$, and up and down the slope farming is practiced. What conservation practice should be adopted if the soil loss is to be reduced to 6.7 Mg/ha (3 t/a)?

5.6. If the soil loss for up and down the slope farming at your present location is 78.4 Mg/ha (35 t/a) from a field having a slope of 6 percent and slope length of 122 m (400 ft), what will be the soil loss if the field is terraced with a horizontal spacing of 26 m (85 ft)? Assume that the cropping-management conditions remain unchanged.

5.7. Compute the average annual soil loss from a terraced field at your present location assuming a slope of 6 percent, $C = 0.2$, $K = 0.3$ t/a, and a slope length of 183 m (600 ft) with terraces at 24-m (80-ft) intervals. Determine the soil lost (1) at the terrace outlet, (2) to the terrace channel (use contouring factor), and (3) to the terrace channel if upslope plowing is practiced.

5.8. Compute the kinetic energy in m-Mg/ha-mm for a 30-min (duration) storm with a return period of 10 years at your present location.

5.9. If the average annual soil loss at Indianapolis, Ind., is 10 Mg/ha, compute the annual erosion for a return period of 10 years assuming all other soil loss factors are the same. Compute the soil loss for a single 10-year return period storm at this location.

5.10. Determine the contour strip width to the nearest complete equipment round for four 76-cm (30-in.) row equipment where the land slope is 6 percent.

5.11. If the soil loss from a field with a 5 percent slope is 44.8 Mg/ha (20 t/a) for up and downslope farming and the cropping-management factor is 0.25, what is the estimated soil loss if the field is contoured and the cropping-management factor is changed to 0.15? What would be the soil loss if strip cropping was substituted for contouring?

CHAPTER 6

Wind Erosion and Control Practices

In the arid and semiarid regions of the United States, large areas are affected by wind erosion. The Great Plains region, an area especially subject to soil movement by wind, represents about 20 percent of the total land area in the United States. Wind erosion not only removes soil but also damages crops, fences, buildings, and highways. Chepil and others (1962) have developed a wind erosion climatic factor that combines annual mean wind velocity with an index of soil moisture conditions to characterize the wind erosion hazard as a function of geographic location. Figure 6.1 shows the distribution of wind erosion hazard in the states west of the Mississippi River. The result of severe wind erosion is evident in Fig. 6.2.

Contrary to popular opinion, many humid regions are also damaged by wind erosion. The areas most subject to damage are the sandy soils along streams, lakes, and coastal plains and the organic soils. Peats and mucks comprise about 10 million ha (25,000,000 ac) located in 34 states. Although the extent of sandy areas is not available, it probably is as much or more than the acreage of peats and mucks.

6.1. Wind Distribution with Height. Air movement of the free atmosphere is retarded near the surface of the soil. In immediate contact with the soil surface the air is nearly at rest because of the drag forces between the air and the soil surface. Internal shear of the air allows an increased velocity with height above the soil until all effect of the soil surface is dissipated.

Wind velocities are normally measured at a given height at major airports and first-order weather stations. For other heights, such as required for predicting evapotranspiration, velocities can be estimated since they vary nearly as the logarithm of the height.

A mean wind velocity-profile equation over stable surfaces is

$$u_z = (u_*/k) \ln[(z - d)/z_0] \tag{6.1}$$

where u_z = wind velocity at z height (L/T),
u_* = friction velocity (L/T) = $(\tau_0/\rho)^{1/2}$,
τ_0 = shear stress at the boundary (F/L^2),

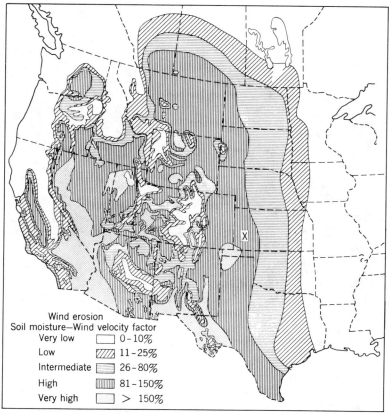

Soil moisture—Wind velocity factor

Very low	0 - 10%
Low	11 - 25%
Intermediate	26 - 80%
High	81 - 150%
Very high	> 150%

Wind erosion

Fig. 6.1. Relative potential soil loss by wind for the western United States and southern Canada as a percentage of that in the vicinity of Garden City, Kansas, marked by X. (From Chepil and others, 1962)

ρ = air density (M/L³),
k = von Karman's constant usually taken as 0.4,
z = height above a reference surface (L),
d = an effective surface roughness height (L),
z_0 = a roughness parameter (L).

The friction velocity u_* is a characteristic velocity in a turbulent boundary layer. The equation is valid only in the first few meters of height above the surface under neutral temperature conditions. These conditions exist when no heat is added or subtracted from the surface (adiabatic). Over short crops and smooth surfaces the effective roughness height d is small or nearly zero. Both d and the roughness parameter z_0 are subject to considerable variation as crops bend and weave with the wind. Below the roughness height turbulent diffusion

Fig. 6.2. Example of severe wind erosion in California. (Photo courtesy U.S. Soil Conservation Service.)

is nearly extinguished and transport occurs primarily by forces of molecular diffusion. This height is a conceptual dimension to make Eq. 6.1 valid over tall crops. For a wide range of crops Stanhill (1969) found the equation

$$\log d = \log h - 0.15$$

applies where d is the effective roughness height and h is the crop height. The roughness parameter z_0 can be estimated from the equation

$$\log z_0 = \log h - 0.09$$

(Tanner and Pelton, 1960). The roughness parameter as defined in Eq. 6.1 is the height where the velocity profile extrapolates to zero.

Wind velocities with height are shown in Fig. 6.3 for three surfaces. Approximate values of d, z_0, and h are shown for wheat. For grass and snow cover the plotted points were computed from Eq. 6.1. Appropriate values of d and z_0 were selected and u_* the friction velocity was computed from the velocities at 4.9-m height. Velocities at other heights were determined using the computed u_* value. As shown, the equation fits the curves reasonably well.

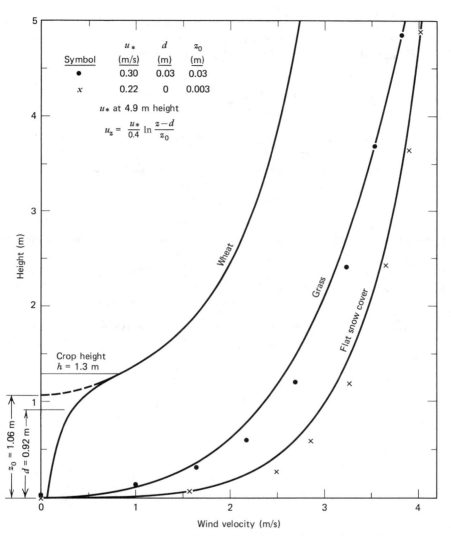

Fig. 6.3. Wind velocity distribution with height over various surfaces. (Curves are at different free wind velocities.)

6.2. Wind Erosiveness. For wind speeds greater than those required to barely move the soil, the rate of soil movement has been found to be directly proportional to the friction velocity cubed. For a specified type of surface and height k, z, and z_0 are constant, therefore u_* is proportional to u_z (Eq. 6.1). Soil movement is thus proportional to u_z cubed. Wind speeds 5.4 m/s (12 mph) or less at 0.3-m height are considered nonerosive.

The sum of the wind erosion energy vectors for all directions gives the wind erosion energy, that is, the relative capacity of the wind to cause soil blowing at a given location. It may be computed by the equation

$$F = \sum_{j=0}^{15} \sum_{i=1}^{n} u_{ij}^3 f_{ij} \tag{6.2}$$

where u_{ij} = average wind speed within the ith wind speed group for all groups above 5.4 m/s,

f_{ij} = percentage of the total observations that occur within the jth direction within the ith speed group,

n = number of wind speed groups.

The sub j's indicate direction (Fig. 6.4) and have values from 0 to 15 inclusive, representing the 16 principal compass directions, numbered counterclockwise starting arbitrarily with East ($j = 0$). Wind data in climatological records are reported by about 12 speed groups (Beaufort numbers). If the average wind speed from all directions was the same and the percentage of observations in each jth direction was equal, the energy would be the same from all directions.

6.3. Wind Erosion Direction. By a procedure described above, components of wind erosion forces parallel and perpendicular to a chosen direction θ can be expressed as

$$F_{\parallel} = \sum_{j=0}^{15} \left[\cos\left(22.5j - \theta\right)\right] \sum_{i=1}^{n} u_{ij}^3 f_{ij} \tag{6.3}$$

and

$$F_{\perp} = \sum_{j=0}^{15} \left[\sin(22.5j - \theta)\right] \sum_{i=1}^{n} u_{ij}^3 f_{ij} \tag{6.4}$$

where $F_{\parallel}$ = forces parallel,

$F_{\perp}$ = forces perpendicular,

θ = the chosen direction measured counterclockwise from East.

These forces are shown graphically in Fig. 6.4 for $j = 2$ as an example. By choosing θ to make $R = F_{\parallel}/F_{\perp}$ a maximum, $\theta = \theta_R$ becomes the prevailing wind erosion direction. For $R = 1.0$, no prevailing wind erosion direction exists and a wind barrier would be equally effective in any direction. An R_m maximum of 2.0 indicates a prevailing wind erosion direction with forces parallel twice as great as those perpendicular to the prevailing wind erosion direction. The R_m value is

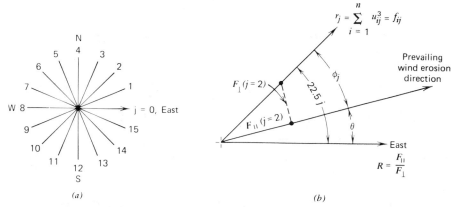

Fig. 6.4. (a) Wind direction numbers and (b) components of wind erosion forces for $j = 2$ relative to the prevailing wind erosion direction.

also called the preponderance of wind erosion forces. Tabulated values of F, θ_R, and R_m for 212 locations in 39 states are presented by Skidmore and Woodruff (1968). For other locations, values could be obtained from climatological data using a digital computer. Determination of potential wind erodibility of farm field and barrier spacing are two examples of practical problems.

WIND EROSION

6.4. Types of Soil Movement. Suspension, saltation, and surface creep in wind erosion are comparable to suspension, saltation, and bed load, respectively, in sediment movement by water. These three distinct types of movement usually occur simultaneously. The major portion of soil movement takes place near the surface at heights not greater than approximately 1 m. Above this height the only movement is normally by suspension, whereas all three types of movement occur near the surface. In laboratory studies 55 to 72 percent of the soil was moved by saltation (particles 0.1- to 0.5-mm dia.), 3 to 38 percent by suspension (typically particles <0.1 mm), and 7 to 25 percent by surface creep (0.5- to 2-mm dia.) according to Chepil (1945).

Movement of soil particles by saltation is caused by the pressure of the wind on the soil particle and its collision with other particles. The characteristic path of a soil particle moved by saltation together with final and initial velocity vectors is shown in Fig. 6.5. As the soil particle leaves the surface, it moves nearly in a vertical direction. The horizontal distance through which the particle continues to rise is about one-fifth to one-fourth of the distance L. As the particle descends to the surface, it travels in a straight line with an angle of descent of about 6 to 12 degrees.

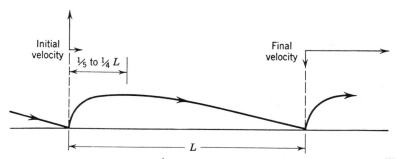

Fig. 6.5. Characteristic path of a soil particle moved in air by saltation. (Redrawn from Bagnold, 1941.)

6.5. Mechanics of Wind Erosion. For a precise understanding of the mechanics of wind erosion, analysis must be made of the nature and magnitude of the forces as they react upon soil particles.

The wind erosion process may be divided into the three simple but distinct phases: (1) initiation of movement, (2) transportation, and (3) deposition.

Initiation of Movement. Soil movement is initiated as a result of turbulence and velocity of the wind. The fluid threshold velocity is defined as the minimum velocity required to produce soil movement by direct action of the wind, whereas the impact threshold velocity is the minimum velocity required to initiate movement from the impact of soil particles carried in saltation. Except very near the surface and at low velocities (less than about 1 m/s), the surface wind is always turbulent. Wind speeds of 5 m/s or less at 30-cm height are usually considered nonerosive for mineral soils.

Transportation. The quantity of soil moved is influenced by the particle size, gradation of particles, wind velocity, and distance across the eroding area. Winds, being quite variable in velocity and direction, produce gusts with eddies and cross currents that lift and transport soil. The quantity of soil moved varies as the cube of the excess wind velocity over and above the constant threshold velocity, directly as the square root of the particle diameter, and increases with the gradation of the soil.

A diagrammatic representation of wind erosion with two different proportions of erodible to nonerodible fractions is shown in Fig. 6.6. The distance between the centers of nonerodible particles is shown as L_x. In the upper drawing the erodible soil between the widely separated nonerodible particles has been removed, but no erosion is evident where the nonerodible particles were close together. For any stabilized surface the ratio of the height of the particles above the surface to the distance between them and the percentage area covered by roughness elements correlated well with the potential for erosion (Lyles et al. 1974).

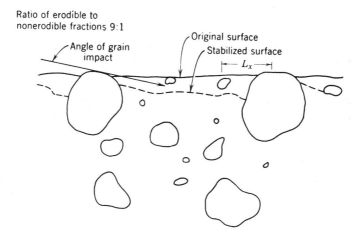

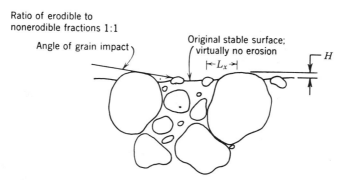

Fig. 6.6. Erosion on soils with different ratios of erodible to nonerodible fractions. (Redrawn from Chepil, 1950.)

The rate of soil movement increases with distance from the windward edge of the field or eroded area. Fine particles drift and accumulate on the leeward side of the area or pile up in dunes. Increased rates of soil movement with distance from the windward edge of the area subject to erosion are the result of increasing amounts of erosive particles, thus causing greater abrasion and a gradual decrease in surface roughness. The rate of erosion varies for different soils, some soils being as much as ten times more erodible than others.

The atmosphere has a tremendous capacity for transporting soil, particularly those soil fractions less than 0.1 mm in diameter. It is estimated that the potential carrying capacity for 1 cubic kilometer of the atmosphere is many thousand tonnes of soil, depending on the wind velocity. As much as 224 kg/ha were deposited in Iowa in 1937 from a dust storm originating in the Texas Panhandle.

Deposition.   Deposition of sediment occurs when the gravitational force is greater than the forces holding the particles in the air. This generally occurs when there is a decrease in wind velocity caused by vegetation or other physical barriers, such as ditches and snow fences. Raindrops may also take dust out of the air.

6.6.   Estimating Wind Erosion Losses.   A wind erosion equation has been developed to indicate the relationships between the amount of wind erosion and the various field and climatic factors that influence erosion. It may be used (1) to determine the potential amount of wind erosion on any field under existing local climatic conditions and (2) as a guide for determining the conditions of surface roughness, soil cloddiness, vegetative cover, sheltering, or width and orientation of field necessary to reduce the potential wind erosion to an acceptable amount. The equation may be expressed as

$$E = f(I', C', K', L', V) \tag{6.5}$$

where E = weight of annual erosion per unit area,
I' = a soil erodibility index, E_1
C' = a climatic factor,
K' = a soil ridge roughness factor,
L' = equivalent field length along the prevailing wind erosion direction,
V = equivalent quantity of vegetative cover.

Relationships among the variables in the equation are extremely complex and cannot be expressed in simple mathematical form. Skidmore and Woodruff (1968) present graphs, tables, maps, and nomographs from which erosion may be determined at 212 locations in 39 states. They are too lengthy and involved to be presented here. A computer solution has been developed by Skidmore et al. (1970) and a slide rule calculator is available commercially.

CONTROLLING SURFACE WIND VELOCITY

The two major types of wind erosion control measures consist of those that reduce surface wind velocities and those that affect soil characteristics. Many measures, such as permanent grass and contouring, provide both types of control. For example, vegetation not only retards surface winds but may also increase soil aggregation in finer textured soils.

The three principal methods of reducing surface wind velocities are vegetative measures, tillage practices, and mechanical methods. In each, the degree of protection is influenced by the height and spacing of obstructions, the breaking effect on the wind, and the resistance of the soil to movement.

Vegetation is generally the most effective means of wind erosion control. Cultivated crops reduce wind velocity and hold soil against the tractive force of the wind. Woody plants, such as shrubs and trees, may be planted to reduce wind velocities over large areas. Wind erosion hazards often exist when fields are bare of growing crops. Cultural methods of control, though temporary, are effective under these conditions. Whenever possible cultural practices should be used before blowing starts, for wind erosion is easier to prevent than to arrest. In general, long periods of drought will obliterate the effects of cultural treatments.

6.7. Cultivated Crops. In general, close-growing crops are more effective for erosion control than are intertilled crops. The effectiveness of crops is dependent upon stage of growth, density of cover, row direction, width of rows, kind of crop, and climatic conditions. Pasture or meadow tends to accumulate soil if there is a good growth of vegetation, the soil coming from neighboring cultivated fields and being deposited by sedimentation. Good management practices such as rotation grazing are important.

Intertilled crops such as corn, cotton, and vegetables offer some protection. Coldwell and others (1942) showed that close-growing crops in the Texas Panhandle gained soil, but intertilled crops lost soil by wind erosion. The best practice is to seed the crop normal to prevailing winds. In the Great Plains region the rows are usually in an east-west direction. A good crop rotation that will maintain soil structure and conserve moisture should be followed. Crops adapted to soil and climatic conditions and providing as much protection against blowing as practical are recommended. For instance, in the Great Plains region forage sorghum and sudan grass are quite resistant to drought and are effective in preventing wind erosion. Stubble mulch farming and cover crops between intertilled crops in more humid regions aid in controlling blowing until the plants become established. In some dry regions emergency crops with low moisture requirements are often planted on summer fallow land before seasons of high intensity winds. In muck soils where vegetable crops are grown, miniature windbreaks consisting of rows of small grain are sometimes planted. Stabilization of sand dunes with vegetation may be accomplished by establishing grasses and then reforesting. Vegetation used should have the ability to grow on sandy soil, the ability to grow in the open, firmness against the wind, and long life. It should also provide a dense cover during critical seasons, provide as uniform an obstruction to the wind as possible, reduce the surface wind velocity, and form an abundance of crop residue.

6.8. Shrubs and Trees. The U.S. Soil Conservation Service (1969) estimated that over 3 240 000 ha (8,000,000 ac) of land were in need of protection by shelterbelts and windbreaks. Although a windbreak is defined as any type of barrier for protection from winds, it is more commonly associated with mechanical or vegetative barriers for buildings, gardens, orchards, and feed lots. A shelterbelt is a longer barrier than a windbreak, usually consisting of shrubs and trees, and is intended for the conservation of soil and moisture and for the protection of field crops. About 450 000 km (280,000 mi) of windbreaks and shelterbelts have been planted in the United States since the middle of the past century. Not only are windbreaks and shelterbelts valuable for wind erosion control but also they save fuel, increase livestock gains, reduce evaporation, prevent firing of crops from hot winds, catch snow during the winter months, provide better fruiting in orchards, and make spraying of trees for insect control more effective as well as provide farm woodlots and wildlife refuge.

The relative wind velocity at a height of 0.4 m above the soil surface near a windbreak is shown in Fig. 6.7. The data were obtained for a slat fence barrier 19 times longer than its height and having an average density of 50 percent but more open in the lower than in the upper half. Similar results could be expected from a tree shelterbelt with an equivalent density.

Shelterbelts should be moderately dense from ground level to treetops if they are to be effective in filtering the wind and lifting it from the surface. Since the

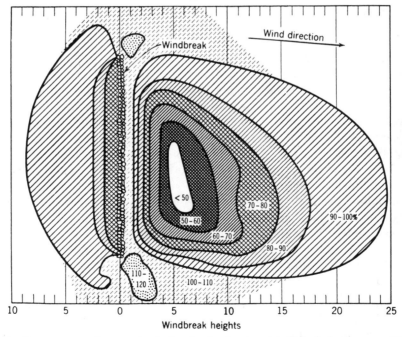

Fig. 6.7. Percentage of normal wind velocity near a windbreak having an average density of 50 percent. (Redrawn from Bates, 1944.)

wind velocity at the ends of the belt as given in Fig. 6.7 is as much as 20 percent greater than velocities in the open, it is evident that long shelterbelts are more effective than short ones. An opening or break in an otherwise continuous belt results in a similar increase in velocity and will reduce the area protected. Roads through shelterbelts should therefore be avoided, and, when essential, they should cross the belt at an angle or they should be curved. In establishing the direction of shelterbelts, records of wind direction and velocity, particularly during vulnerable seasons, should be considered, and the barrier should be oriented as nearly as possible at right angles to the prevailing direction of winds.

From wind-tunnel tests Woodruff and Zingg (1952) found that the distance of full protection from a windbreak is

$$d = 17h \ (v_m/v) \cos \theta \qquad (6.6)$$

where d = distance of full protection (L),
  h = height of the barrier in the same units as d (L),
  v_m = minimum wind velocity at 15-m height required to move the most erodible soil fraction (LT^{-1}),
  v = actual wind velocity at 15-m height (LT^{-1}),
  θ = the angle of deviation of prevailing wind direction from the perpendicular to the windbreak.

Chepil (1959) reported that v_m for a smooth bare surface after erosion has been initiated and before wetting by rainfall and subsequent surface crusting was about 9.6 m/s. Equation 6.6 is valid only for wind velocities below 18 m/s. It may also be adapted for estimating the width of strips by using the crop height in the adjoining strip in the equation.

Typical shelterbelts for the Great Plains are shown in Fig. 6.8. A tight row of shrubs on the windward side is desirable, and, when combined with conifers, and low, medium, and tall deciduous trees, the shelterbelt provides a compact and rather dense barrier. Such an extensive shelterbelt may not always be required. Single-row belts are preferred in many areas because fewer trees and less land are needed, and the shelterbelt is easier to cultivate and maintain. Studies in North Dakota indicated that shelterbelts with a density of much less than 50 percent were effective in trapping snow for distances up to 90 m with a 5-m single row of trees. Local recommendations should be followed for varieties, spacings, and other practices.

Shelterbelts are designed principally to control erosion, but crop production may be increased even though some land is occupied by the trees. Under most situations in the Great Plains, Stoeckeler (1965) reported that where shelterbelts occupy 1 to 5 percent of the gross land area, optimum crop gains are most likely to be obtained. Among the crops most responsive to shelterbelt protec-

Fig. 6.8.   Typical shelterbelts. (a) A half-mile long windbreak of evergreens, deciduous trees, and shrubs protects a Nebraska farmstead and fields. (b) A three-row evergreen and broadleaf tree windbreak. (c) A three-row broadleaf windbreak in winter. (Photographs courtesy U.S. Soil Conservation Service.)

tion are garden crops, flower bulbs, citrus, and fruit trees. Those with the least response are drought-hardy crops.

6.9.   Field and Contour Strip Cropping.   Field and contour strip cropping consists of growing alternate strips of clean-cultivated and close-growing crops in the same field. Field strip cropping is laid out parallel to a field boundary or other guideline. In some of the plains states, strips of fallow and grain crops are alternated. The chief advantages of strip cropping are (1) physical protection against blowing, provided by the vegetation, (2) soil erosion limited for a distance equal to the width of strip, (3) greater conservation of moisture, particularly from snowfall, and (4) the possibility of earlier harvest. The chief disadvantages are (1) machines problems in farming narrow strips and (2) greater number of edges to protect in case of insect infestation.

The strips should be of sufficient width to be convenient to farm, yet not so wide as to permit excessive erosion. The width of strips depends on the inten-

(b)

(c)

135

sity of wind, row direction, crops, and erodibility of the soil and can be determined from Eq. 6.5.

The required width of field strips as determined by Eq. 6.5 can be increased to the extent that each strip is fully protected from wind by a standing crop stubble or some other barrier on its windward side as determined from Eq. 6.6. The zone of protection is about 10 times the barrier height so for a crop stubble it is essentially negligible.

6.10. Primary and Secondary Tillage. The objective of primary and secondary tillage for wind erosion control is to produce a rough, cloddy surface with some plant residue exposed on the surface. In order to obtain maximum roughness the land normally should be cultivated as soon after a rain as possible. Large clods as well as a high percentage of large aggregates are desirable.

Small ridges normal to the direction of prevailing winds are effective in wind erosion control. The effect of such ridges on wind movement is shown in Fig. 6.9. Very low velocities were observed between the ridges and approximately 2 cm below the crest. In the furrow the direction of movement was opposite to that of the wind. The decrease in wind velocity and change in direction between

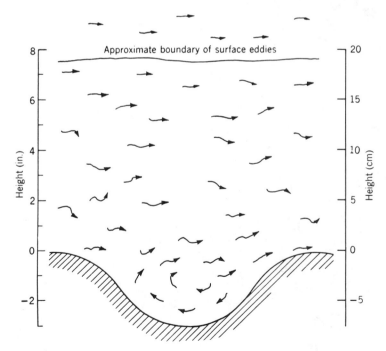

Fig. 6.9. Diagrammatic representation of wind structure across ridges. (Redrawn from Chepil and Milne, 1941.)

the ridges causes soil deposition. For a height of about 15 cm above the ridges there is an area of considerable turbulence.

Tillage may be quite effective as an emergency control measure. Soil blowing usually starts in a small area where the soil is less stable or is more exposed than in other parts of the field. Where the entire field starts to blow, the surface should be put in a rough and cloddy condition as soon as practicable. This tillage should begin at the windward side of the field and continue by making widely spaced trips across the field. When the field has been stripped, the areas between the strips may then be cultivated.

Crop residues exposed on the surface are an effective means of control, especially when combined with a rough soil surface. This practice is usually called stubble mulch tillage. The effectiveness of various soil treatments on wind velocity and soil erosion are shown in Fig. 6.10. These experiments were carried out in a wind tunnel. All surfaces were exposed to the same wind velocity as measured at a height of 30 cm. The reduction in wind velocity was greatest for the ridges plus straw. The effect of the straw mulch in reducing soil

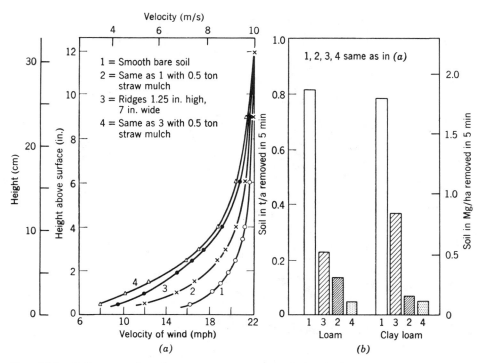

Fig. 6.10. Influence of surface treatment on (a) wind velocity and (b) soil erosion. (Redrawn from Chepil, 1944.)

erosion was greater than for ridges alone. The combined effect of straw and ridges was appreciably greater than the effect of either straw or ridges. As compared to bare soil, 1120 kg/ha of wheat straw reduced the amount of erosion by 83 and 88 percent for loam and clay loam soils, respectively.

Crop residues act in two ways: they reduce wind velocity and trap eroding soil. Short stubble is generally less effective than long stubble. A mixture of straw and stubble on the surface provides more protection against erosion than equivalent amounts of straw or stubble alone. The higher the wind velocity the greater the quantity of crop residue required.

The effect of various quantities of crop residue on soil loss for the soil aggregate fraction less than 0.42 mm in diameter is shown in Fig. 6.11. The data show that the soil loss varies inversely as the 0.8 power of the weight of plant residue. With 60 percent of the soil fractions less than 0.42 mm, the soil loss may be reduced from 12 to 2 Mg/ha by increasing the residue from 280 to 2240 kg/ha, respectively.

6.11. Mechanical Methods. Mechanical barriers such as windbreaks are of little importance for field crops, but they are frequently employed for the pro-

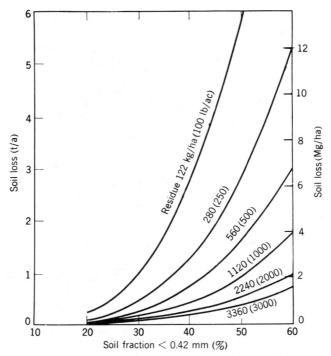

Fig. 6.11. Effect of crop residues on soil loss for soils with varying amounts of erodible fractions. (Redrawn from Englehorn and others, 1952.)

tection of farmsteads and small areas. Some of the mechanical methods of control include slat or brush fences, board walls, and vertical burlap or paper strips, as well as surface protection, such as brush matting, rock, and gravel. These barriers may be classed as semipermeable or impermeable. Semipermeable windbreaks are usually more effective than impermeable structures because of diffusion and eddying effects on the leeward side of the barrier. A slat snow fence is a good example of this action. Slat fences, picket fences, and vertical burlap or paper strips are sometimes used for the protection of vegetable crops in organic soils. Brush matting, debris, rock, and gravel may be suitable in stabilizing sand dune areas.

Terraces have some effect on wind erosion. In the Texas Panhandle, Coldwell and others (1942) found that terraces lost less soil than the interterraced area and in some instances gained soil. Most of the soil that was lost from the interterraced area was collected in the terraces. Where vegetation was growing on the ridge and in the channel, terraces were still more effective.

CONTROLLING SOIL FACTORS

The principal soil factors influencing wind erosion control include the conservation of moisture to improve vegetative growth and conditioning of the surface soil to improve aggregation.

6.12. Conserving Soil Moisture. The conservation of moisture, particularly in arid and semiarid regions, is of utmost importance for wind erosion control as well as for crop production. The means of conserving moisture fall into three categories: increasing infiltration, reducing evaporation, and preventing unnecessary plant growth. In practice these can be accomplished by such practices as level terracing, contouring, mulching, and selecting suitable crops.

The greater the amount of mulch on the surface the greater the quantity of moisture conserved. The effect of crop residue and tillage practices on the conservation of moisture is shown in Fig. 6.12. These data show the percentage of rainfall conserved from April to September. Where the straw was disked or plowed in, some of the residue was left on the surface, and, where the soil was plowed and disked, no straw was present. Although there was no runoff for the basin-listed area, the quantity of moisture conserved was about half of that for the treatment with 4.5 Mg/ha of straw on the surface, probably owing to greater evaporation. For the same reason disking and plowing were less effective than any of the mulch treatments.

The time of seedbed preparation in some regions has considerable effect on the conservation of moisture. This effect is shown in Fig. 6.13 for winter wheat at Hays, Kansas. Late plowing (September–October) resulted in a decrease in available moisture as compared to early plowing (July) or summer fallow.

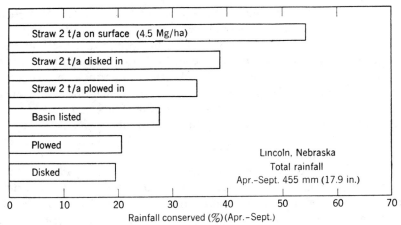

Fig. 6.12. Effect of straw and tillage practices on storage of soil moisture. (From data by Duley and Russel, 1939.)

Wheat yields for late and early plowing were 37 and 65 percent, respectively, of that for summer fallow.

Organic soils do not blow appreciably if the soil is moist. Where the subsoil is wet, rolling the soil with a heavy roller will increase capillary movement and moisten the surface layer. Controlled drainage where the water level is main-

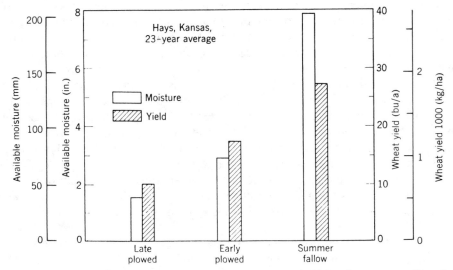

Fig. 6.13. Effect of time of seedbed preparation on available moisture at seeding time and on wheat yields. (From data by Hallsted and Mathews, 1936.)

tained at a specified depth may also reduce blowing. In irrigated areas the surface is often wet down by overhead sprinkling at critical periods.

Terracing is a good moisture conservation practice where level or conservation bench terraces are suitable and where the slopes are gentle enough so that the water can be spread over a relatively large area. Such practices as contouring, strip cropping, and mulching are effective in increasing the total infiltration and thereby the total soil moisture available to crops. Field strip cropping generally does not conserve as much moisture as contour strip cropping, but it is somewhat more effective in reducing surface wind velocities.

6.13. Conditioning Topsoil. Since wind erosion is influenced to a large extent by the size and apparent density of aggregates, an effective method of conditioning the soil against wind erosion is by adopting practices that produce nonerosive aggregates (>1-mm dia.) or large clods. During the periods of the year when the soil is bare or has a limited amount of crop residue, control of erosion may depend on the degree and stability of soil aggregation.

Tillage may or may not be beneficial to soil structure, depending on the moisture content of the soil, type of tillage, and number of operations. For optimum resistance to wind erosion in semiarid regions it is desirable to perform primary tillage as soon as practical after a rain. The number of operations should be a minimum because tillage has a tendency to reduce soil aggregate size and to pulverize the soil. Secondary tillage for seedbed preparation should be delayed as long as practical.

Good crop and soil management practices are necessary to maintain good soil structure. In the Great Plains a typical recommended rotation may include a row crop, fallow, and winter wheat. The organic matter in the soil should be held to a high level and lime and fertilizers applied where necessary.

Soil structure is affected by the climatic influences of the season, rainfall, and temperature. Freezing and thawing generally have a beneficial effect in improving soil structure where sufficient moisture is present. However, in dry regions the soil is more susceptible to erosion because of rapid breakdown of the clods into smaller aggregates.

REFERENCES

Bagnold, R. A. (1941). *The Physics of Blown Sand and Desert Dunes*. Methuen, London.

Bates, C. G. (1944). "The Windbreak as a Farm Asset." *U.S. Dept. Agr. Farmers Bull.*, **1045** (revised).

Chepil, W. S. (1945). "Dynamics of Wind Erosion: I. Nature of Movement of Soil by Wind." *Soil Sci.* **60**, 305–320.

———— (1950). "Properties of Soil Which Influence Wind Erosion: I. The Governing Principle of Surface Roughness." *Soil Sci.* **69**, 149–162.

—— (1944). "Utilization of Crop Residues for Wind Erosion Control." *Sci. Agr.* **24**, 307–319.

—— (1959). "Wind Erodibility of Farm Fields." *J. Soil Water Cons.* **14** (5), 214–219.

—— (1960). "How to Determine Required Width of Field Strips to Control Wind Erosion." *J. Soil Water Cons.* **15** (2), 72–75.

Chepil, W. S., and R. A. Milne (1941). "Wind Erosion of Soil in Relation to Roughness of Surface." *Soil Sci.* **52**, 417–431.

Chepil, W. S., F. H. Siddoway, and D. V. Armbrust (1962). "Climatic Factor for Estimating Wind Erodibility of Farm Fields." *J. Soil Water Cons.* **17** (4), 162–165.

Chepil, W. S., and N. P. Woodruff (1963). "The Physics of Wind Erosion and Its Control." *Advan. Agron.* **15**, 211–302; Academic Press, New York.

Coldwell, A. E., et al. (1942). "The Relation of Various Types of Vegetative Cover to Soil Drift." *Am. Soc. Agron. J.* **34**, 702–710.

Duley, F. L., and J. C. Russel (1939). "The Use of Crop Residues for Soil and Moisture Conservation." *Am. Soc. Agron. J.* **31**, 703–709.

Engelhorn, C. L., et al. (1952). "The Effects of Plant Residue Cover and Clod Structure on Soil Losses by Wind." *Soil Sci. Soc. Am. Proc.* **16**, 29–33.

Hallsted, A. L., and O. R. Mathews (1936). "Soil Moisture and Winter Wheat with Suggestions on Abandonment." *Kansas Agr. Expt. Sta. Bull.* **273**.

Lyles, L., R. L. Schrandt, and N. F. Schneidler (1974). "How Aerodynamic Roughness Elements Control Sand Movement." *Trans. Am. Soc. Agr. Eng.* **17** (1), 134–139.

Rosenberg, N. J. (1974). *Microclimate: The Biological Environment.* John Wiley, New York.

Skidmore, E. L., P. S. Fisher, and N. P. Woodruff (1970). "Wind Erosion Equation: Computer Solution and Application." *Soil Sci. Soc. Am. Proc.* **34** (5), 931–935.

Skidmore, E. L. and N. P. Woodruff (1968). "Wind Erosion Forces in the United States and Their Use in Predicting Soil Loss." *U.S. Agr. Res. Serv. Agr. Hbk.* **346**.

Stanhill, G. (1969). "A Simple Instrument for Field Measurement of Turbulent Diffusion Flux." *J. Appl. Meteorol.* **8**, 509–513.

Stoeckeler, J. H. (1965). "The Design of Shelterbelts in Relation to Crop and Yield Improvement." *World Crops,* 3–8, Grampian Press Ltd., England.

Tanner, C. B. and W. L. Pelton (1960). "Potential Evapotranspiration Estimates by the Approximate Energy Balance Method of Penman." *J. Geophys. Res.* **65**, 3391–3413.

U.S. Soil Conservation Service (1969). "Soil and Water Conservation Needs Inventory." Preliminary report, base year 1967. *Soil Conservation* **35** (5), 99–109. (See USDA Stat. Bull. 461, 1971.)

Woodruff, N. P., and A. W. Zingg (1952). "Wind-Tunnel Studies of Fundamental Problems Related to Windbreaks." U.S. Soil Conservation Service, SCS-TP-112.

PROBLEMS

6.1. Determine the spacing between windbreaks that are 15 m (50 ft) high if the 5-year return period wind velocity at 15-m (50-ft) height is 15.6 m/s (35 mph) and the wind direction deviates 10 degrees from the perpendicular to the field strip. Assume a smooth, bare soil surface and a fully protected field.

6.2. Determine the full protection strip width for field strip cropping if the crop in the adjacent strip is wheat 0.9 m (3 ft) tall and the wind velocity at 15-m (50-ft) height is 8.9 m/s (20 mph) at 90 degrees with the field strip.

CHAPTER 7

Vegetated Waterways

The design of vegetated waterways is more complex than the design of channels lined with concrete or other stable material because of variation in the roughness coefficient with depth of flow, stage of vegetal growth, hydraulic radius, and velocity.

7.1. Uses of Vegetated Waterways. Runoff from sloping land must flow to lower lands in a controlled manner that will not result in gully formation. Flow may be concentrated by the natural topography or by contour furrows, terraces, or other works of man. In any event considerable amounts of energy are dissipated as flow proceeds down a slope. A flow of 1.4 m³/s for 30 m down a 5 percent slope releases energy at the rate of over 20 000 watts. If this energy acts upon bare soil, considerable quantities of soil particles will be detached and transported by the water. The resultant gullies may divide a field into several parts. Fields thus become smaller and more numerous, rows are shortened, movement from field to field is obstructed, and the farm value is decreased. Roads, bridges, buildings, and fences frequently are jeopardized by gully development. Soil carried from gullied areas contributes to costly downstream sedimentation damage and pollution.

The basic approach to the control of such gullies involves (1) reduction of peak flow rates through the gully by full utilization of field protection practices, and (2) provision of a stable channel that can handle the flow that remains. This stabilization is best accomplished by providing vegetal protection for the channel together with modifying the cross section and grade of the channel so as to limit the flow velocities to a level that the vegetation can withstand.

Providing properly proportioned channels protected by vegetation is frequently a complete solution to the problem of gully formation. For large runoff volumes or steep channels it may be necessary to supplement the vegetated watercourse by permanent gully control structures. Vegetated waterways should be used to handle natural concentrations of runoff or to carry the discharge from terrace systems, contour furrows, diversion channels, or emergency spillways for farm ponds or other structures. Vegetated waterways should not be used for continuing flows, such as may discharge from tile drains, as prolonged wetness in the waterway will result in poor vegetal protection.

Special measures must be taken in the planning and execution of control measures on large gullies.

DESIGN

7.2. Determination of Runoff. In the design of a vegetated watercourse, the functional requirements should be determined and then the channel proportioned to meet these requirements. The capacity of the waterway should be based on the estimated runoff from the contributing drainage area. The 10-year return period storm is a sound basis for vegetated waterway design. For exceptionally long watercourses it may be desirable to estimate the flow for each of several reaches of the channel to account for changing drainage area. For short channels the estimated flow at the waterway outlet is the practical design value.

7.3. Shape of Waterway. The cross-sectional shape of the channel as it is constructed may be parabolic, trapezoidal, or triangular. The parabolic cross section approximates that of natural channels. Under the normal action of channel flow, deposition, and bank erosion, the trapezoidal and triangular sections tend to become parabolic. In some channels no earthwork is necessary; the natural drainageway or meadow outlet is adequate, and only the width and location of the waterway need be defined.

A number of factors influence the choice of the shape of cross section. Channels built with a blade-type machine may be trapezoidal if the bottom width of the channel is greater than the minimum width of the cut. Triangular channels may also be readily constructed with blade equipment. Trapezoidal channels having bottom widths less than a mower swath are difficult to mow. Flat triangular or parabolic channels with side slopes of 4:1 (4 horizontal to 1 vertical) or flatter may be easily maintained by mowing. Side slopes of 4:1 or flatter are also desirable to facilitate crossing with farm equipment.

Broad-bottom trapezoidal channels require less depth of excavation than do parabolic or triangular shapes. During low flow periods, sediment may be deposited in trapezoidal channels with wide, flat bottoms. Uneven sediment deposition may result in meandering of higher flows and development of turbulence and eddies that will cause local damage to vegetation. Triangular channels reduce meandering, but high velocities may damage the bottom of the waterway.

Parabolic cross sections should usually be selected for natural waterways. A trapezoidal section with a slight V bottom is most easily constructed where the waterway is artificially located as in a terrace outlet along a fence line.

The geometric characteristics of the three shapes of cross sections are given in Fig. 7.1. This figure defines the three types of cross sections and gives formulas necessary for computing the hydraulic characteristics of each.

10 year storm
4:1 slopes   Z = 4

Cross-Sectional Area a	Wetted Perimeter, p	Hydraulic Radius $R=\frac{a}{p}$	Top Width
$bd + Zd^2$	$b + 2d\sqrt{Z^2+1}$	$\dfrac{bd + Zd^2}{b + 2d\sqrt{Z^2+1}}$	$t = b + 2dZ$ $T = b + 2DZ$

Trapezoidal cross section

Zd^2	$2d\sqrt{Z^2+1}$	$\dfrac{Zd}{2\sqrt{Z^2+1}}$ or $\dfrac{d}{2}$ approx.	$t = 2dZ$ $T = \dfrac{D}{d}\,t$

Triangular cross section

$\dfrac{2}{3}\,td$	$t + \dfrac{8d^2}{3t}$	$\dfrac{t^2 d}{1.5t^2 + 4d^2}$ or $\dfrac{2d}{3}$ approx.	$t = \dfrac{a}{0.67d}$ $T = t\left(\dfrac{D}{d}\right)^{\frac{1}{2}}$

Parabolic cross section

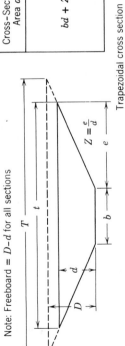

Note: Freeboard = $D-d$ for all sections

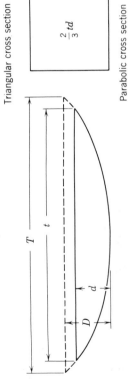

Fig. 7.1. Channel cross section, wetted perimeter, hydraulic radius, and top width formulas.

7.4. Selection of Suitable Vegetation.   Soil and climatic conditions are primary factors in the selection of vegetation. Vegetation recommended for various regions of the United States is indicated in Table 7.1. Other factors to be considered are duration, quantity, and velocity of runoff, ease of establishment of vegetation, time required to develop a good protective cover, suitability to the farmer in regard to utilization of the vegetation as a seed or hay crop, spreading of vegetation to adjoining fields, cost and availability of seed, and retardance to shallow flows in relation to sedimentation.

7.5. Design Velocity.   The ability of vegetation to resist erosion is limited. The permissible velocity in the channel is dependent upon the type, condition, and density of vegetation and the erosive characteristics of the soil. Uniformity of cover is very important, as the stability of the most sparsely vegetated area controls the stability of the channel. Permissible velocities for bunch grasses or other nonuniform covers are lower than those for sod-forming grasses. Bunch grasses produce nonuniform flows with high localized erosion. Their open roots do not bind the soil firmly against erosion.

Permissible velocities are also influenced by bed slope. Steeper channels produce increasing turbulence with intense localized erosion. Suggested design values for velocity are given in Table 7.2. It should be recognized that the design velocity is an average velocity rather than the actual velocity in contact with the vegetation or with the channel bed. Figure 7.2 shows the velocity distribution in a grass-lined channel and illustrates this point. Though the average velocity in the cross section is about 0.8 m/s, the velocity in contact with the vegetation and bed is less than 0.3 m/s. Design of vegetated waterways is based upon the Manning formula (Eq. 7.1) for velocity.

Table 7.1   Vegetation Recommended for Grasses Waterways

Geographical Area of U.S.	Vegetation
Northeastern—L, N, R, S[a]	Kentucky bluegrass, red top, tall fescue, white clover
Southeastern—N, O. P, T, U	Kentucky bluegrass, tall fescue, Bermuda, brome, Reed canary
Upper Mississippi—K, L, M, N	Brome, Reed canary, tall fescue, Kentucky bluegrass
Western Gulf—H, I, J, P, T	Bermuda, King Ranch bluestem, native grass mixture, tall fescue
Southwestern—C, D, G	Intermediate wheatgrass, western wheatgrass, smooth brome, tall wheatgrass
Northern Great Plains—G, F, H	Smooth brome, western wheatgrass, red top switchgrass, native bluestem mixture

[a] Letters refer to land resource regions shown in Chapter 1, but recommended vegetation does not necessarily apply to all areas in the region.

VELOCITY DEPENDS ON
① TYPE, CONDITION, DENSITY OF VEG
② EROSIVE CHARACT OF SOIL

Table 7.2 Permissible Velocities for Vegetated Channels

Cover	Permissible Velocity For Erosion Resistant Soils (m/s) (fps)		
	% Slope in Channel		
	0–5	5–10	Over 10
Bermuda grass	2.4 (8)[a]	2.1 (7)[a]	1.8 (6)[a]
Blue grama Buffalo grass Kentucky bluegrass Smooth brome Tall fescue	2.1 (7)[a]	1.8 (6)[a]	1.5 (5)[a]
Annual crops for temporary protection Alfalfa Crabgrass Kudzu Lespedeza sericea Weeping lovegrass	1.1 (3.5)[b]	NR[c]	NR
Grass mixture	1.5 (5)[b]	1.2 (4)[b]	NR

[a] Moderately resistant soils reduce velocities 0.3 m/s (1 fps). Easily eroded soils reduce velocities 0.6 m/s (2 fps).

[b] Easily eroded soils reduce velocities 0.3 m/s (1 fps).

[c] Not recommended.

Source: Modified from Ree (1949).

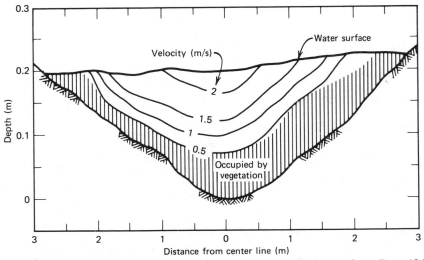

Fig. 7.2. Velocity distribution in a grass-lined channel. (Redrawn from Ree, 1949.)

7.6. Roughness Coefficient. Slope and hydraulic radius are evaluated readily from the geometry of the channel. However, the roughness coefficient is more difficult to evaluate. Extensive tests by Ree (1949, 1958, 1976, 1977) and others have provided techniques and data to determine roughness coefficients for various types of vegetation. Figure 7.3 illustrates the complexity of the problem. The roughness coefficient varies tremendously with the depth of flow. Flows at very shallow depths encounter a maximum resistance because the vegetation is upright in the flow. The slight increase in resistance in the low flow range apparently is due to the greater bulk of vegetation encountered with increasing depth. Intermediate flows bend over and submerge some of the grass, and resistance drops off sharply, as more and more vegetation is submerged.

Resistance to flow is also influenced by the gradient of the channel. Decreasing resistance apparently results from higher velocities on steeper slopes with an accompanying greater flattening of the vegetation. Type and condition of vegetation have a great influence on the retardance. Newly mowed grass offers less resistance than rank growth. Long plants, stems, and leaves tend to whip and vibrate in the flow, thus introducing and maintaining considerable turbulence. The cross-sectional shape of the channel has only minor influence upon the roughness coefficient in the range of cross sections commonly used.

The product vR, velocity multipled by hydraulic radius, has been found to be a satisfactory index of channel retardance for design purposes. Vegetation has

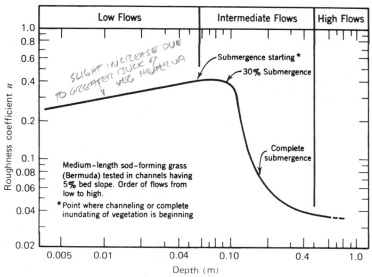

Fig. 7.3. Hydraulic behavior of a medium-length sod forming grass. (Redrawn from Ree, 1949.)

STEEPER SCOPES ⇨ HIGHER VELOCITY → GREATER FATTENING ? VEG

∴ STEEPER ⇨ DECREASED RESISTANCE

been grouped into five retardance categories designated A through E. Table 7.3 gives a portion of this classification of vegetation by degree of retardance, and Fig. 7.4 shows the n-vR curves for five retardance categories. In the past it has been common practice to use $n = 0.04$ for vegetated waterways. In many channels this may be satisfactory, but careful consideration should be given to the vegetation and flow conditions especially for long, large channels where refinements in design may result in lower construction costs.

7.7. Channel Capacity. The channel must be proportioned to carry the design runoff at average velocities less than or equal to the permissible velocity. This is accomplished by application of the Manning formula,

$$V = \frac{1.486\,R^{2/3}\,s^{1/2}}{n}$$

$$v = R^{2/3}s^{1/2}/n \tag{7.1}$$

$R = \dfrac{A}{WP}$

$S = \dfrac{6}{6^+}$

where v = average velocity of flow in m/s,
  n = roughness coefficient of the channel,
  R = a/p, the cross-sectional area divided by the wetted perimeter in m,
  s = hydraulic gradient (slope of channel).

The dimensions of the channel must be so selected that

$$v = q/a \tag{7.2}$$

where q = flow rate to be carried in m³/s.

Nomographs in Figs. 7.5, 7.6, and 7.7 give solutions to the Manning equation for retardance classes B, C, and D, respectively. For trapezoidal channels with 4:1 side slopes the depth of flow can be determined for the required cross sectional area, bottom width, and hydraulic radius. Larson and Manbeck (1960) found that for small parabolic cross sections $d = 1.5R$ and for triangular cross sections $d = 2R$ with less than 1 percent error in velocity. These approximations greatly simplify the calculations, but when many solutions are required, these equations could easily be solved on a computer. The channel should be designed to carry the runoff at a permissible velocity under conditions of minimum retardance (short grass) that may be encountered during the runoff season. This condition establishes the basic proportions of the channel, for example, the bottom width of a trapezoidal channel. Additional depth should then be added to the channel to provide adequate capacity under conditions of maximum retardance (long grass). A freeboard of 0.1 to 0.15 m should be added to the design depth. Examples 7.1 and 7.2 illustrate design procedure.

Example 7.1. Design a trapezoidal grassed waterway to carry 5.66 m³/s (200 cfs) on a 3 percent slope on erosion-resistant soil. The vegetation is to be Bermuda grass, and the channel should have 4:1 side slopes.

Table 7.3 Classification of Vegetal Cover According to Retardance

Retardance Class	Cover	Condition	Avg. Height (cm) (in.)
B	Alfalfa	Good stand, uncut	28 (11)
	Bermuda grass	Good stand, tall	30 (12)
	Blue grama	Good stand, uncut	33 (13)
	Brome, tall fescue	Long	
	Kudzu	Dense or very dense, uncut	
	Lespedeza sericea	Good stand, not woody, tall	48 (19)
	Reed canary	Long	
	Weeping lovegrass	Good stand, tall or mowed	33–61 (13–24)
	Wheat	Mature, good, $vR \geq 1$	
C	Bermuda grass	Good stand, unmowed	15 (6)
	Brome, tall fescue	Mowed	
	Centipede grass	Very dense cover	15 (6)
	Common lespedeza	Good stand, uncut	28 (11)
	Crabgrass	Fair stand, uncut	25–122 (10–48)
	Grass mixture (orchard grass, red top, Italian ryegrass, common lespedeza)	Good stand, uncut	15–20 (6–8)
	Kentucky bluegrass	Good stand, headed (6–12 in.)	15–30 (6–12)
	Reed canary	Mowed	
	Wheat	Mature, poor, $vR \geq 0.5$	
D	Bermuda grass	Good stand, cut	6 (2.5)
	Common lespedeza	Excellent stand, uncut	11 (4.5)
	Buffalo grass	Good stand, uncut	8–15 (3–6)
	Grass mixture (as above)	Good stand, uncut	10–13 (4–5)
	Lespedeza sericea	Good stand, cut	5 (2)

Source: Modified from Ree (1949, 1958).

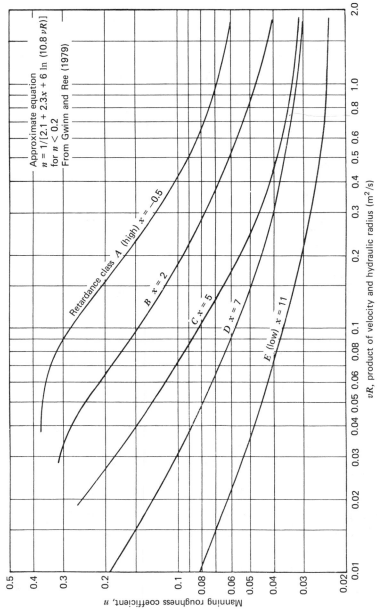

Fig. 7.4. Roughness coefficient as a function of vR for various retardance classes of vegetation. (Redrawn from U.S. SCS, 1966.)

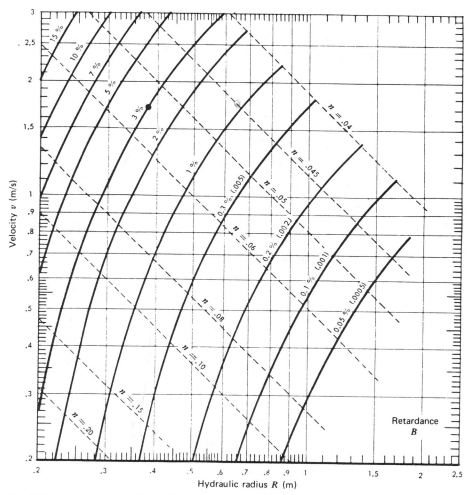

Fig. 7.5. Nomograph of the Manning equation for vegetated channels of retardance class B (high vegetal retardance). (From U.S. SCS, 1966.)

Solution. From Table 7.2 the permissible velocity is 2.4 m/s (8 fps). Table 7.3 shows Bermuda grass in retardance class D when mowed and in class B when long. To design for stability against erosion, the mowed condition is the more critical. Entering Fig. 7.7 for class D retardance with $v = 2.4$ m/s (8 fps) and a slope of 3 percent, $R = 0.31$ m (1.02 ft). A trial and error solution could be obtained from the curve in Fig. 7.4. The trapezoidal cross-sectional area must be $(5.66/2.4) = 2.36$ m² (25 ft²) and $R = 0.31$ m (1.02 ft). From Fig. 7.8 for $R = 0.31$ m and $a = 2.36$ m², read $b = 4.0$ m and the depth of flow $d = 0.41$ m (1.3 ft). These dimensions will provide a stable channel with $v = 2.4$ m/s (8 fps).

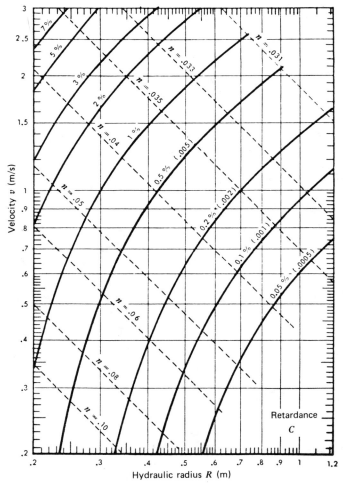

Fig. 7.6. Nomograph of the Manning equation for vegetated channels of retardance class C (moderate vegetal retardance). (From U.S. SCS, 1966.)

The design depth must now be increased when the grass is long with retardance class B because the velocity is reduced. The previous bottom width of 4.0 m must be retained. By trial and error select a depth of 0.53 m (1.7 ft) that will have $a = 3.24$ m^2 (34.9 ft^2) and $R = 0.39$ m (1.28 ft). Entering Fig. 7.5 with $R = 0.39$ m and a slope of 3 percent gives $v = 1.75$ m/s (5.7 fps). At the 0.53-m depth, $q = 3.24 \times 1.75 = 5.67$ m^3/s (200 cfs), which checks, or within 10 percent is adequate. An alternate solution is to select the approximate additional depth increment, 0.16 m (0.13 + 0.03) from Table 7.4 for the increased roughness. The depth of flow is 0.41 + 0.16 = 0.57 m (1.9 ft).

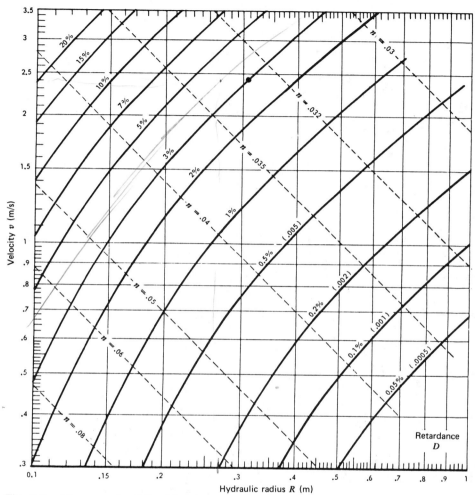

Fig. 7.7. Nomograph of the Manning equation for vegetated channels of retardance class D (low vegetal retardance). (From U.S. SCS, 1966.)

An additional freeboard or safety factor of 0.1 m (0.3 ft) may be included; this would increase the depth to 0.67 m (2.2 ft) making the constructed top width T = 4.0 + (2 × 0.67 × 4) = 9.4 m (30.7 ft).

The example shows that the bottom width is determined by the need not to exceed the permissible velocity under the mowed condition of minimum retardance and that the depth is determined by the need to provide capacity under conditions of maximum retardance.

Example 7.2 Design a parabolic grassed waterway for the same conditions as in Example 7.1.

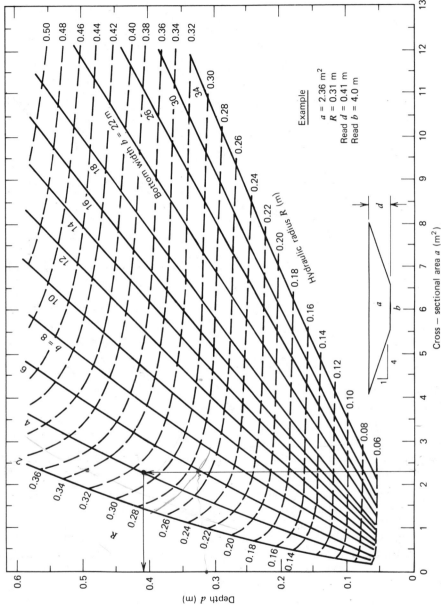

Fig. 7.8. Dimensions of trapezoidal cross sections with 4:1 side slopes. (Redrawn in SI units from Larson and Manbeck, 1960.)

Table 7.4 Approximate Channel Depth Increment for Grassed Waterways with Retardance Increase from Class C (short) to Class B (long)

Slope Range (%)	*Channel Cross Section and Width*		
	Triangular t = 2 to 6 m *Trapezoidal* b = 2 to 6 m	*Parabolic* t = 6 to 27 m *Trapezoidal* b = 6 to 27 m	*Trapezoidal* b = 27 m or more
1–2	0.17 m (0.6 ft)	0.15 m (0.5 ft)	0.12 m (0.4 ft)
2–5	0.13 (0.4)	0.12 (0.4)	0.09 (0.3)
5 or more	0.10 (0.3)	0.09 (0.3)	0.06 (0.2)

Note: For change from retardance D to B add 0.03 m (0.1 ft) to the above values for all slopes and cross sections. For change from retardance D to C use 0.03 m (0.1 ft) as the depth increment for all slopes and cross sections.
Source: Larson and Manbeck (1960) as computed by C. L. Larson.

Solution.   From Example 7.1 the hydraulic requirements for a stable channel for class D retardance are $a = 2.36$ m² (25 ft²) and $R = 0.31$ m (1.02 ft). Using the approximate relationship for parabolic channels of $d = 1.5R$, $d = 1.5 \times 0.31 = 0.46$ m (1.52 ft). From Fig. 7.1, $t = 3a/2d = (3 \times 2.36)/(2 \times 0.46) = 7.70$ m (25.2 ft). These dimensions provide for a stable channel.

The additional increase in depth when in class B retardance (long grass) is 0.15 m (0.5 ft) from Table 7.4. The design depth of flow is $0.46 + 0.15 = 0.61$ m (2.0 ft), and the constructed depth including 0.1 m freeboard is $0.61 + 0.1 = 0.71$ m (2.3 ft). Top width $T = 7.70(0.71/0.46)^{1/2} = 9.59$ m (31.4 ft).

A more accurate method for obtaining the flow depth with long grass is to assume a new depth of 0.57 m (1.9 ft) (trial and error), which results in $R = 0.57/1.5 = 0.38$ m (1.25 ft). From Fig. 7.5 read $v = 1.70$ m/s (5.6 fps). Since the top width of a parabolic cross section by definition varies as the square root of the depth, the new $t = 7.70(0.57/0.46)^{1/2} = 8.57$ m (28.1 ft). Then $q = (2/3)8.57 \times 0.57 \times 1.70 = 5.56$ m³/s (196 cfs), which is adequate. Adding 0.10 m (0.3 ft) freeboard, the constructed dimensions of the channel are $d = 0.67$ m (2.2 ft) and $T = 9.29$ m (30.5 ft), which are similar to the trapezoidal cross section in Example 7.1.

Waterways are often designed on the basis of tables, charts, or rules of thumb developed for a particular area. An engineer or technician working in a given region gains confidence in such shortcuts particularly adapted to local conditions.

Immediately after construction the channel may be called upon to carry runoff under conditions of little or no vegetation. It is not practical to design for this extreme condition. In many channels it may be practical and desirable to divert flow from the channel until vegetation is established. In others the possibility that high runoff will occur before vegetation is established is accepted as a calculated risk.

7.8. Drainage. Waterways that are located in seepy draws or below seeps, springs, or tile outlets will be continually wet for long periods of time. The wet condition will inhibit the development and maintenance of a good vegetal cover and will maintain the soil in a soft, erosive condition. Subsurface (pipe) drainage or diversion of such flow is essential to the success of the waterway.

Low continuous flow of surface water entering at some point may be intercepted by a catch basin and carried off by a pipe drain. A concrete or asphalt trickle channel of about 0.2 square meters cross section is sometimes placed in the bottom of a waterway to carry prolonged low flows. Seepage along the sides or upper end of the waterway may be intercepted by pipe drains. Pipe should be placed to one side of the center of the waterway to prevent washing out of pipe in case of failure of the waterway.

WATERWAY CONSTRUCTION

7.9. Shaping Waterways. The procedure and amount of work involved in shaping a waterway depends upon the topographic situation and the equipment available. If the watercourse is to be located in a natural draw or meadow outlet where there is little gullying, only smoothing and normal seedbed preparation are required. Some improvement in alignment of the channel may be desired to remove sharp bends. This will improve the hydraulic characteristics, facilitate farming operations, and reduce channel maintenance. If the waterway is to reclaim an established gully, considerable earthwork is required. The gully must be filled and the waterway cross section properly shaped.

Small waterways may be easily shaped with regular farm equipment. Large gullies, however, can be most satisfactorily handled by a bulldozer or other heavy earth-moving equipment.

ESTABLISHMENT OF VEGETATION

7.10. Seedbed Preparation. Soil in the waterway should be brought to a high fertility level and limed in accordance with the soil and plant requirements. Manure worked into the seedbed provides needed plant nutrients and furnishes organic material which will help the soil to resist erosion. Specific lime and fertilizer needs depend upon local crop and soil conditions.

7.11. Seeding. Waterway seeding mixtures should include some quick-growing annual for temporary control as well as a mixture of hardy perennials for permanent protection. Seed should either be broadcast or drilled nonparallel to the direction of flow. Mulching after seeding helps to secure a good stand.

Fig. 7.9. Paper net being installed to secure straw mulch on a newly seeded sod waterway. (Courtesy U.S. Soil Conservation Service.)

Where high flows must be turned into a channel before seedings can become established, sodding may be justified.

Many new materials have been under test to determine their effectiveness in controlling infiltration and runoff as a means of facilitating revegetation. Such materials include chemical soil stabilizers, plastic, fiber or other mesh or net covers (see Fig. 7.9), asphalt mulches, and plastic or other surface covers.

The primary purpose of such materials is to stabilize the soil surface during the establishment of grass seedlings. The materials serve (1) as a mulch to reduce rate of drying and crusting, (2) as a barrier to absorb the energy of raindrops or wind, thus protecting the surface structure of the soil, and (3) as a retarding barrier to reduce the velocity of flow of runoff water.

WATERWAY MAINTENANCE

7.12. Causes of Failure. Failure of vegetated waterways may result from insufficient capacity, excessive velocity, or inadequate vegetal cover. The first two of these are largely a matter of design. The condition of the vegetation, however, is influenced not only by the initial preparation of the waterway but also by the subsequent management. Use of a waterway especially in wet weather, as a lane, stock trail, or pasture injures the vegetation and often results in failure. Terraces that empty into a waterway at too steep a grade may cut back into the terrace channel, injuring both terrace and waterway. Careless handling of machinery in crossing a waterway may injure the sod. When land adjacent to the waterway is being plowed, the ends of furrows abutting against the vegetated strip should be staggered to prevent flow concentration down the edges of the watercourse.

7.13. Controlling Vegetation. Waterways should be mowed and raked several times a season to stimulate new growth and control weeds. A rotary-type mower cuts the grass fine enough to make raking unnecessary. Annual application of manure and fertilizer maintains a dense sod. Any breaks in the sod should be repaired. Rodents that are damaging waterways should be controlled.

7.14. Sediment Accumulation. Good conservation practice on the watershed is the most effective means of controlling sedimentation. Accumulated sediments smother vegetation and restrict the capacity of the waterway. Extending vegetal cover well up the side slopes of the waterway and into the outlets of terrace channels helps to prevent sediment from being deposited in the watercourse. Control of vegetation to prevent a rank, matted growth reduces the accumulation of sediment. High allowable design velocities also decrease sedimentation.

VEGETATION OF LARGE GULLIES

The discussions to this point have been mainly applicable to waterways and to the stabilization of small gullies. Larger gullies may be controlled by reduction of the surface inflow, or by shaping and intensive natural or artificial revegetation.

7.15. Control of Inflow. Where large gullies exist, there is generally found a contributing watershed area either of considerable size or one having a high runoff potential. To facilitate the vegetation and control of the gully, the normal conservation practices that will protect small gullies must be replaced with more effective methods.

Diversion terraces, constructed above the gully head and carefully laid out on a grade that will resist channel erosion, can be used to intercept runoff from the watershed area above the gully and then convey it to a safely stabilized outlet area. Palmer (1963) reports that in some river terrace areas a system of dikes may be used to hold runoff back from the gully head until it can normally infiltrate into the permeable river terrace soils.

7.16. Sloping Gully Banks. Bank sloping should be done only to the extent required for establishment of vegetation or for facilitating tillage operations. Where trees and shrubs are to be established, rough sloping of the banks to about 1:1 should be sufficient. Where gullies are to be reclaimed as grassed waterways, sloping of banks to 4:1 or flatter usually is desired.

Sloping banks of small gullies may be accomplished with plows, disks, and other farm tools. Large gullies must be shaped with heavy earth-moving equipment.

7.17. Natural Revegetation. If the runoff that has caused the gully is diverted, and livestock is fenced from the gullied area, plants will begin to come in naturally. A gradual succession of plant species eventually will protect the gullied area with grasses, vines, shrubs, or trees native to the area in question. In some gullied areas, the development of vegetation may be stimulated by fertilizing and by spreading a mulch to conserve moisture and protect young volunteer plants. Vertical gully banks may be roughly sloped to prevent caving and provide improved conditions for natural seeding. The opportunity to provide protective cover by natural revegetation frequently is overlooked, and unnecessary expenditures are made for structures and plantings.

7.18. Artificial Revegetation. Selection of vegetation to be established artificially in a reclaimed gully should be governed by the use intended for the planted area. Grasses and legumes may be planted if the vegetation is to be used for a hay or pasture crop. Where gullies are reclaimed as drainageways in cultivated fields, sod-forming vegetation should be selected to permit crossing of the drainageway with farm machines.

In some areas trees and shrubs are easier to establish in gullies than are grasses, particularly if the gully is not to be shaped to permit operation of farm implements. Locally adapted trees planted on gullied areas may be utilized for fence posts or rough timber. Shrubs, such as dogwoods and lespedezas, are desirable for establishing the gullied area as a wildlife refuge. Trees and shrubs should be planted in accordance with local recommendations for control of erosion and establishment of wildlife refuges.

REFERENCES

Chepil, W. S. et al. (1963). "Vegetative and Non-Vegetative Materials to Control Wind and Water Erosion." *Soil Sci. Soc. Am. Proc.* **27**, 86–89.

Chow, V. T. (1959). *Open Channel Hydraulics*. McGraw-Hill, New York.

Cowan, W. L. (1956). "Estimating Hydraulic Roughness Coefficients." *Agr. Eng.* **37**, 473–475.

Fenzl, R. N. and J. R. Davis (1962). "Hydraulic Resistance Relationships for Surface Flows in Vegetated Channels." *Am. Soc. Agr. Eng. Trans.* **7** (1), 46–51, 55.

Gwinn, W. R. and W. O. Ree (1979). "Maintenance Effects on the Hydraulic Properties of a Vegetation-Lined Channel." ASAE Paper No. 79-2063.

Larson, C. L., and D. M. Manbeck (1960). "Improved Procedures in Grassed Waterway Design." *Agr. Eng.* **41**, 694–696.

Palmer, R. S. (1963). "River-Terrace Diking for Gully Control." *Am. Soc. Agr. Eng. Trans.* **6** (4), 325–328.

Ree, W. O. (1976). "Effect of Seepage Flow on Reed Canarygrass and Its Ability to Protect Waterways." U.S. Dept. Agr., Agr. Res. Serv., ARS-S-154.

———— (1949). "Hydraulic Characteristics of Vegetation for Vegetated Waterways." *Agr. Eng.* **30**, 184–187, 189.

———— (1977). "Performance Characteristics of a Grassed-Waterway Transition." U.S. Dept. Agr., Agr. Res. Serv., ARS-S-158.

———— (1958). "Retardation Coefficients for Row Crops in Diversion Terraces." *Am. Soc. Agr. Eng. Trans.* **1** (1), 78–80.

Ree, W. O., and V. J. Palmer (1949). "Flow of Water In Channels Protected by Vegetative Linings." U.S. Dept. Agr. Tech. Bull. 967.

U.S. Soil Conservation Service (1966). *Handbook of Channel Design for Soil and Water Conservation.* SCS-TP-61. Revised 1954. Converted to SI units.

PROBLEMS

7.1. Determine the velocity of flow in a parabolic-, a triangular-, and a trapezoidal-shaped waterway, all having a cross-sectional area of 1.86 m² (20 ft²), a depth of flow of 0.3 m (1.0 ft), a channel slope of 4 percent, and a roughness coefficient of 0.04. Assume 4:1 side slopes for the trapezoidal cross section.

7.2. Design a parabolic-shaped grassed waterway to carry 1.42 m³/s (50 cfs). The soil is easily eroded; the channel has a slope of 4 percent; and a good stand of Bermuda grass, cut to 60 mm (2½ in), is to be maintained in the waterway.

7.3. Design a trapezoidal-shaped waterway with 4:1 side slopes to carry 0.57 m³/s (20 cfs) where the soil is resistant to erosion and the channel has a slope of 12 per cent. Brome grass in the channel may be either mowed or long when maximum flow is expected.

7.4. Design a parabolic-shaped waterway to carry 1.70 m³/s (60 cfs) from a terraced field where the soil is resistant to erosion and the channel slope is 6 percent. Tall fescue is normally mowed, but the channel should have adequate capacity when the grass is long.

7.5. Design a trapezoidal-shaped emergency spillway for a farm pond to carry 1.70 m³/s (60 cfs) with retardance class C condition. The maximum velocity in the spillway is 1.52 m/s (5 fps). Around the end of the dam the channel slope is to be 2 percent, but it is increased to 7 percent from this section to stream channel below. The depth of flow should not exceed 0.3 m (1.0 ft). Determine the bottom width, design depth, and recommend design slope for both the 2 and 7 percent sections.

7.6. If the velocity of flow is 2.2 m/s (7.2 fps) for complete submergence ($n = 0.035$) of Bermuda grass described in Fig. 7.3, what is the velocity at 30 percent submergence considering only the change in roughness coefficient? If the hydraulic radius was also reduced from 0.24 to 0.12 m (0.8 to 0.4 feet), what is the expected velocity?

7.7. At low flow in a channel the velocity is 0.61 m/s (2 fps) and the hydraulic radius is 0.15 m (0.5 ft) for a short grass with retardance class D. From Fig. 7.4 determine the relative change in the roughness coefficient for long grass with retardance class B for the same conditions. If the velocity is increased to 3.05 m/s (10 fps) and the hydraulic radius to 0.3 m (1.0 ft) for the long grass (class B), what is the change in the roughness coefficient?

7.8. A diversion terrace is needed for a flow of 0.85 m³/s (30 cfs). Assume a poor stand of grass and use an allowable velocity of 0.9 m/s (3.0 fps). Determine d, D, and the required slope using a 2.44-m (8-ft) bottom width. Allow 0.1 m (0.3 ft) of freeboard with long grass (class D). Use a trapezoidal cross section with 4:1 side slopes.

CHAPTER 8

Terracing

Terracing is a method of erosion control accomplished by constructing broad channels across the slope of rolling land. It has been estimated that over 36 million hectares (90 million acres) of cropland in the United States could be more effectively protected from runoff and erosion damage through the use of well designed and maintained terrace systems. The use of diversions, a form of terrace, to protect bottomlands, buildings, and special areas from destructive runoff and erosion from hillsides is becoming increasingly important. The first terraces consisted of large steps or level benches as compared to the broadbase terraces now common. For several thousand years, bench terraces have been widely adopted over the world, particularly in Europe, Australia, and Asia. In the United States, ditches that functioned as terraces were constructed across the slopes of cultivated fields by farmers in the southern states during the latter part of the eighteenth century. As shown in Fig. 8.1, most terraces are found in the central and south central states with a considerable number in the southeast.

As technology has advanced, terrace design has been scientifically adapted to the hydrologic and erosion control needs of the treated areas. Channel cross sections have been modified to become more nearly compatible with modern mechanization.

8.1. Function of Terraces. The function of terraces is to decrease the length of the hillside slope, thereby reducing sheet and rill erosion, preventing the formation of gullies, and retaining runoff in areas of inadequate precipitation. In dry regions such conservation of moisture is important in the control of wind erosion. In most areas graded terraces are more effective in reducing erosion than runoff, whereas level terraces are effective in reducing runoff as well as controlling erosion.

Since crop rows are always made parallel to the terrace channel, terracing also includes contouring as a conservation practice. Since terracing requires additional investment and causes some inconvenience in farming, it should be considered only where other cropping and soil management practices, singly or in combination, will not provide adequate erosion control.

8.2. Terrace Classification. The two major types of terraces are the bench terrace, which reduces land slope, and the broadbase terrace, which removes

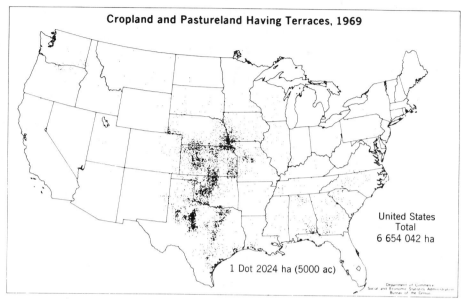

Fig. 8.1. Geographic distribution of cropland and pastureland having terraces in U.S. (1969). (Courtesy U.S. Bureau of the Census.)

or retains water on sloping land (see Fig. 8.2). The original bench terraces were adapted to slopes of 25 to 30 percent, were costly to construct, and were not always suited to modern cultivation equipment. The modern conservation bench terrace is adapted to slopes up to 6 to 8 percent and may be changed to meet cultivation machinery needs.

The broadbase terrace has wide use throughout the United States for control of runoff, for erosion control, and for moisture conservation. The long process of development of the broadbase terrace has led to a variety of types, terms, and classifications. On the basis of alignment they are classified as either parallel or nonparallel. Terraces may be constructed from the uphill or downhill side or both, and side slopes may be steep or flat. Grassed-backslope terraces are common in some areas. Some are made with only an excavated channel, using the cut soil to fill interterrace area depressions.

TYPES OF TERRACES

A <u>broadbase terrace</u> is a broad surface channel or embankment constructed across the slope of rolling land. On the basis of primary function the broadbase terrace is classified as graded or level. The distinguishing characteristic of this terrace is farmability $(4:1 \leftarrow 8:1)$

In addition to usual factors affecting runoff and erosion, soil and water losses from terraced areas are influenced by both the spacing and the length of the terrace, as well as the velocity of flow in the channel.

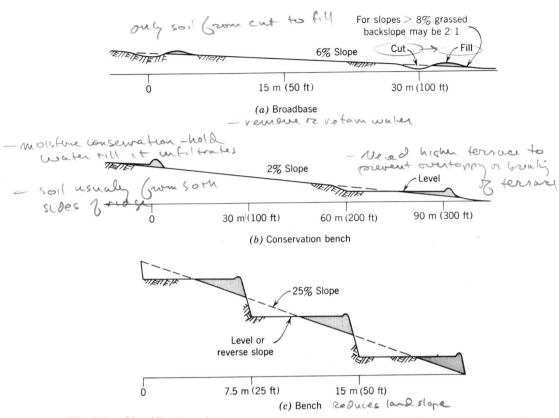

only soil from cut to fill (handwritten)

For slopes > 8% grassed backslope may be 2:1

6% Slope

Cut → Fill

0   15 m (50 ft)   30 m (100 ft)

(a) Broadbase

— remove or retain water (handwritten)

— moisture conservation —hold water till it unfiltrates (handwritten)

2% Slope

— Need higher terrace to prevent overtopping or breaking of terrace (handwritten)

Level

— soil usually from soth slopes of ridge (handwritten)

0   30 m (100 ft)   60 m (200 ft)   90 m (300 ft)

(b) Conservation bench

25% Slope

Level or reverse slope

0   7.5 m (25 ft)   15 m (50 ft)

(c) Bench   *Reduces land slope* (handwritten)

Fig. 8.2. Classification of terraces. (a) Broadbase. (b) Conservation bench. (c) Bench.

8.3. Graded or Channel-Type Terrace. The primary purpose of this type of terrace is to remove excess water in such a way as to minimize erosion. Terraces control erosion by reducing the slope length and conducting the intercepted runoff to a safe outlet at a nonerosive velocity.

In general, this type of terrace is constructed, as shown in Fig. 8.2a, by cutting a shallow channel on the uphill side and using only this soil to build the embankment. The side slopes of both the channel and the ridge are kept as flat as possible to facilitate farming operations. For land slopes over 8 percent the ridge backslope may be a 2:1 slope, if grassed. Because of the importance of constructing and maintaining a satisfactory channel, this type of terrace should not be built on deep sands or on soils that are too stony, steep, or shallow to permit adequate construction.

8.4. Level or Ridge-Type Terrace. The primary purpose of this type of terrace is moisture conservation; erosion control is a secondary objective. In

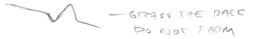

STEEP BACKSLOPE (handwritten)   *— GRASS THE BACK DO NOT FARM* (handwritten)

low-to-moderate rainfall regions they trap and hold rainfall for infiltration into the soil profile. On permeable soils they may be used for this same purpose in high rainfall areas.

The embankment for this type of terrace is usually constructed of soil taken from both sides of the ridge. This is necessary to obtain a sufficiently high embankment to prevent overtopping and breaking through by the entrapped runoff water. The channel is level and is sometimes closed at both ends to assure maximum water retention. On slopes of over 2 percent, water in the channel is spread over a relatively small area, thus limiting the effect upon crop yield. The conservation bench terrace, designed to correct this deficiency, is shown in Fig. 8.2b.

8.5. Bench Terrace. The early conventional bench terrace system consisted of a series of flattened shelf-like areas that converted a steep slope of 20 to 30 percent to a series of level, or nearly level, benches as shown in Fig. 8.2c. The use of such terraces, while important in many ancient civilizations, is limited mainly to providing for more efficient distribution of irrigation water where extremely rough terrain exists.

The conservation bench terrace is designed for semiarid regions where maximum moisture conservation is needed. It consists of an earthen embankment and a very broad flat channel that resembles a level bench. The sloping area above this bench and extending upslope to the next terrace ridge, is the runoff contributing area to the benched area. The relationship of this type terrace to the conventional level terrace, both in terms of cross section and in relative moisture storage pattern, is shown in Fig. 8.3.

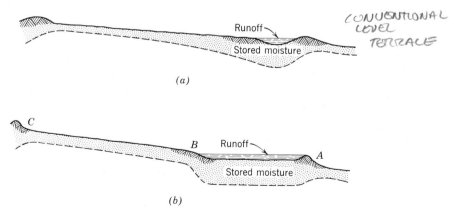

Fig. 8.3. Comparison of cross sections of (a) conventional level terrace and (b) conservation bench terrace showing available soil moisture stored. Ratio of storage area *AB* to runoff area *BC* may be varied to suit soil, cover, and topographic conditions. (After Hauser et al., 1962.)

TERRACE SPACING — VERTICAL DISTANCE (VI)
  — TOP TERRACE; FROM TOP OF HILL TO
                  BOTTOM OF CHANNEL
  — OTHERS — BETWEEN CHANNELS

168   TERRACING

TERRACE DESIGN

The design of a terrace system involves the proper spacing and location of terraces, the design of a channel with adequate capacity, and development of a farmable cross section. For the graded terrace, runoff must be removed at nonerosive velocities in both the channel and the outlet. Soil characteristics, cropping and soil management practices, and climatic conditions are the most important considerations in terrace design.

8.6. Terrace Spacing. Spacing is expressed as the vertical distance between the channels of successive terraces. For the top terrace the spacing is the vertical distance from the top of the hill to the bottom of the channel. This vertical distance is commonly known as the vertical interval or V.I. For a given slope the vertical spacing and the horizontal spacing are directly related. The horizontal spacing for parallel terraces is usually selected as an even number of rounds for row crop equipment. The vertical interval is more convenient for terrace layout and construction with surveying equipment.

Graded. Under given conditions, graded terrace spacing is often expressed as a function of land slope by the empirical formula

$$V.I. = 0.3(XS + Y) \tag{8.1}$$

where $V.I.$ = vertical interval between corresponding points on consecutive terraces or from the top of the slope to the bottom of the first terrace in meters,

X = constant for geographical location as given in Fig. 8.4,

Y = constant for soil erodibility and cover conditions during critical erosion periods (see Fig. 8.4),

S = average land slope above the terrace in percent.

Spacings thus computed may be varied as much as 25 percent to allow for soil, climatic, and tillage conditions. Terraces are seldom recommended on slopes over 20 percent, and in many regions slopes from 10 to 12 percent are considered the maximum.

Where soil loss data are available, spacings should be based on slope lengths using contouring and the appropriate cropping-management factor which will result in soil losses within the permissible loss as outlined in Chapter 5. Estimation of terracing spacing is illustrated in the following examples.

Example 8.1. If the soil loss was 16.8 Mg/ha (7.5 t/a) at Memphis, Tenn., for K = 0.1 t/a, l = 122 m (400 ft), S = 8 percent, C = 0.2, and P = 0.6 (contouring) in the USLE, what is the maximum slope length and corresponding terrace spacing to reduce the soil loss to the terrace channel to 6.7 Mg/ha (3 t/a)?

Handwritten annotations at top:

$VI \left\{ \dfrac{\%S}{100} \right\}$

$\tan \theta = \dfrac{\%S}{100}$

$\sin \theta = \dfrac{VI}{L}$

$VI = 0.3 (x5\% + Y)$ — May vary as much as 25% for tillage etc.

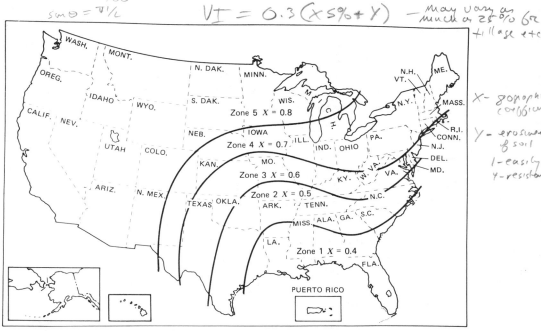

Handwritten annotations at right:
X - geographic condition
Y - erosion of soil
1 - easily
4 - resistant

Fig. 8.4. Values of X and Y in the terrace spacing equation (Eq. 8.1). For Y of "1," low intake rates and little cover. For Y of "4," average or above intake rates and good cover. Interpolate between 1 and 4 for other conditions. (Modified from U.S. SCS, 1969.)

Solution.   From Fig. 5.7 read $LS = 2.0$. The maximum LS value to reduce soil loss to 6.7 Mg/ha (3 t/a) is

$$LS = 2.0 \times 6.7/16.8 = 0.80.$$

From Fig. 5.7 read from the 8 percent curve a slope length of 21.3 m (70 ft). By similar triangles,

$$V.I. = 8 \times 21.3/100 = 1.7 \text{ m (5.6 ft)}$$

Example 8.2.   Compute the terrace spacing for Example 8.1 from Eq. 8.1 assuming that the soil has a low intake rate and that good cover conditions exist.

Solution.   From Fig. 8.4 read $X = 0.5$ and assume $Y = 2$. Substituting in Eq. 8.1,

$$V.I. = 0.3 [(0.5 \times 8) + 2] = 1.8 \text{ m (5.9 ft)}$$

Since the permissible spacing based on soil loss takes into account more of the erosion variables, a vertical interval of 1.7 m (5.6 ft) as computed in Example 8.1 is a better estimate.

Terrace spacing should not be so wide as to cause excessive rilling and the resultant movement of large amounts of soil into the terrace channel. On sandy soils excessive interterrace erosion makes the soils progressively sandier as the fines are carried away and the sand grains are left behind.

Level. The spacing for level terraces is a function of channel infiltration and runoff; however, in more humid areas where erosion control is important, the slope length may limit the spacing. The runoff from the terraced area should not cause overtopping of the terrace, and the infiltration rate in the channel should be sufficiently high to prevent serious damage to crops. Spacings vary widely in different parts of the country.

8.7. Terrace Grades. Gradient in the channel must be sufficient to provide good drainage and to remove runoff at nonerosive velocities. The minimum slope is desirable from the standpoint of soil loss. With reference to slope in the channel, terraces are constructed with uniform or variable grades. Level terraces have zero grades.

In the uniform-graded terrace the slope remains constant throughout its entire length. Because of higher velocities (greater slope) in the upper portions of the channel of a uniform graded terrace, sediment is not as likely to be deposited as in the upper portions of the variable-graded terrace. A grade of 0.4 percent is common in many regions; however, grades may range from 0.1 to 0.6 percent, depending on soil and climatic factors. Generally, the steeper grades are recommended for impervious soils and short terraces.

ASAE (1972) recommended maximum velocities of 0.46 m/s for extremely erosive soils and 0.61 m/s for most other soils, when the roughness coefficient in the Manning formula is taken as 0.03. Recommended minimum and maximum grades are given in Table 8.1. As will be discussed later, higher allowable grades with short terraces greatly facilitate the layout of parallel terraces.

The variable-graded terrace is more effective because the capacity increases toward the outlet with a corresponding increase in runoff. The grade usually varies from a minimum at the upper portion to a maximum at the outlet end. The resulting reduced velocity in the upper reaches provides for greater absorption of runoff. Variable gradient makes possible greater flexibility in design; for instance, either constant velocity or constant capacity could be provided by varying the grade in the channel. Such designs are sometimes required, particularly in large diversion terraces.

8.8. Terrace Length. Size and shape of the field, outlet possibilities, rate of runoff as affected by rainfall and soil infiltration, and channel capacity are

Table 8.1   Maximum and Minimum Terrace Grades

Terrace Length or Length from Upper End of Long Terraces		Maximum slope (%)
Meters	*Feet*	
30 or less	99 or less	2.0
31 to 60	100 to 199	1.2
61 to 150	200 to 499	0.5
151 to 365	500 to 1199	0.35
366 or more	1200 or more	0.3
		Minimum Slope (%)
Soils with slow internal drainage		0.2
Soils with good internal drainage		0.0

Source: Beasley (1963).

factors that influence terrace length. The number of outlets should be a minimum consistent with good layout and design. Extremely long graded terraces are to be avoided; however, long lengths may be reduced in some terraces by dividing the flow midway in the terrace length and draining the runoff to outlets at both ends of the terrace (see Fig. 8.5). The length should be such that erosive velocities and large cross sections are not required. On permeable soils longer terraces may be permitted than on impermeable soils. The maximum length for graded terraces generally ranges from about 300 to 500 m, depending on local conditions. The maximum applies only to that portion of the terrace that drains toward one of the outlets.

There is no maximum length for level terraces, particularly where blocks or dams are placed in the channel about every 150 m. These dams prevent total loss of water from the entire terrace and reduce gully damage should a break occur. The ends of the level terrace may be left partially or completely open to prevent overtopping in case of excessive runoff.

8.9.   Terrace Cross Section.   The terrace cross section should provide adequate capacity, have broad farmable side slopes, and be economical to construct with available equipment. The cross section of a broad-base terrace for design purposes can be considered a triangular channel as shown in Fig. 8.6*a*. The flow depth (d) is the height to the top of ridge (h) less a freeboard of about 8 cm. After smoothing, the ridge and bottom widths will be about 1 m, which will give a cross section that approximates the shape of a terrace after 10 year's of farming as shown in Fig. 8.6*b*.

In designing the cross section the front slope width (W_f) is chosen equal to the machinery width ordinarily used for row-crop operations. The depth of flow is

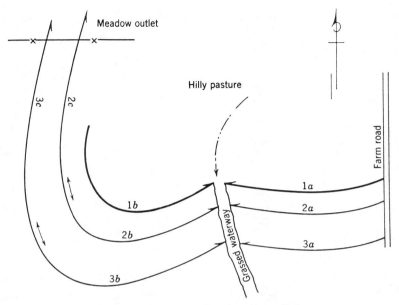

Fig. 8.5. Typical layout for broadbase graded terraces.

determined from the runoff rate for a 10-year return period storm or for the required runoff volume for storage-type terraces. When the side slope widths are equal ($W_c = W_f = W_b = W$), cuts and fills from the geometry are

$$c + f = h + sW \tag{8.2}$$

where c = cut (L),
        f = fill (L),
        h = depth of channel including freeboard (L),
        s = original land slope (L/L),
        W = width of side slope (L).

For a balanced cross section with cut equal to fill, Table 8.2 was obtained from Eq. 8.2. The front side slope ratio was also computed for W = 427 cm (14 ft). This ratio is the same for all slopes and varies only with the depth of flow. Larson (1969) has developed other similar equations for grassed backslope terraces shown in Fig. 8.6*c*. These terraces have backslopes too steep to farm and are planted to permanent grass to stabilize the slope. Fill soil may be obtained from the lower side of the terrace, which tends to reduce the land slope between terraces.

Example 8.3. For a channel depth (h) of 33 cm and for 8-row (76-cm or 30-in. row width) equipment 6.1-m (20-ft) width on 7 percent land slope, compute the

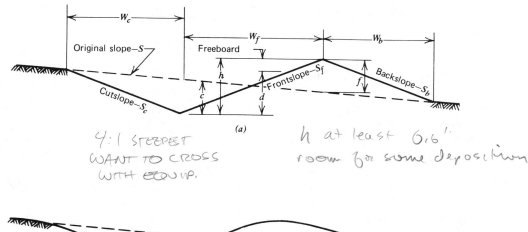

4:1 STEEPEST
WANT TO CROSS
WITH EQUIP.

h at least 6.6'
room for some deposition

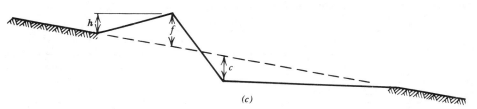

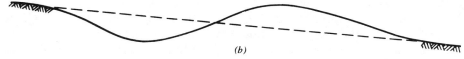

Fig. 8.6. (a) Design of a broadbase cross section. (b) Broadbase cross section after 10 years of farming. (c) Grassed backslope cross section.

cut and fill heights and the side slope ratio for the frontslope and the backslope assuming a balanced cross section.

Solution. From Eq. 8.2 and letting cuts equal fills,

$$c = f = (h + sW)/2$$
$$= (0.33 + 0.07 \times 6.1)/2$$
$$= 0.38 \text{ m (1.2 ft)}$$

By geometry for the frontslope

$$S_f = 6.1/0.33 = 18.5 \text{ or } 18{:}1 \text{ side slope ratio.}$$

Similarily for the backslope

$$S_b = 6.1/(0.38 + 0.07 \times 6.1) = 7.5$$

or about 8:1 side slope ratio.

Pipe outlet terraces shown in Fig. 8.7 were known in the early 1900s, but the present modern version was developed in Iowa in the 1960s. These terraces eliminate the need for grassed waterways. By straightening the terrace at natural channels with an earth fill, it is easier to make adjacent terraces parallel. The pipe outlet as shown in Fig. 8.7b has an orifice plate to restrict the outflow and thus with storage provided in the terrace channel some degree of flood control is achieved. To provide this storage the top of the terrace ridge may be constructed at the same elevation along its length even though the bottom of the channel may have a slope to the pipe inlet. During this temporary storage of water in the terrace much of the sediment will settle out in the terrace especially near the pipe inlet. In this way considerable pollution control is obtained while filling in the natural channel and making it more farmable. These terraces may also be constructed with a grassed backslope as shown in Fig. 8.7a.

For practical reasons a terrace is usually constructed with a uniform cross

Table 8.2   Depth of Channel Cut for Broad Base Terraces
(Balanced cross section, cut equals fill)

Land Slope (%)	Depth of Cut or Fill [a] (cm)						
	Depth of Flow (cm)						
	21	*24*	*27*	*30*	*34*	*37*	*43*
2	19	20	22	23	25	27	30
4	23	25	26	28	30	31	34
6	27	29	30	32	34	35	38
8	32	33	35	36	38	40	43
10	36	37	39	40	42	44	47
12	40	42	43	45	47	48	51
14	45	46	47	49	51	52	55
Front Side Slope Ratio[b] = 427/(d + 8)	(Same front side slope for all land slopes)						
	15	13	12	11	10	9	8

[a] Depth of cut or fill computed from Eq. 8.2.
[b] Slope width $W_c = W_f = W_b = 427$ cm (14 ft) and total depth, $h = d + 8$ cm freeboard. (See Fig. 8.6a.)

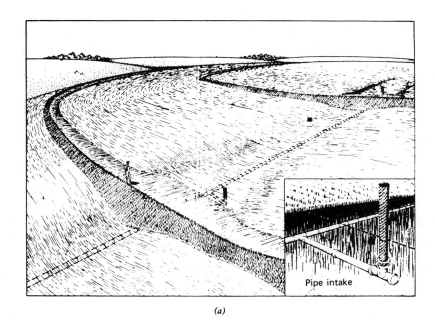

(a)

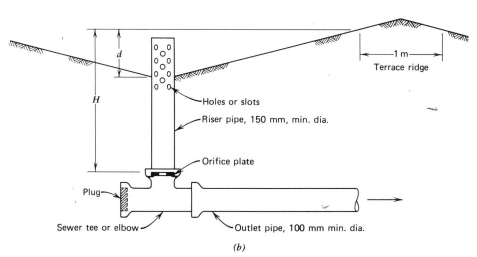

(b)

Fig. 8.7. (a) Grassed backslope and pipe outlet graded terraces and (b) details of controlled-flow pipe intake. (Redrawn from U.S. SCS, 1969.)

section from the outlet to the upper end, although this construction results in the upper portion of the channel being overdesigned. On the conservation bench terrace the ridge is generally built up to provide a settled height of 0.3 m above the level of the bench. The ends of the bench are blocked to retain 0.15 m of water on each bench before they overtop into a vegetated waterway. The ridge is generally constructed with a 5:1 back slope and planted with grass.

8.10. Terrace Channel Capacity. With graded terraces the rate of runoff is more important than total runoff, whereas both rate and total runoff influence the design of level, pipe outlet, and conservation bench terraces. Graded terraces are designed as drainage channels or waterways, and level terraces function as storage reservoirs. The terrace channel acts as a temporary storage reservoir subjected to unequal rates of inflow and outflow. For graded terraces the Manning velocity equation given in Chapter 7 is suitable for design. A roughness coefficient of 0.04 is usually selected for tilled soil and design is based on this condition because overtopping would cause the worst damage. The maximum design velocity will vary with the erosiveness of the soil but should rarely exceed 0.6 m/s for soil devoid of vegetation. The channel depth should permit a freeboard of about 20 percent of the total depth after allowing for settlement of the ridge (fill).

For graded terraces the design peak runoff rate should be based on a return period of 10 years (see Chapter 4). Some authorities recommend a 24-hour duration storm, but this is done because their runoff procedures are based on 24-hour rainfall rather than the time of concentration. The runoff volume for level, pipe outlet, and conservation bench terraces should be based on a 10-year, 24-hour duration storm. The orifice for pipe outlet terraces is usually selected so that the design storm will be removed in a 48-hour period. Orifice flow is discussed in Chapter 9.

PLANNING THE TERRACE SYSTEM

The terrace system should be coordinated with the complete water-disposal system for the farm, giving adequate consideration for proper land use. Terrace systems should be planned by watershed areas and should include all terraces that may be constructed at a later date. Where practicable, adjacent farms having fields in the same drainage area may have joint terracing systems. Factors such as fence and road location must be considered.

8.11. Selection of Outlets. One of the first steps in planning is the selection of outlets or disposal areas. Since level terraces generally do not require outlets, their location and layout are greatly simplified. The design, construction, and maintenance of vegetated outlets and watercourses as discussed in Chapter 7 are applicable for terrace outlets.

Outlets are of many types, such as natural draws, constructed channels, sod flumes, permanent pasture or meadow, road ditches, waste land, concrete or stabilized channels, pipe drains, and stabilized gullies. Natural draws, where properly vegetated, provided a desirable and economical outlet. Where these draws do not permit adequate field size, constructed waterways along field boundaries may be considered, as well as pipe outlets previously described. Terrace outlets on to pasture land should be staggered by increasing the length of each terrace a few meters, starting with the lowest terrace. Sod flumes and concrete channels are to be avoided because of excessive cost. Road ditches and active gullies may scour or enlarge if terrace runoff is added. Permission should be obtained from the appropriate highway agency before outletting into road ditches.

8.12. Terrace Location. After a suitable outlet is located, the next step is the location of the terraces. Factors that influence terrace location include (1) land slope, (2) soil conditions, such as degree and extent of erosion, (3) proposed land use, (4) boulders, trees, gullies, and other impediments to cultivation, (5) farm roads, (6) fences, (7) row layout, (8) type of terrace, and (9) outlet. Minimum maintenance, ease of farming, and adequate control of erosion are the criteria for good terrace location. Better alignment of terraces can usually be obtained by placing the terrace ridge just above eroded spots, gullies, and abrupt changes in slope. Satisfactory locations for roads and fences are on the ridge, on the contour, or on the spoil to the side of the outlet.

Unless there are obvious reasons for doing otherwise, the top terrace is laid out first, starting from the outlet end. It is important that the top terrace be located in the proper place, so that it will not overtop and cause failure of other terraces below. Some general rules for the location of the top terrace are: (1) the drainage area above the top terrace ordinarily should not exceed 1 hectare; (2) if the top of the hill comes to a point, the interval may be increased to 1½ times the regular vertical interval; (3) on long ridges, where the terrace approximately parallels the ridge, the regular vertical interval should be used; and (4) if short abrupt changes in slope occur, the terrace should be placed just above the break.

Obstructions or topographic features below the top terrace or the need to make terraces parallel may necessitate locating a terrace at some other point downslope. This terrace is called the key terrace because other terraces are located from it. Terraces above and below the key terrace are located by using the normal vertical interval.

A typical terrace layout is shown in Fig. 8.5. The top terrace (1a and 1b) is a diversion that intercepts runoff from the pasture and prevents overflow on the cultivated land below. Since the slope below terrace 1a is uniform, terraces 2a and 3a are laid out parallel to it. Because terraces 2bc and 3bc are each longer than 500 m, outlets are provided at each end.

Level terraces are located in much the same manner as graded terraces. On flat slopes in the Great Plains, level terraces are sometimes constructed so that runoff is allowed to flow from one terrace to the next by opening alternate ends of the terraces.

8.13. Layout Procedure. A good tripod level and the application of surveying techniques along with field experience are sufficient for terrace layout. When available, topographic maps are especially helpful in planning parallel terraces.

The first step is to determine the predominant slope above the terrace and then to determine a suitable vertical interval. Stakes are normally set along the proposed terrace every 15 m, although intervals are shorter if turns in the line are sharp. The grade in the channel is provided by placing the stakes on the desired grade, allowance being made at the outlet to compensate for the difference in the elevation of the constructed terrace channel and the stake line. Additional terraces are staked in the same manner. Terraces may be made parallel by adjusting the location of the stakes set for uniform graded terraces as shown in Fig. 8.8. Realignment of these stakes should be limited to provide a cut of not more than 30 cm below the bottom of the normal channel or a ridge height not in excess of 90 cm. For parallel pipe outlet terraces these heights are more variable. Beasley (1963), Larson (1969), and U.S. SCS (1969) describe other procedures for locating parallel, variable-cut terraces, for balancing the cut and fill volumes, and for computing water storage requirements. These procedures involve adding many short segments, suited to computer solutions.

The staking procedure for graded terraces that drain into an established grass waterway is illustrated in Fig. 8.9. The difference in elevation of the stake line and the bottom of the constructed channel will depend on the land slope, shape of cross section, and location of the staked line in relation to the terrace channel. Normally, the center line of the ridge is staked as shown. On slopes more than about 6 percent the staked line may be lower than the bottom of the channel. In Fig. 8.9 the vertical scale has been exaggerated in order to show y more clearly. As shown, the rod reading at $0 + 15$ assuming $y = 15$ cm, and 6 cm slope in 15 m is 186 cm. Note that the remaining rod readings decrease by 6 cm for each 15-m station to establish the desired grade of 0.4 percent.

Use of 4-, 6-, and 8-row farm equipment makes the elimination of sharp turns and point rows increasingly important. In many instances, with a relatively small amount of land forming, the topography of a field can be sufficiently changed to permit establishment of a system of parallel terraces. Figure 8.10 illustrates the layout adjustment and point row elimination that can be achieved

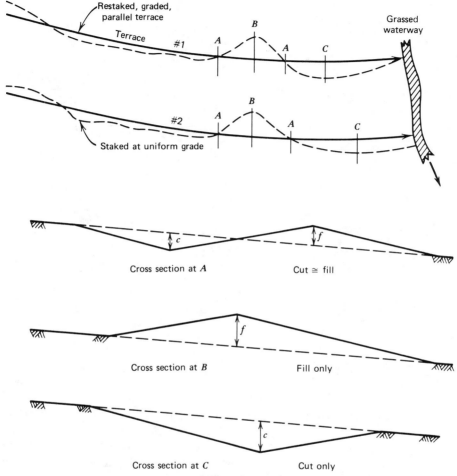

Fig. 8.8. Layout of parallel terraces with varying cuts and fills.

by parallel terrace construction. Pipe outlet terraces facilitate making terraces parallel without land forming.

TERRACE CONSTRUCTION

8.14. Construction Equipment. A variety of equipment is available for terrace construction. Terracing machines include the bulldozer, pan or rotary scraper, motor patrol, and elevating grader. Smaller equipment, such as moldboard and disk plows are suitable for slopes less than about 8 percent, but the rate of construction is much less than with heavier machines.

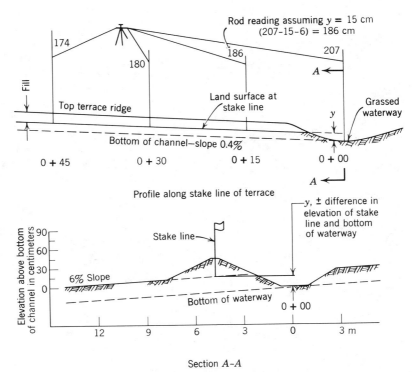

Fig. 8.9. Staking procedure for graded terraces with a grassed waterway outlet.

8.15. Factors Affecting Rate of Construction. The rate of construction of terraces is affected chiefly by the following factors: (1) equipment, (2) soil moisture, (3) crops and crop residues, (4) degree and regularity of land slope, (5) soil tilth, (6) gullies and other obstructions, (7) terrace length, (8) terrace cross section, and (9) experience and skill of the operator. Soil and crop conditions are likely to be most suitable in the spring and fall.

8.16. Settlement of Terrace Ridges. The amount of settlement in a newly constructed terrace ridge depends largely on soil and moisture conditions, type of equipment, construction procedure, and amount of vegetation or crop residue. The percentage of settlement based on unsettled height will vary as follows (1) moldboard plow or bulldozer, 10 to 20 percent; (2) elevating grader, 15 to 25 percent; and (3) blade grader (motor patrol) 0 to 5 percent. These data are applicable for soils in good tillable condition with little or no vegetation or residue and for normal construction procedure. In general, those machines that move over the loose fill during construction provide greater compaction than those that carry the soil on to the ridge, such as the elevating grader.

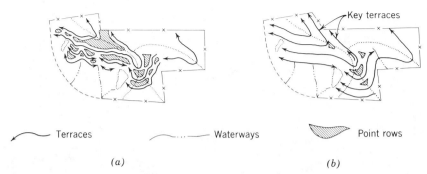

Terraces   Waterways   Point rows

(a)   (b)

Fig. 8.10. Terrace layout to improve alignment and reduce point rows. (a) Conventional layout with many point rows. (b) Parallel terraces with few point rows. (Redrawn from Jacobson, 1961.)

TERRACE MAINTENANCE

Proper maintenance is as important as the original construction of the terrace. However, it need not be expensive since normal farming operations will usually suffice. The terrace should be watched more carefully during the first year after construction.

8.17. Tillage Practices. In a terraced field all farming operations should be carried out as nearly parallel to the terrace as possible. The most evident effect of tillage operations, after several years, is the increase in the base width of the terrace. Procedure for plowing out point rows is similar to that for contoured areas given in Chapter 5. Other tillage practices such as stubble-mulch operations, disking, and harrowing as well as planting can be performed parallel to the plow furrows. With parallel terraces all crop rows are parallel to the terraces.

The effect of one-way plowing, in which the furrow slice is moved up the slope and the deadfurrow placed in the channel, is shown in Fig. 8.11. Except for a soil loss of about 7 percent from the channel, the soil has been transferred from the interterraced area to the upper slope of the ridge below.

DIVERSIONS

Most effective control of gullies is by complete elimination of runoff into the gully or the gullied area. This may often be accomplished by diverting runoff from above the gully and causing it to flow in a controlled manner to some suitably protected outlet. A diversion, referred to in some areas as a diversion terrace, is particularly effective for this purpose. It also effectively protects bottomland from hillside runoff and diverts water from uncontrolled areas away

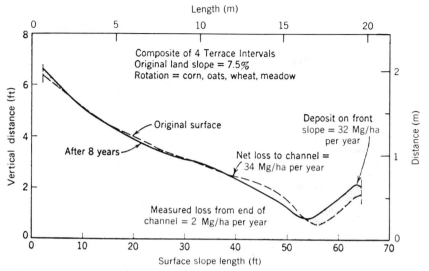

Fig. 8.11. Effect of one-way plowing for 8 years on soil movement between terraces at Bethany, Missouri. (Redrawn from Zingg, 1942.)

from buildings, strip cropped fields, and other special purpose areas. A typical application is shown in Fig. 8.12.

8.18. Design of Diversions. A diversion is a channel constructed around the slope and given a slight gradient to cause water to flow to the desired outlet. The capacity of diversion channels should be based upon estimates of peak runoff for the 10-year return period if it is to empty into a vegetated waterway. If the diversion is to outlet into a permanent structure, the design should be the same as for the structure. The design procedures for diversions are the same as for vegetated waterways discussed in Chapter 7.

Cross-section design may vary to fit soil, land slope, and maintenance needs. Side slopes of 4 to 1 with bottom widths that permit mowing are frequently used. Since sediment deposition is often a problem in diversions, the designed velocity of water flow should be kept as high as the channel protection will permit. In the event that the channel cross section has been designed to permit cultivation, the velocity of flow must be based on bare soil conditions, that is, a maximum of about 0.5 m/s.

REFERENCES

American Society Agricultural Engineers (ASAE). Terrace and Related Slope Modification Committee (1972). ASAE Yearbook, R268.1.

Beasley, R. P. (1963). "A New Method of Terracing." *Missouri Agr. Expt. Sta. Bull.* 699 (revised July).

Fig. 8.12. Typical diversion to protect cultivated bottomland from damaging hillside runoff. (U.S. SCS, 1973.)

Hauser, V. L., et al. (1962). "A Comparison of Level and Graded Terraces in the Southern High Plains." *Am. Soc. Agr. Eng. Trans.,* **5** (1), 75–77.

Hauser, V. L., and M. B. Cox (1962). "Evaluation of Zingg Conservation Bench Terrace." *Agr., Eng.* **43**, 462–464, 467.

Jacobson, P. (1962). "A New Method for bench Terracing Steep Slopes." *Am. Soc. Agr. Eng. Trans.* **6** (3), 257–258.

Jacobson, P. (1961). "A Field Method for Staking Cut and Fill Terraces." *Agr. Eng.* **42**, 684–687.

Larson, C. L. (1969). "Geometry of Broad-Based and Grassed-Backslope Terrace Cross Sections." *Am. Soc. Agr. Eng. Trans.* **12** (4), 509–511.

McCool, D. K., R. P. Beasley, and I. L. Berry (1962). "Terrace Channel Design Using Spatially Varied Flow and Tractive Force Theories." *Am. Soc. Agr. Eng. Trans.* **5**, 190–191, 196.

Smith, D. D. (1956). "Time Study of Parallel Terraces." *Agr. Eng.* **37**, 342–345.

Zingg, A. W. (1942). "Movement within the Surface Profile of Terraced Lands." *Agr. Eng.* **23**, 93–94.

U.S. Soil Conservation Service (SCS) (1973). "Drainage of Agricultural Lands." Water Information Center, Port Washington, New York.

U.S. Soil Conservation Service (1969). Engineering Field Manual for Conservation Practices. Litho.

PROBLEMS

8.1. On one graph, plot two curves with the slope in percent (0 to 10) as the abscissa, and the vertical interval and the horizontal spacing for graded terraces as ordinates. Follow recommendations specified for your area and for low intake soil with good cover.

8.2. Determine the time required for the flow to travel a distance of 100 m (328 ft) in a terrace channel having a slope of 0.3 percent, $n = 0.04$, depth of flow of 30 cm (1 ft), and side slopes for $W_c = W_f = 4.3$ m (14 ft).

8.3. Determine rod readings for the first four 15-m (50-ft) stations for a uniform-graded terrace having a slope of 0.3 percent if the rod reading at the outlet is 170 cm (5.6 ft). If the stake line is 15 cm (0.5 ft) higher than the bottom of the finished terrace channel, what should the rod readings be at these four stations?

8.4. If the rod reading at the outlet of the top terrace is 146 cm (4.8 ft) and the vertical interval (V.I.) for the second terrace is 138 cm (4.5 ft), what should be the rod reading at the outlet of the second terrace assuming the survey instrument has not been moved?

8.5. Determine the slopes for three sections of a 300-m (984-ft) length terrace to give a constant velocity of about 0.6 m/s (2 fps). Assume that 0.085 m³/s (3 cfs) of surface runoff enters the channel at the upper end of each 100-m (300-ft) section, $n = 0.04$, and the cutslope and frontslope are 8:1.

8.6. Compute the cut volume and the fill volume for a 200-m (656-ft) terrace on a 4 percent slope with a depth of flow of 30 cm. Assume the three slope widths are 4.3 m (14 ft), a freeboard of 8 cm, and a balanced cross section (cut = fill).

8.7. Determine the vertical interval from Eq. 8.1 for graded terraces on average intake soil and good cover with a slope of 6 percent at your present location. What is the peak runoff from the second terrace if the terrace length is 366 m (1200 ft), $C = 0.5$, and $T_c = 10$ min? Use the rational method for a return period of 10 years (see Chapter 4).

8.8. The runoff from a 457-m (1500-ft) length graded terrace with a channel grade of 0.4 percent is 0.31 m³/s (11 cfs). Assuming a roughness coefficient of 0.03, and the slope widths are 4.3 m (14 ft), compute the depth of flow using the Manning formula and determine the total depth of the channel allowing a freeboard of 10 cm.

8.9. A system of graded terraces for your present location is to be made parallel and to be farmed with four-row equipment 76-cm (30-in.) rows. To provide the nearest complete rounds of travel for this equipment, what slope distance between terraces is needed if the soil has a low intake rate and poor cover and the land slope is 2 percent?

8.10. If a pipe outlet terrace is to store 50 mm (2 in.) of runoff from a 10-year return period 24-hour duration storm, determine the opening size in mm (in.) for the orifice to remove this volume in 48 hours. Assume an average pressure head of 1.5 m (4.9 ft) on the orifice, orifice discharge coefficient of 0.6, and the runoff area above the terrace of 1 ha (2.47 ac).

Tile Outlet

— set terrace spacing

— set storage requirement + available storage

— 2" of runoff — up to 24 hr

— set tile size

— orifice size

— USE MANNINGS — MAKE SURE YOU COMPENSATE FOR CHANGE of SLOPE

CHAPTER 9

Conservation Structures

Major channels, whether they are designed to convey irrigation water, drainage flow or flood runoff, often require stabilization by structures of concrete, metal, or wood. This chapter discusses hydraulic principles in relation to the design of major channel stabilization structures and some uniquely related to drainage or irrigation.

Provision of a stable channel frequently involves reducing the gradient of the channel to maintain velocities below an erosive level. Much of the fall in the channel is taken up by structures which are designed to dissipate the energy of the falling water. The gradient of the channel reaches between structures is designed to maintain nonsilting and nonscouring velocities.

9.1. Temporary and Permanent Structures. *Temporary Structures.* Temporary structures can only be recommended in situations where cheap labor and materials can be utilized. Increasing mechanization and high labor costs have resulted in a great decline in the popularity of temporary channel stabilization structures. For gully control, shaping the channel and establishing vegetation in accordance with the principles discussed in Chapter 7, provide more efficient and effective control. Temporary structures (described by Jepson, 1939) may be constructed of creosoted planks, rocks, logs, brush, woven wire, sod, or earth.

Smith (1952) reported on the performance of 50 temporary structures that had been used on the Soil Conservation Service experimental farm at Bethany, Missouri; only 5 percent of the structures were found to have functioned as intended. It was concluded that vegetal protection was established as easily without temporary structures as with them.

Permanent Structures. Structures constructed of permanent materials may be required to control the overfall at the head of a large gully, to drop the discharge from a vegetated waterway into a drainage ditch, to take up the fall at various points in any channel, or to provide for discharge through earth fills. Figure 9.1 shows the profile of a gully that has been reclaimed by methods involving the use of several types of permanent structures. Standard designs are available from references of the Soil Conservation Service and the Bureau of Reclamation.

9.2. Functional Requirements of Control Structures. Not only must a control structure have sufficient capacity to pass the design discharge, but the kinetic energy of the discharge must be dissipated within the confines of the structure in a manner and to a degree that will protect both the structure and the downstream channel from damage. The two primary causes of failure of permanent control structures are (1) insufficient hydraulic capacity and (2) insufficient provision for energy dissipation.

9.3. Design Features. The basic components of a hydraulic structure are the inlet, the conduit, and the outlet. Structures are classified and named in accordance with the form that these three features take. Figure 9.2 identifies the various types of inlets, conduits, and outlets that are commonly used. In addition to these hydraulic features, the structure must include suitable wing walls, side walls, head wall extensions, and toe walls to prevent seepage under or around the structure and to prevent damage from such local erosion as may occur. These structural components are identified in Fig. 9.3 for one common type of structure. It is important that a firm foundation be secured for permanent structures. Wet foundations should be avoided or provided with adequate artificial drainage. Surface soil and organic material should be removed from the site to allow a good bond between the structure and the foundation material.

Models. The design criteria for conservation structures have been developed from intensive observation of the behavior of small-scale laboratory models. The results of such laboratory studies have been summarized in empirical formulas, graphs, or tables that relate certain critical dimensions of the structure to characteristics of the flow.

Model studies of open channel structures are based on the Froude law, which requires that the Froude number of flow through the model must be equal to the Froude number of flow through the prototype. Application of this law assumes that the force of gravity is the only force producing motion. Other forces, such as fluid friction and surface tension are neglected. The Froude number is defined as the ratio of the inertial force to the gravitational force or by the dimensionless quantity, $F = v/(gd)^{1/2}$.

Most control structures include a section at which flow at critical depth, explained below, occurs. The Froude law will be satisfied if the critical depth of flow at such a section in the prototype (the full-scale structure) is equal to the scale factor times the critical depth of flow at that section in the model. Thus, design equations for certain components of structures are often expressed as functions of critical depth.

In the operation of models the flow rate and the various dimensions of the model are varied, and the operation of the model is observed. The influence of the variables on erosion of the downstream channel is checked by observation of the scour pattern produced in a sand bed at the outlet section of the model.

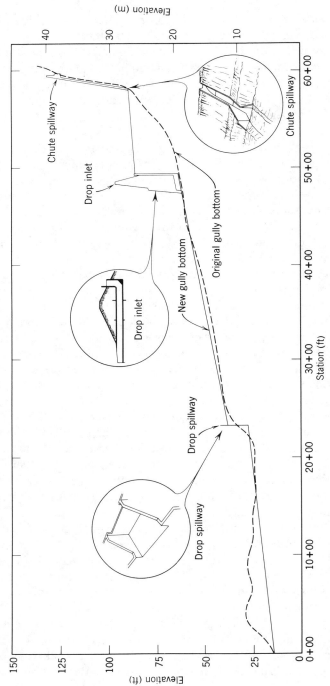

Fig. 9.1 Profile of a gully showing the application of several types of permanent structures.

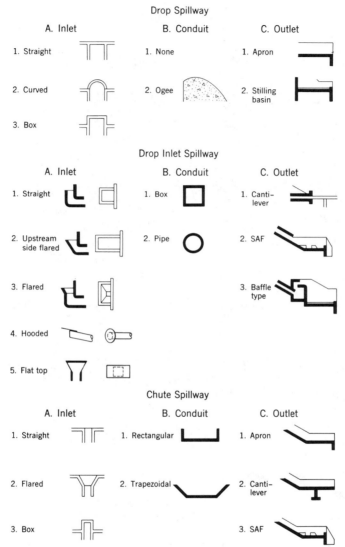

Fig. 9.2. Classification of components of hydraulic structures.

Additional information on models may be found in *Similitude in Engineering* by Murphy (1950) and in the ASCE Manual of Engineering Practice No. 25, *Hydraulic Models* (1942).

Critical Depth. A given quantity of water in an open conduit may flow at two depths having the same energy head. When these depths coincide, the energy head is a minimum and the corresponding depth is termed the critical depth.

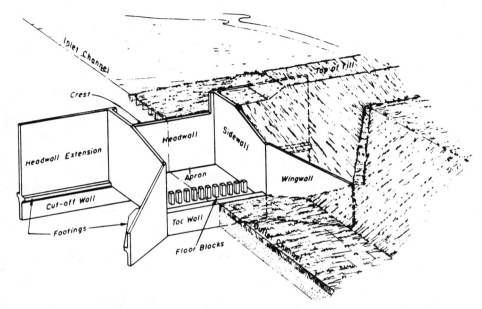

Fig. 9.3. Straight drop spillway showing structural components. (From U.S. Soil Conservation Service, Engineering Field Manual, 1969.)

This is illustrated by Fig. 9.4. The expression for critical depth at a rectangular section may be developed as follows. The specific energy head at a section with reference to the channel bed is

$$H_e = y + \frac{v^2}{2g} = y + \frac{q^2}{2a^2g} = y + \frac{q^2}{2b^2y^2g} \qquad (9.1a)$$

Differentiating with respect to y,

$$\frac{dH_e}{dy} = 1 - \frac{q^2}{b^2y^3g} \qquad (9.1b)$$

Setting $dH_e/dy = 0$ to determine y when H_e is a minimum, and letting this value of y be d_c, the critical depth,

$$1 - \frac{q^2}{b^2y^3g} = 0$$

$$d_c = \sqrt[3]{\frac{q^2}{b^2g}} \qquad (9.2)$$

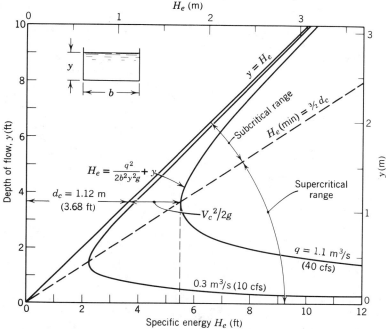

Fig. 9.4. Depth of flow and specific energy for two flow rates in a rectangular channel. (After U.S. Bureau of Reclamation, 1977.)

Equation 9.1*a* may also be solved to determine the depth at which maximum discharge will occur for a given energy, H_e. Solving for q in Eq. 9.1*a* yields

$$q^2 = 2y^2b^2g(H_e - y) \tag{9.3}$$

Assuming H_e is constant and setting $dq/dy = 0$,

$$\frac{dq}{dy} = \frac{2b^2g(2yH_e - 3y^2)}{2[2y^2b^2g(H_e - y)]^{1/2}} = 0 \tag{9.4}$$

If $y = d_c$, then by solving Eq. 9.4

$$d_c = \tfrac{2}{3}H_e \tag{9.5}$$

Thus, for a given H_e, maximum flow occurs at the critical depth or when $H_e = \tfrac{3}{2}y$. Equation 9.3 is plotted in Fig. 9.5 for $H_e = 1.68$ m showing a maximum q of 1.13 m³/s (40 cfs) at a depth of 1.12 m, which is $\tfrac{2}{3}(1.68)$. The curve shows that the discharge would be less at any other depth either above or below the critical depth of 1.12 m (3.68 ft).

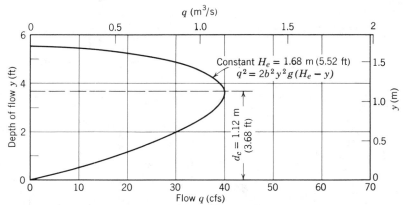

Fig. 9.5. Flow rate and depth of flow at a constant level of specific energy in a rectangular channel.

Hydraulic Jump. When a given flow changes from a flow depth, y, less than critical depth to a flow depth greater than critical depth, the phenomenon is referred to as a hydraulic jump. The profile and depth-energy relationships of a hydraulic jump are shown in Fig. 9.6. Energy due to velocity is converted to energy of elevation and some energy is lost to friction through turbulence in the process. Control structures are often designed so that a hydraulic jump forms within the downstream portion of the structure and velocity is reduced to a nonerosive level in the subcritical range.

When the inflow is at critical depth the Froude number is unity and the two depths in Fig. 9.6 are equal, Thus, no jump would occur.

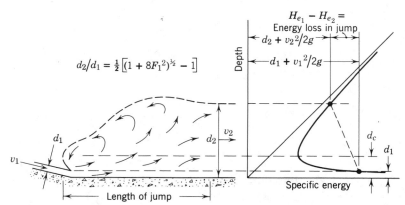

Fig. 9.6. Energy relationships in a hydraulic jump. (After U.S. Bureau of Reclamation, 1977.)

- flow capacity (25-50 yr)
- Elevation change req'd
- Dimensions of Approach channel
- Structure stability

DROP SPILLWAYS

A typical drop spillway is shown in Fig. 9.3. Drop spillways may have a straight, arched, or box-type inlet. The energy dissipator may be a straight apron or some type of stilling basin.

9.4. Function and Limitations. Drop spillways are installed in channels to establish permanent control elevations below which an eroding stream cannot lower the channel floor. These structures control the stream grade not only at the spillway crest itself but also through the ponded reach upstream. Drop structures placed at intervals along a channel can stabilize it by changing its profile from a continuous steep gradient to a series of more gently sloping reaches. Where relatively large volumes of water must flow through a narrow structure at low head, the box-type inlet is preferred. The curved inlet serves a similar purpose and also gives the advantage of arch strength where masonry construction is used. Drop spillways are usually limited to drops of 3 m, flumes or drop-inlet pipe spillways being used for greater drops.

9.5. Design Features. *Capacity.* The free flow capacity for drop spillways is given by the weir formula

$$q = 0.552CLh^{3/2} \qquad (9.6)$$

3.2

where q = the discharge in m³/s,
 C = weir coefficient (English units),
 L = weir length in m, — *all sides*
 h = depth of flow in m.

$q = 3.2Lh^{3/2}$
cfs

— SOL'N — START w/ H = 2, SOLVE FOR L

The length L is the sum of the lengths of the three sides of a box inlet, the circumference of an arch inlet, or the crest length of a straight inlet. The value of C varies considerably with entrance conditions. Blaisdell and Donnelly (1951) have prepared correction charts to modify C for a wide range of conditions of entrance and crest geometry for box inlets. Where the ratio of head to box width is 0.2 or greater, the ratio of the width of the approach channel to the total length L is greater than 1.5, and no dikes or other obstacles are within $3h$ of the crest, a value of $C = 3.2$ may be used with an accuracy of ±20 percent. Discharge characteristics of a typical box-inlet drop spillway are given in Fig. 9.7. A value of $C = 3.2$ will also give satisfactory results for the straight inlet. The inlet should have a freeboard of 0.15 m above h, the height of the water surface.

Whenever the tailwater is nearly up to or above the crest of the inlet section, submergence decreases the capacity of the structure. When such conditions occur, other design equations apply.

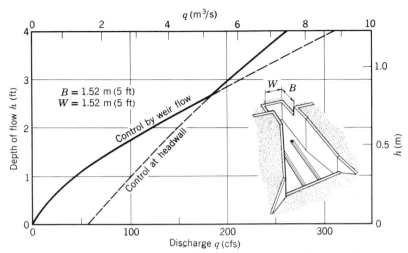

Fig. 9.7. Discharge-head relationships of a box-inlet drop spillway. (From Blaisdell and Donnelly, 1951).

Apron Protection.   The kinetic energy gained by the water as it falls from the crest must be dissipated and/or converted to potential energy before the flow is discharged from the structure. For straight-inlet drop structures the dissipation and conversion of energy are accomplished in either a straight apron or a Morris and Johnson stilling basin. Dimensions for the Morris and Johnson stilling basin are given in Fig. 9.8. For larger structures the Morris and Johnson outlet is preferred, as it results in a shorter apron and the transverse sill induces a hydraulic jump at the toe of the structure. The longitudinal sills serve to straighten the flow and prevent transverse components of velocity from eroding the side slopes of the downstream channel. The flow pattern through a Morris and Johnson stilling basin is shown in dimensionless form in Fig. 9.9.

The stilling basin design for the box-inlet drop spillway is given in Fig. 9.10.

CHUTES

Flumes or chutes carry flow down steep slopes through a concrete-lined channel rather than by dropping the water in a free overfall.

9.6.   Function and Limitations.   Chutes may be used for the control of heads up to 6 m. They usually require less concrete than do drop-inlet structures of the same capacity and drop. However, there is considerable danger of undermining of the structure by rodents, and, in poorly drained locations, seepage may threaten foundations. Where there is no opportunity to provide temporary storage above the structure, the flume with its inherent high capacity is preferred over the drop-inlet pipe spillway. The capacity of a chute is not decreased by sedimentation at the outlet.

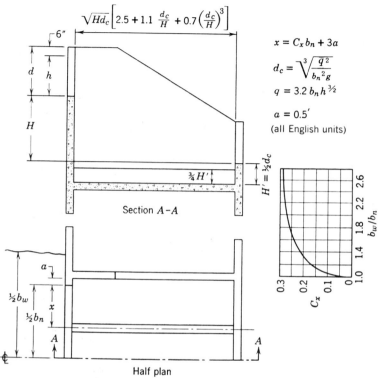

$$\sqrt{Hd_c}\left[2.5 + 1.1\,\frac{d_c}{H} + 0.7\left(\frac{d_c}{H}\right)^3\right]$$

$x = C_x b_n + 3a$

$d_c = \sqrt[3]{\dfrac{q^2}{b_n{}^2 g}}$

$q = 3.2\, b_n h^{3/2}$

$a = 0.5'$

(all English units)

Section A–A

$\frac{3}{4}H'$

$H' = \frac{1}{2}d_c$

$\frac{1}{2}b_w$

$\frac{1}{2}b_n$

a

x

Half plan

Fig. 9.8. Design dimensions for drop spillway with straight inlet and Morris and Johnson outlet. (From Morris and Johnson, 1942.)

9.7. Design Features. *Capacity.* Flume capacity normally is controlled by the inlet section. Inlets may be similar to those for straight-inlet or box-inlet drop spillways, and in such inlets capacity formulas already discussed will apply. Blaisdell and Huff (1948) have investigated the performance of other types of flume entrances.

Outlet Protection. The cantilever-type outlet should be used where the channel grade below the structure is unstable. In other situations, either the straight-apron or SAF outlet is used. The straight apron is applicable to small structures. Figure 9.11 shows dimensions of this type of outlet protection.

FORMLESS FLUME

9.8. Function and Limitations. This structure has the advantage of low-cost construction. It may be used to replace drop spillways where the fall does not exceed 2 m and the width of notch required does not exceed 7 m. The flume is constructed by shaping the soil to conform to the shape of the flume and applying a 13-cm layer of concrete reinforced with woven wire mesh. No forms

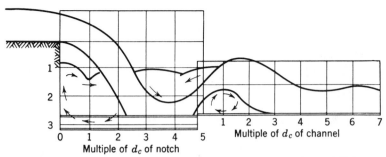

Fig. 9.9. Flow pattern through a drop spillway with a Morris and Johnson stilling basin. (From Morris and Johnson, 1942.)

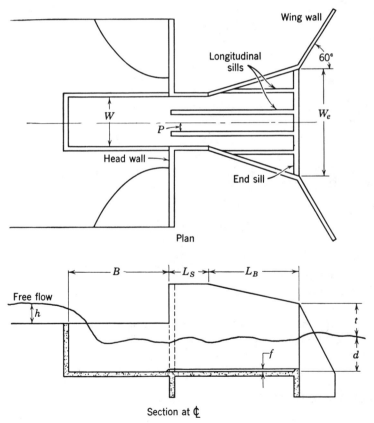

Fig. 9.10. The box-inlet drop spillway. (See reference for design formulas.) (From Blaisdell and Donnelly, 1951.)

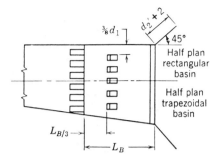

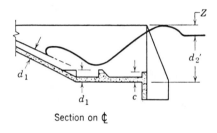
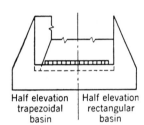

Fig. 9.11. The Saint Anthony Falls (SAF) stilling basin. (See reference for design formulas.) (From Blaisdell, 1948.)

are needed, thus, the construction is simple and inexpensive. The formless flume should not be used where water is impounded upstream (danger of undermining the structure by seepage) or where freezing occurs at great depth.

9.9.   Design Features.   Figure 9.12 shows the design features and dimensions of the formless flume. The capacity is given by

$$q = 2.13Lh^{3/2} \tag{9.7}$$

where q = discharge in m³/s,
   L = notch width in m,
   h = flow depth in m.

The depth of the notch, D, is h plus a freeboard of 0.15 m.

PIPE SPILLWAYS

Pipe spillways may take the form of a simple conduit under a fill, or they may have a riser on the inlet end and some type of structure for outlet protection. These conditions are illustrated in Fig. 9.13. The pipe in Fig. 9.13c, called an

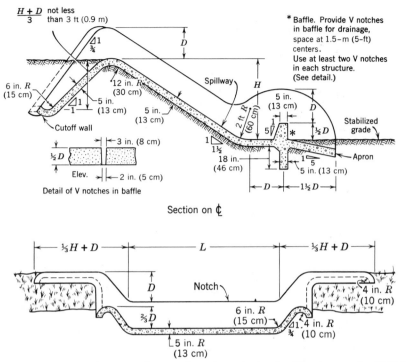

Fig. 9.12. The formless flume. (Based on design by Wooley et al., 1941.)

inverted siphon, is often used when water in an irrigation canal must be conveyed under a natural or artificial drainage channel. Inverted siphons must withstand hydraulic pressures much higher than those encountered in other pipe spillways and therefore require special attention to structural design.

9.10. Function and Limitations. The pipe spillway used as a culvert has the simple function of providing for passage of water under an embankment. When combined with a riser or drop inlet, the pipe spillway serves to lower water through considerable drop in elevation and to dissipate the energy of the falling water. Drop-inlet pipe spillways are thus frequently used as gully control structures. This application is usually made where water may pond behind the inlet to provide temporary storage. The hydraulic capacity of pipe spillways is related to the square root of the head, and hence they are relatively low-capacity structures. This characteristic is used to advantage where discharge from the structure is to be restricted.

9.11. Design Features. *Culverts.* Culvert capacity may be controlled either by the inlet section or by the conduit. The headwater elevation may be above or below the top of the inlet section. The several possible flow conditions

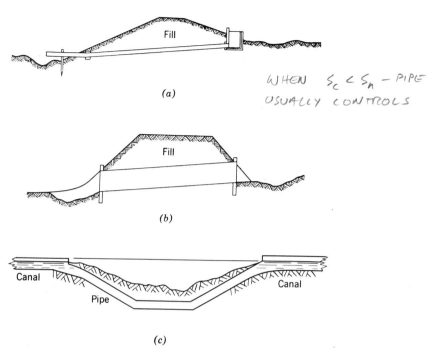

WHEN $S_c < S_n$ — PIPE
USUALLY CONTROLS

(a)

(b)

(c)

Fig. 9.13. Types of pipe spillways. (a) Drop-inlet pipe spillway with cantilever outlet. (b) Simple culvert. (c) Inverted siphon.

are represented in Fig. 9.14. Solution of a culvert problem is primarily the determination of the type of flow that will occur under given headwater and tailwater conditions. Consider a culvert as shown in Figs. 9.14a and 9.14b. Pipe flow (conduit controlling capacity) will occur under most conditions when the slope of the culvert is less than the neutral slope, which for small angles of θ is

$$s_n = \tan \theta = \sin \theta = \frac{H_f}{L} = K_c \frac{v^2}{2g} \qquad (9.8)$$

where θ = slope angle of conduit,
H_f = friction loss in conduit length L in m,
L = length of conduit in m,
K_c = friction loss coefficient,
v = velocity of flow in m/s,
g = gravitational constant in m/s².

$$S_n = \frac{k_c \left(\frac{v^2}{2g}\right)}{\sqrt{1 - \left(\frac{k_c v^2}{2g}\right)^2}}$$

In some situations inlet losses may be so great that pipe flow will not occur even though the culvert is on less than neutral slope. Therefore, capacity of the

(handwritten top) S_c – SLOPE AS IT IS INSTALLED

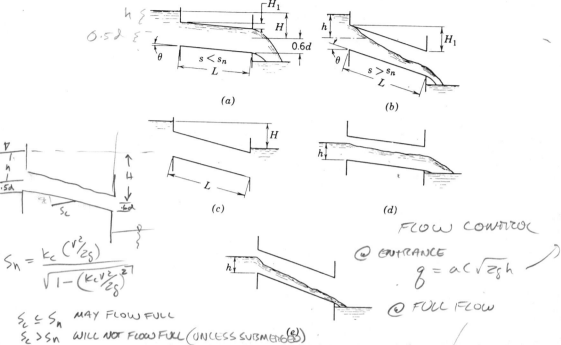

(handwritten annotations around figure)
$h \{$
$0.5d \; \{$
θ

$\triangledown$
$\uparrow$ 1.
$\;$ h
$\uparrow$ 4
.5d
S_c $\downarrow$ .6d

$S_n = \dfrac{k_c \left(\dfrac{v^2}{2g}\right)}{\sqrt{1 - \left(k_c \dfrac{v^2}{2g}\right)^2}}$

$S_c \le S_n$ MAY FLOW FULL
$S_c > S_n$ WILL NOT FLOW FULL (UNLESS SUBMERGED)

FLOW CONTROL
@ ENTRANCE
$q = ac\sqrt{2gh}$
@ FULL FLOW

Fig. 9.14. Possible conditions of flow through culverts. (a) Full: free outfall, pipe flow. (b) Part full: free outfall, orifice flow. (c) Full: outfall submerged, pipe flow. (d) Inlet not submerged: conduit controls, open channel flows. (e) Inlet not submerged: inlet controls, weir flow. (Modified from Mavis, 1943.)

entrance should be checked and compared with pipe flow capacity to determine which is limiting. The capacity of a culvert under conditions of pipe flow is given by

$$q = \frac{a\sqrt{2gH}}{\sqrt{1 + K_e + K_b + K_c L}} \tag{9.9}$$

where q = flow capacity in (L^3T^{-1}),
$\quad a$ = conduit cross-sectional area (L^2),
$\quad H$ = head causing flow (L),
$\quad K_e$ = entrance loss coefficient,
$\quad K_b$ = loss coefficient for bends in the culvert. *(handwritten)* } Appendix C

Values of K_b, K_c, and K_e are given in Appendix C. Since most culverts do not have bends, K_b can often be omitted.

If the conduit is at greater than neutral slope and the outlet is not submerged, the flow will be controlled by the inlet section on short culverts, and orifice flow prevails. Capacity is then given by

$C = 0.6$

$$q = aC\sqrt{2gh} \qquad a = Area \qquad (9.10)$$

where a = cross-sectional area (L^2),

h = head to the center of the orifice (L) (either SI or English units).

The coefficient, C, for a sharp-edged orifice is 0.6. For more detailed values and for other orifices, consult King and Brater (1963) or other hydraulics books. Examples 9.1 and 9.2 serve to clarify the above discussions.

In situations where the headwater elevation does not reach the elevation of the top of the inlet section, there is again the possibility of control of flow by either the conduit or the inlet section. In Figs. 9.14d and 9.14e, the conduit controls if the slope of the conduit is too flat to carry the maximum possible inlet flow at the required depth. This depth is equal to the headwater depth above the inlet invert minus the static head loss due to entrance losses and acceleration. Conditions of control at the entrance section occur when the slope of the conduit is greater than that required to move the possible flow through the inlet. For conditions of control by the entrance section, solution for circular culverts may be made from Fig. 9.15. This figure will also apply when the inlet is submerged. Examples 9.3 and 9.4 clarify the solution for conditions of an unsubmerged inlet. Values of the roughness coefficient, n, for conduits may be found in Appendix B.

Example 9.1. Determine the capacity of a 762-mm (30-in.) diameter corrugated culvert 18.29 m (60 ft) long with a square-edged entrance. Elevation of the inlet invert is 127.92 m (419.7 ft), and the elevation of the outlet invert is 127.71

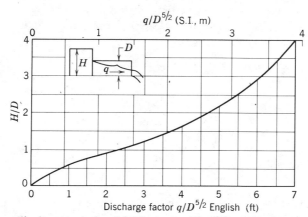

Fig. 9.15. Stage-discharge relationship for control by a square-edge entrance inlet to a circular pipe. (Redrawn from Mavis, 1943.)

m (419.0 ft). Headwater elevation is 129.54 m (425.0 ft), and tailwater elevation is 126.80 m (416.0 ft).

Solution. Assume pipe flow prevails, for which Eq. 9.9 applies. From Appendix C, find $K_e = 0.5$ and $K_c = 0.112$. Head loss h through the structure is from the headwater elevation to an elevation of $0.6d$ above the elevation of the outlet invert, which is $129.54 - (127.71 + 0.6 \times 0.762) = 1.37$ m.

$$q = \frac{3.14 \times 0.762^2}{4} \left[\frac{2 \times 9.8 \times 1.37}{1 + 0.5 + 0.112 \times 18.29} \right]^{1/2}$$

$$= 1.255 \text{ m}^3/\text{s} \ (44.4 \text{ cfs})$$

From Eq. 9.8 determine the neutral slope of the culvert for a discharge of 1.255 m³/s:

$$s_n = \frac{0.112 \times 1.255^2}{2 \times 9.8 \times 0.45^2} = 0.043 = 4.3\%$$

Actual slope of the culvert is $(127.92 - 127.71)/18.29$, or 0.0115. Since the culvert slope is less than the neutral slope, pipe flow prevails and 1.255 m³/s is the discharge. Checking for orifice control at the entrance ($h = 1.24$ m) gives a discharge of 1.34 m³/s using Eq. 9.10. Since this flow is more than pipe flow capacity of 1.255 m³/s, pipe flow rather than orifice flow controls.

Example 9.2. Determine the capacity of the 762-mm (30-in.) culvert of Example 9.1 if the elevation of the outlet invert is 125.15 m (410.6 ft), and the tailwater elevation is 124.36 m (408.0 ft).

Solution. Assume pipe flow and calculate the discharge by Eq. 9.9 with $h = 129.54 - (125.15 + 0.46) = 3.93$ m (12.9 ft). Note $0.6d = 0.46$.

$$q = 0.456 \left[\frac{2 \times 9.8 \times 3.93}{1 + 0.5 + (3.28 \times 0.0341 \times 18.288)} \right]^{1/2}$$

$$= 2.12 \text{ m}^3/\text{s} \ (75.0 \text{ cfs})$$

$$\text{Neutral slope, } s_n = \frac{0.112 \times 2.12^2}{2 \times 9.8 \times 0.456^2} = 0.123$$

$$\text{Actual slope, } s = (127.92 - 125.15)/18.29 = 0.15$$

Since the culvert is at greater than neutral slope, pipe flow will not exist. Entrance conditions will prevail, and the problem is solved by application of the orifice flow formula, Eq. 9.10. From Example 9.1 $q = 1.34$ m³/s (47.3 cfs).

An alternate solution may be obtained from the curve in Fig. 9.15 with head $H = 129.54 - 127.92 = 1.62$ m. For $H/D = 1.62/0.762 = 2.13$, read $q/D^{5/2} = 2.65$ (S.I. units). Solving for q, which is $2.65 \times 0.762^{5/2} = 1.34$ m³/s, the discharge is thus the same as computed from Eq. 9.10.

Example 9.3. Determine the capacity of a 1.52-m (5-ft) diameter concrete culvert 30.5 m (100 ft) long. The culvert entrance is square-edged. Elevation of the inlet invert is 157.76 m (517.6 ft), and the elevation of the outlet invert is 155.60 m (510.5 ft). Headwater elevation is 158.80 m (521.0 ft), and tailwater elevation is 152.40 m (500.0 ft).

Solution. Assume that the conduit controls and entrance conditions are not limiting. Neglect for the moment the loss of static head at the culvert entrance due to acceleration of the flow entering the culvert. Under these assumptions, the depth of flow in the culvert would be 1.04 m (3.4 ft). Calculating the flow by the Manning formula, $a = 1.33$ m² (14.3 ft²), $n = 0.015$, $R = 0.445$ m (1.46 ft), and $s = 0.071$, for which $q = 13.76$ m³/s (486 cfs). Checking in Fig. 9.15, $H/D = 0.68$ and $q = 1.98$ m³/s (70 cfs). Since only this flow rate can enter the culvert, a flow of 13.76 m³/s could not possibly occur; thus entrance conditions prevail and the capacity is 1.98 m³/s.

Example 9.4. Determine the capacity of a culvert as in Example 9.3, but having the outlet invert at an elevation of 157.75 m (517.55 ft).

Solution. As before, we note that the maximum possible flow through the entrance inlet is 1.98 m³/s. However, in this case the culvert is on a flat slope, and conduit flow conditions may limit the flow. Assume a flow depth in the conduit of 0.76 m (2.5 ft). Then $a = 0.91$ m² (9.8 ft²), $n = 0.015$, $R = 0.38$ m (1.25 ft), and $s = 0.0005$. Substituting in the Manning formula $v = 0.78$ m/s (2.56 fps) and $q = 0.71$ m³/s (24.9 cfs). Assume the approach velocity is negligible; then the loss of static head at the culvert entrance due to acceleration is $v^2/2g = 0.78^2/19.6 = 0.03$ m (0.102 ft). Depth of water at the entrance is 1.04 m (3.4 ft), and a loss of 0.03 m (0.102 ft) would give 1.01 m (3.3 ft), which does not correspond with our assumption of 0.76 m (2.5 ft). Thus, the first assumption of flow depth was in error. Now assume a flow depth in the culvert of 1.00 m (3.28 ft), for which $a = 1.26$ m² (13.5 ft²), $n = 0.015$, $R = 0.44$ m (1.44 ft), and $s = 0.0005$. Substituting in the Manning formula $v = 0.86$ m/s (2.83 fps) and $v^2/2g = 0.04$ m (0.12 ft). Subtracting the 0.04 m from the entrance depth of 1.04 m leaves a flow depth of 1.00 m, which agrees with the original assumption and is the correct depth of flow. Thus, flow is limited by the conduit, and the discharge capacity is

$$q = 1.26 \times 0.44^{2/3} \times 0.0005^{1/2}/0.015 = 1.08 \text{ m}^3/\text{s (38 cfs)}$$

Drop Inlets.   The discharge characteristics of a drop-inlet pipe spillway are given in Fig. 9.16. At low heads, the crest of the riser controls the flow, and discharge is proportional to $h^{3/2}$. Under this condition, the discharge should be calculated as outlined in Section 9.5. When this type of flow equals the capacity of the conduit or conduit inlet section, the flow becomes proportional to the square root of the total head loss through the structure or the head on the conduit inlet.

Hood Inlets.   For farm pond mechanical spillways and similar small structures the hood inlet has largely replaced the drop inlet entrance. For slopes up to 30 percent the hood inlet when provided with a suitable antivortex device will cause the pipe to prime and flow full. Hood inlets shown in Figs. 9.17a and 9.17b have been developed by Blaisdell and Donnelly (1958). Beasley et al. (1960) reported that model and field tests of a hood inlet with an end plate shown in 9.17c gave satisfactory performance although the entrance loss was somewhat higher than the other two. The discharge characteristics of these three inlets are shown in Fig. 9.17d for a pipe length of 110D. For H/D less than 1, weir flow occurs, and up to H/D of about 1.4 the flow is rather erratic. At H/D of 1.4 the vortex is eliminated and pipe flow controls. The entrance coefficient for thin-wall pipe (ratios of wall thickness to pipe diameter below 0.04) is slightly higher than for heavier pipe.

The design capacity is determined from the pipe-flow equation as described for culverts. Entrance-loss coefficients are given in Appendix C. Although the approach conditions have little effect on spillway performance, the presence of the face of the dam will reduce somewhat the entrance loss coefficient. High velocities near the hood inlet may erode the dam, but such a hole is small and becomes stabilized in a short time. For example, Blaisdell and Donnelly (1958) report that for $q/D^{5/2} = 15$ (English units, ft) and the size of the bed material equal to 0.001D, the scour hole diameter is only 6D. Hood inlets are simple and easy to install and for small pipe diameters are much more economical than the reinforced concrete drop inlet. They are also available commercially.

The hood drop inlet shown in Fig. 9.18 is a hood inlet on the pipe of a drop-inlet pipe spillway shown in Fig. 9.16. Because the hood inlet must have a head of 1.1D to prime, the pipe spillway on large structures will require a specific and significant rise in the reservoir level before the spillway will flow at design capacity. The hood drop inlet will thus reduce the height and cost of the dam compared to that for a hood inlet. Extensive model tests have been conducted by Yalamanchili and Blaisdell (1975).

Outlet Protection.   For small culverts or drop-inlet pipe spillways, a cantilever-type outlet is usually satisfactory. The straight-apron outlet may be used in some instances. Large drop-inlet pipe spillways may be provided with the SAF stilling basin discussed in Section 9.7.

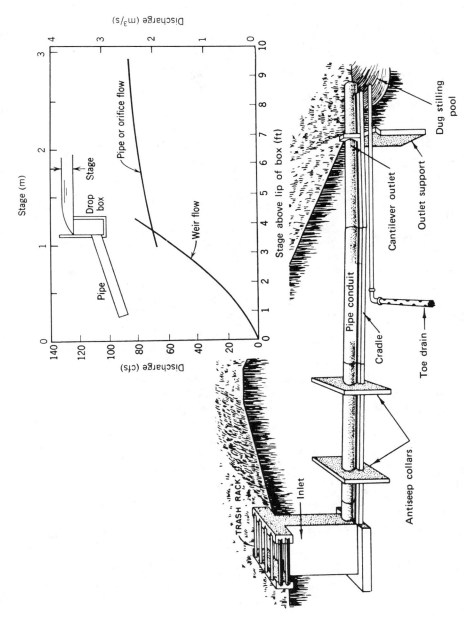

Fig. 9.16. Discharge characteristics of a drop-inlet pipe spillway.

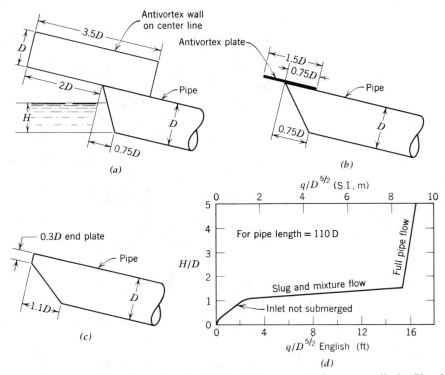

Fig. 9.17. Types of antivortex pipe inlets. (a) Splitter-type antivortex wall. (b) Circular or square antivortex top plate. (c) End antivortex plate. (d) Hood inlet discharge. (From Blaisdell and Donnelly, 1958 and Beasley and others, 1960.)

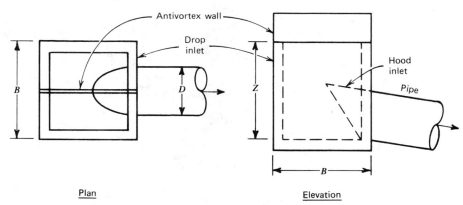

Fig. 9.18. The hood drop-inlet spillway. (Redrawn from Yalamanchili and Blaisdell, 1975.)

IRRIGATION AND DRAINAGE STRUCTURES

Many types of permanent structures are needed to control irrigation water. Most of these are placed in canals or farm distribution channels. Figure 9.19a shows an adjustable gate for controlling the flow into a farm ditch from a larger canal. A concrete division box for distributing the flow from a well into several small channels is shown in Fig. 9.19b. A control box for maintaining the water level in peat, muck, or other permeable soil is shown in Fig. 9.20. Flow through these structures can be determined by application of the principles given in this chapter.

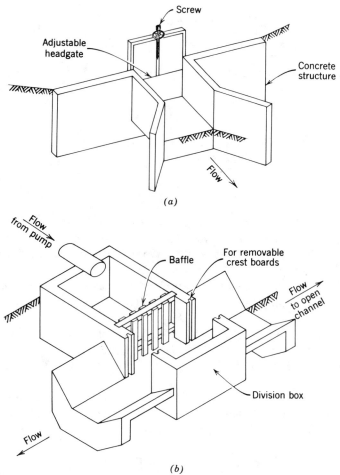

Fig. 9.19. Concrete irrigation structures. (a) Canal flow control gate. (b) Pump outlet and division-box. (Redrawn from U.S. Soil Conservation Service, 1969.)

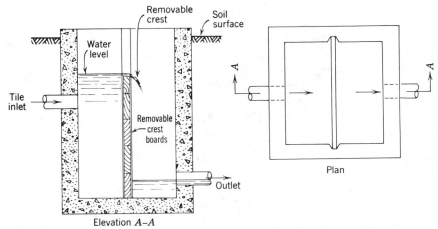

Fig. 9.20. Drainage structure for water level control with removable crest boards (also called stop logs).

REFERENCES

American Society of Civil Engineers, Committee of the Hydraulic Division on Hydraulic Research (1942). "Hydraulic Models." *ASCE Manuals of Engineering Practice,* no. 25.

Beasley, R. P., et al. (1960). "Canopy Inlet for Closed Conduits." *Agr. Eng.* **41,** 226–228.

Blaisdell, F. W. (1948). "Development and Hydraulic Design, Saint Anthony Falls Stilling Basin." *Trans. Am. Soc. Civil Eng.* **113,** 483–520.

Blaisdell, F. W., and C. A. Donnelly (1951). "Hydraulic Design of the Box Inlet Drop Spillway." *U.S. Dept. Agr. SCS-TP-106.*

——— (1958). "Hydraulics of Closed Conduit Spillways, Pt X: The Hood Inlet." *Univ. of Minn. St. Anthony Falls Hydraulic Lab. Tech. Paper no. 8, Series B.*

Blaisdell, F. W., and A. N. Huff (1948). "Report of Tests Made on Three Types of Flume Entrances." *U.S. Dept. Agr. SCS-TP-70.*

Jepson, H. G. (1939). "Prevention and Control of Gullies." *U.S. Dept. Agr. Farmers' Bull. 1813.*

King, H. W., and E. F. Brater (1963). *Handbook of Hydraulics* (5th ed.). McGraw-Hill, New York.

Mavis, F. T. (1943). "The Hydraulics of Culverts." *Pa. Eng. Expt. Sta. Bull, 56.*

Morris, B. T., and D. C. Johnson (1942). "Hydraulic Design of Drop Structures for Gully Control." *Proc. Am. Soc. Civil Engrs.* **68,** 17–48.

Murphy, G. (1950). *Similitude in Engineering.* Ronald Press, New York.

Smith, D. D. (1952). "A 20-Year Appraisal of Engineering Practices in Soil and Water Conservation. *Agr. Eng.* **33**, 553, 556.

U.S. Bureau of Reclamation (1977). *Design of Small Dams,* 2nd ed. U.S. GPO, Washington, D.C.

U.S. Soil Conservation Service (1969). "Engineering Field Manual for Conservation Practices" (lithograph). Washington, D.C.

Wooley, J. C., et al. (1941). "The Missouri Soil Saving Dam." Missouri Agr. Expt. Sta. Bull. 431.

Yalamanchili, K., and F. W. Blaisdell (1975). "Hydraulics of Closed Conduit Spillways, Part XIII. The Hood Drop Inlet." U.S. Dept. Agr. ARS-NC-23.

PROBLEMS

9.1. What is the maximum capacity of a straight-drop spillway having a crest length of 3 m (10 ft) and a depth of flow of 0.9 m (3 ft)?

9.2. Determine the design dimensions for the drop spillway in Problem 9.1, using the Morris and Johnson outlet if the drop in elevation is 1.5 m (5 ft). The waterway is 4.6 m (15 ft) wide.

9.3. Determine the crest length for a straight-inlet drop spillway to carry 7.5 m^3/s (265 cfs) if the depth of flow in not to exceed 0.9 m (3 ft). What should be the dimensions of a square-box inlet for the same conditions?

9.4. Determine the design dimensions for a 1.5 $\times$ 1.5-m (5 $\times$ 5 ft) box-inlet drop spillway to carry 7.1 m^3/s (250 cfs) if the end sill is 3 m (10 ft) in length.

9.5. What is the capacity of a formless flume 1.8 m (6 ft) wide when the flow depth is 0.6 m?

9.6. Determine the discharge of a 30-m (100-ft) 0.9 $\times$ 0.9-m (3 $\times$ 3-ft) concrete box culvert having a square entrance, $n = 0.013$, $s = 0.003$, and elevations of 10.97 m (36.0 ft) at the center of the conduit at the outlet, 12.89 m (42.3 ft) for the headwater, and 12.34 m (40.5 ft) for the tailwater.

9.7. Tabulate and plot the head-discharge up to a 1.5-m (5-ft) depth above the crest for a 0.9 $\times$ 0.9-m (3 $\times$ 3-ft) drop inlet attached to an 457-mm (18-in.) diameter concrete pipe ($n = 0.015$) 45.7 m (150 ft) in length. Assume pipe flow controls in the conduit; tailwater height is not higher than the center of the pipe at the outlet; radius of curvature of the pipe entrance is 0.09 m (0.3 ft); and the difference in elevation between crest and center of pipe at outlet is 5.49 m (18 ft).

9.8. Compute the critical depth of flow in a rectangular chute 1.82 m (6 ft) in width if the flow is 2.83 m^3/s (100 cfs). If the roughness coefficient is 0.015, what is the slope of the chute?

9.9. Determine the critical depth for a triangular-shaped channel having side slopes of 1:1 if the flow is 2.83 m^3/s (100 cfs).

9.10. If the discharge as controlled by pipe flow through a drop inlet pipe spillway is 1.30 m³/s (46 cfs) with a depth of flow of 0.6 m (2 ft) over the crest of the inlet box, what is the minimum size of a square box inlet to keep the pipe flowing full. At this condition weir flow equal pipe flow. Assume the weir coefficient for the box is 3.2, and one side of the box serves as an antivortex wall.

9.11. Determine the capacity of a 610-mm (24-in.) diameter culvert 30 m (100 ft) in length. $K_e = 0.25$, $K_c = 0.049$ (SI units), $s = 0.032$, headwater is 2.13 m (7 ft) above inlet invert, and tailwater is 0.37 m (1.2 ft) above outlet invert. What is the neutral slope assuming full pipe flow?

9.12. A 457-mm (18-in.) diameter culvert 15 m (50 ft) in length is to be placed under a farm road to carry an estimated peak runoff of 0.28 m³/s (10 cfs). The invert elevation of the pipe inlet is 15.24 m (50 ft) and the outlet invert is 14.94 m (49 ft). Maximum tailwater elevation is 15.09 m (49.5 ft). The entrance is square with a concrete headwall flush with the end of the pipe. Corrugated metal pipe with $n = 0.025$ is to be installed. Assume ponding of water above the culvert has no effect on runoff. To what elevation would the road have to be built so that the peak flow would not overtop the road?

9.13. Determine the discharge of a circular-plate hood inlet pipe spillway through a pond dam if the conduit is a 305-mm (12-in.) diameter corrugated metal pipe, 30 m (100 ft) long on a slope of 16 percent. The depth of water above the inlet invert is 0.91 m (3 ft).

9.14. Determine the minimum dimensions of a square drop inlet and pipe spillway to cause the pipe just to flow full with a 0.3-m (1-ft) depth of flow above the spillway crest. The pipe is 610 mm (24 in.) in diameter and 30 m (100 ft) in length, and the total head causing pipe flow is 4.88 m (16 ft). Assume $K_e = 0.15$, $K_c = 0.066$ (SI units), and the weir coefficient is 3.2.

CHAPTER 10

Earth Embankments and Farm Ponds

In all land-use programs the availability of water for crops, livestock, and many miscellaneous purposes is of primary importance. Farm ponds and reservoirs provide a logical source of such water, for they may be designed and adjusted to fit the individual land-use plans and conditions that exist on the farm. Conservation and protection of land also depend upon the control of excess waters. Earth embankments in the form of dikes, levees, and detention dams are important protective structures.

The design of earth embankments that are effective and safe requires thorough integration of the principles of soil physics and soil mechanics with sound engineering design and construction principles.

EARTH EMBANKMENTS

Regardless of the structure size, the basic design principles apply equally to all conservation structures, such as dikes, levees, farm ponds, and reservoirs having a total height above ground level not to exceed 15 m. While structures in excess of 15 m in height may be found in upstream watershed projects, such structures have specialized design requirements not covered in this text.

10.1. Types of Earth Embankments. The discussions in this chapter are limited to the rolled-fill type of earth embankment in which the soil material has been spread in uniform layers and then compacted at optimum moisture until maximum density is achieved. The selection and design of embankments for water control is predicated upon (1) the foundation properties, that is, stability, depth to impervious strata, relative permeability, and drainage conditions, and (2) the nature and availability of the construction materials.

There are three major types of earth fills. The *simple embankment* type is constructed of relatively homogeneous soil material and is either keyed into an impervious foundation stratum, as shown in Fig. 10.1a, or is constructed with an upstream blanket of impervious material, as shown in Fig. 10.1b. This type is limited to low fills and to sites having sufficient volumes of satisfactory fill materials available. The *core* or *zoned type* of design utilizes within the dam a central section of highly impermeable or puddled soil materials extending from

211

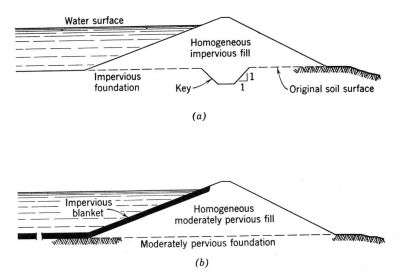

Fig. 10.1. (a) Simple embankment utilizing "key" construction. (b) Simple embankment with an impervious "blanket" seal.

above the water line to an impermeable stratum in the foundation. In some instances an upstream blanket is used in conjunction with this design. These designs, shown in Fig. 10.2, reduce the percentage of high-grade fill materials needed for construction. The *diaphragm* type uses a thin wall of plastic, butyl, concrete, steel, or wood to form a barrier against seepage through the fill. A "full-diaphragm" cutoff extends from above the water line down to and sealed into an impervious foundation stratum as shown in Fig. 10.3*a*. The "partial diaphragm," Fig. 10.3*b*, does not extend through this full range and is sometimes referred to as a cutoff wall. The internal or buried diaphragm, particularly when constructed of rigid materials, has the disadvantage of being unavailable for inspection or repair if broken or cracked due to settlement in the foundation or the fill. The more recent use of flexible film diaphragms of plastic and butyl rubber have partially overcome this problem.

10.2. Foundation Requirements. Earth dams and embankments may be built upon a wide range of foundation conditions provided these conditions are properly considered in design.

On small dams these investigations may be limited to auger borings. On larger structures the subsurface exploration should be more thorough. Wash borings, test pits, and other standard procedures should be employed to determine the underlying soil and geologic conditions.

In evaluating the physical and engineering properties of the foundation soils and the embankment materials, a number of standard soil mechanics tests, such

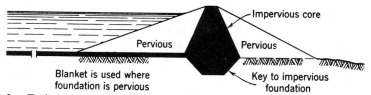

Fig. 10.2.   Embankment utilizing a central core and key of impermeable material.

as particle size distribution, liquid limit, plasticity index, shear strength, compressibility, and permeability can be utilized. A Unified Soil Classification System has been developed to provide a logical groupings of soils based upon these physical and mechanical properties. A full discussion of exploration procedures, soil tests, classification, and interpretation of the data in terms of their use appears in several references: USBR (1977), Richey (1961), Taylor (1948), Creager et al. (1945), Spangler and Handy (1973) and similar references. Foundation materials have been classified by Creager et al. (1945) in the following way: (1) *Ledge rocks.* Under earth-filled dams ledge rocks present a potential permeability hazard and frequently need grouting. (2) *Fine uniform sands.* If below ''critical density'' (void ratio at which a soil can undergo deformation without change of volume) fine uniform sands must be consolidated to prevent flow when saturated under load. (3) *Coarse sands and gravel.* From the stability standpoint they will consolidate under load. An upstream blanket may be required to prevent seepage losses. (4) *Plastic clays.* They require careful analysis to assure that shear stress imposed by the weight of the dam is less than the shear strength of the foundation material; flattened side slopes may be required to reduce shear stress.

Knowledge of porous strata, preglacial gorges, geologic faults, and other hazardous conditions will be of value in design of the structure.

10.3. Design to Suit Available Materials.   The design of a dam or embankment should be based upon the most economical use of the available materials immediately adjacent to the site. For example, if satisfactory core materials are unavailable and must be hauled some distance, the hauling cost should be compared to the cost of a thin-section diaphragm of concrete or steel.

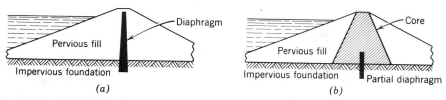

Fig. 10.3.   (a) Full-diaphragm and (b) Partial-diaphragm construction for seepage control.

Cross-section design depends on both the foundation conditions and the fill material available. Where depth to an impervious foundation is not too great and where supplies of quality core materials can be found, designs shown in Figs. 10.2 and 10.3*b* can be used. The combination core and blanket design shown in Fig. 10.2 is adapted to sites having extremely deep pervious foundations. Other designs may be developed to utilize diaphragms alone and in combination with cores and other construction features to meet specific conditions.

For optimum compaction and water-holding capacity, experiments and experiences have shown that preparations of gravel, sand, silt, and clay that will compact to maximum density are, generally, not over 20 percent gravel, 20 to 50 percent sand, not over 30 percent silt, and 15 to 25 percent clay. Soils having high shrinkage and swelling characteristics and ungraded soils, when their use cannot be avoided, should be placed in the downstream interior of the embankment. Here they are subject to less moisture change and because of overburden weight have less volume change than if placed elsewhere in the embankment. Soils having higher percentages of graded sandy materials resist changes in moisture, temperature, and internal stresses. Organic soils should be entirely eliminated from the fill.

In levee construction, the selection of material is usually limited to that found adjacent to the structure. As these materials change in their characteristics along the route of the levee, it will be necessary to adjust the cross-sectional design of the embankment in compensation.

10.4. Seepage Through the Embankment. While the characteristics of the soil materials in both the foundation and the embankment may have a marked effect upon the seepage losses through and beneath the embankment, many design factors may also have an influence. An understanding and knowledge of the position of the seepage line permits adjustments to be made in the design that will permit improved seepage control.

The seepage line is the upper line of seepage. It is the "line above which there is no hydrostatic pressure and below which there is hydrostatic pressure." Some refer to it as the "phreatic line." It has been found that the seepage line is affected by (1) the permeability of the fill materials and of the foundation, (2) the position and flow of ground water at the site, (3) the type and design of any core wall or cut-off within the embankment, and (4) the use of drainage devices to collect seepage in the downstream portion of the structure.

It is important to know at what point the seepage line intersects the downstream slope so that adequate toe drainage can be developed. If the intersection is well up the face of the dam and the rate of seepage is sufficient to move the soil, serious sloughing will result.

Where a dam is homogeneous and is located on an impervious foundation, the seepage line will cut the downstream face above the base of the dam. Its location is dependent on the cross section of the dam and is not affected by the permeability of the dam unless the permeability changes. The basic shape of the seepage line is that of a parabola varying at the ends as the intake and outflow conditions vary.

By making certain assumptions, an approximate seepage line may be located as illustrated for the composite dam shown in Fig. 10.4a. The upstream and downstream portions of this dam are relatively pervious while the center portion is impervious. When wide differences in the permeability of these two portions exist, it is usually sufficient to determine the position of the seepage line through the most impervious portion of the dam. The upstream shell will have little effect on the position of the line and the downstream portion will act as a drain, thus placing the line slightly above the tail water.

In this rough solution, it is assumed that where the discharge slopes are flatter than 1:1

$$e = h/3 \qquad (10.1)$$

where e = distance from the impervious base of the dam up to the intersection of the seepage line and the downstream face of the dam,

   h = distance from the impervious base to the level of the water in the reservoir.

By applying Darcy's law to a unit width of dam and assuming that the mean discharge area will be $(h + e)/2$, the formula for expressing the discharge through the dam is

$$q = \frac{K(h - e)}{L} \frac{(h + e)}{2} = \frac{K}{2} \frac{(h^2 - e^2)}{L} \qquad (10.2)$$

where q = discharge rate per unit length (L^3/T),

   K = hydraulic conductivity of the material comprising the least permeable section of the dam (L/T),

   L = length of flow (L).

By trial and error or by differentiation, a value of e for which q is maximum may be computed. As indicated above, this is approximately $h/3$ for slopes flatter than 1:1. Therefore, by substituting $h/3$ for e in Eq. 10.2,

$$q = \frac{4Kh^2}{9L} \qquad (10.3)$$

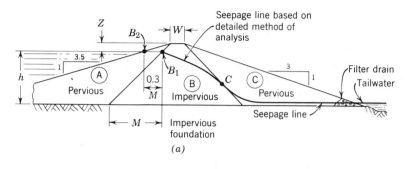

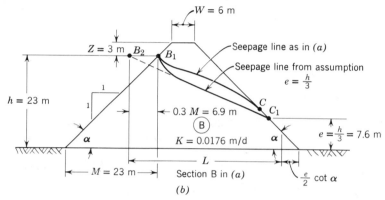

Fig. 10.4. Seepage line determination for a composite dam by (a) detailed method of analysis and (b) approximation method. (Redrawn from Creager et al., 1945.)

Example 10.1. Construct the approximate seepage line and estimate the seepage rate through the dam shown in Fig. 10.4.

Solution. Considering only the impervious section B, $e = 23/3 = 7.6$ m (25 ft) from Eq. 10.1. The intersection of the seepage line and the downstream face, point C_1, can then be plotted in Fig. 10.4b using $e = 7.6$ m (25 ft). Then, from point B_2 on the water surface upstream from the face of the dam the seepage line is drawn to point C_1. Point B_2 on the water surface is equal to $0.3M$, where M is the horizontal projection of the wetted upstream slope. As shown in Fig. 10.4b, B_2 to $B_1 = 0.3 \times 23 = 6.9$ m (22.5 ft) $= 0.3M$. The upstream or ingress end of the seepage line may be sketched in as shown in Fig. 10.4. The mean length of the seepage line,

$$L = (2Z + h - e/2) \cot \alpha + W + 0.3M$$

using symbols shown in Fig. 10.4.

By substituting the values computed above in the equation,

$$L = (2 \times 3 + 23 - 7.6/2) \times 1 + 6 + 6.9 = 38.1 \text{ m (125 ft)}$$

By substitution in Eq. 10.3, the discharge through the dam per unit length is

$$q = \frac{4 \times 0.0176 \times 23 \times 23}{9 \times 38.1} = 0.1086 \text{ m}^3/\text{d per lineal meter of length}$$

(1.169 cfd per lineal feet of length)

The determination of the probable position of the seepage line in levees follows the same principles as those outlined above. Maximum flood stage is substituted for the water surface elevation. Determination of the seepage line in levees is important where sustained peaks are expected and where construction materials are particularly permeable. Levees controlling short duration flood peaks rarely become thoroughly enough saturated to create a hazardous condition.

Frequently farm ponds are needed at sites where the storage area will not hold water. In such instances it is necessary to resort to a "bag-type" construction in which an impervious blanket is constructed over the entire impounding area, as illustrated in Fig. 10.1b.

Blankets should not be applied to the upstream face of a dam where frequent, total, or rapid drawdown is anticipated because of the danger of slumping, due to the internal hydrostatic pressure in the saturated portion of the dam, and of cracking upon drying. Uncontrolled stock watering or wading also causes disturbance of the blanket.

For earth embankments of the size covered in this text, a blanket thickness equal to 10 percent of the depth of the reservoir above the blanket but with a minimum of 1 m is usually satisfactory if constructed from the same impervious materials as in the dam. Where blankets are used on small structures, they usually extend out from the toe of the dam 8 to 10 times the depth of the water at the dam.

Where the soil materials on the pond bottom approach the proportions of 70 percent sharp well-graded sand, 20 to 25 percent clay, and sufficient silt to provide good gradation of particle size, the seal can be accomplished by ripping the soil to a depth of about 0.3 m and then recompacting at optimum moisture with a sheepsfoot roller until maximum density is attained. When proper soils are not available at the site, materials must be selected from adjacent borrow pits, mixed, spread, and compacted.

Support of the head of stored water is essential; therefore, consideration must be given to the necessary mantle thickness needed to prevent the water pressure from forcing the soil out through underlying rock crevices. Under most conditions, a minimum of 0.6 m of soil should exist between the compacted blanket and the rock formations. Soils with the particle size gradation outlined above have sufficient structural resistance to support the weight of the water, thus preventing slippage and flow of the soil mantle into underlying rock crevices. Soils having high clay contents do not have this resistance to superimposed loads. Therefore, to provide the needed supporting power, high clay content soil mantles must have greater thickness.

Where satisfactory construction materials are unavailable, seals may sometimes be accomplished by incorporating swelling clays, such as bentonite, with the soil. This material swells to fill the interstices between the soil particles, thereby decreasing the permeability. Bentonites alone will not support the weight of the stored water; therefore, they should be used only with soils containing a minimum of 10 to 15 percent sand, which will provide the supporting capacity. Rates of application must be adjusted to individual soil conditions.

Similarly, the use of polyphosphates and other chemical additives that tend to deflocculate the soil, thus reducing its porosity, must be carefully selected in relation to the soil type and its chemical characteristics.

Where special site or construction conditions dictate maximum precautions against seepage through the reservoir sides and bottom or where water costs dictate maximum water storage efficiency, consideration may be given to use of total lining techniques. Plastic and butyl films sealed into one-piece reservoir liners may be practical.

10.5. Foundation Treatment. A site examination and the results of exploratory foundation borings will determine whether the embankment is to be placed on a foundation of (1) rock, (2) fine-grained material (silt and clay), or (3) coarse-grained material (sand and gravel). Exposed rock foundation should be examined for joints, fractures, and permeable strata that might contribute to seepage and eventual structural damage through piping. The usual treatment is pressure grouting to a depth in the bedrock equal to the design head of water above the rock surface by techniques described in USBR (1977).

Where fine-grained impermeable material constitutes the foundation, minimum treatment should include removal of vegetative cover, topsoil having a high organic matter content, and soil that has been loosened through frost action or of recent deposition. If no cutoff trench is necessary because of the impermeability of the foundation material, other steps should be taken to bond the impervious zone of the embankment to the foundation. This can be done with a simple *key trench* cut 0.6 to 1 m into the foundation and having a bottom width of 4 to 6 m (see Fig. 10.2).

Where the foundation consists of porous material, seepage should be stopped by construction of a *cutoff trench* extending to bedrock or to an impervious soil stratum. In general, a minimum bottom width of 3−6 m should be maintained.

The minimum height of a cutoff seal should be one half the height of the dam; its depth below original ground surface will depend upon the position of the impermeable stratum. Side slopes of the trench should not be less than 1:1. Where the cutoff seal is extended as a full core, it should be designed with a minimum top width of 1.2 m and the side slopes should not be steeper than ½:1.

Where satisfactory soil materials for cores are unavailable or where there is an excessive distance through the pervious foundation material to impeding strata, cutoff or core walls of steel sheet piles or reinforced concrete may be used. The effectiveness of these devices is based on the care taken in installation and in keying the cutoff into the soil materials. Cement-bound curtain cutoffs and chemical grouting have been used with success where there is adequate economic justification for their use.

10.6. Drainage. In relatively impervious and homogeneous dams the seepage line frequently appears high on the downstream slope, resulting in a saturated unstable soil condition. The construction of a system of toe filters or drains will lower the seepage line until it intersects with the drain. The capacity of toe filters or drains should be at least twice the computed maximum seepage discharge through the dam. Figure 10.5 illustrates several types of effective drainage systems.

Where a saturated pervious foundation underlies an impervious top layer, consideration must be given to the possibility of piping due to uplift pressure that exceeds the pressure of the impervious mantle and the reservoir head. Where the overlying mantle is too thick to permit use of drainage ditches, pressure-relief wells may be used to intercept the seepage and reduce the uplift pressure.

Drainage blankets and filters are designed to be (1) pervious so that seepage removal can be accomplished, (2) operational as a filter so as to prevent movement of soil particles from the foundation or embankment into the seepage discharge, and (3) of sufficient weight to overbalance the high upward seepage forces that might lead to downstream blowouts or related piping failures.

Adequate drainage is equally important in levee construction, for a saturated impounding structure is always potentially dangerous. A tile line or drainage ditch located near the land side toe at a depth that will affect the lower end of the saturation line is a standard design. These drains are carried away from the levee and outletted through controlled relief wells.

10.7. Side Slopes and Berms. The side slope of an earth dam or levee is dependent upon the height of the structure, the shearing resistance of the foundation soil, and upon the duration of inundation.

On structures less than 15 m high with average materials, the side slopes should be no steeper than 3:1 on the upstream face and 2:1 on the downstream face. Coarse, uncompactable soils may need side slopes of 3:1 or 4:1 to assure

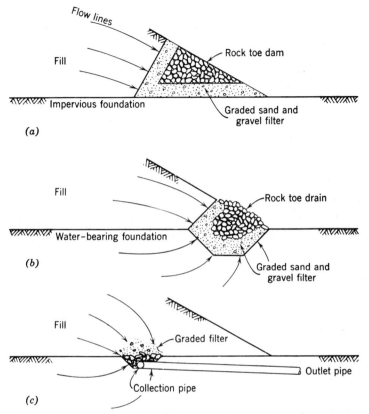

Fig. 10.5. (a) Simple rock-fill toe drain. (b) Deep rock toe designed to drain both the fill and the water-bearing foundation. (c) Pipe-type drain with collection pipe.

stability. Where more precise designs are required and on dams exceeding 15 m in height, a complete soil analysis should be made to determine the horizontal shear strength of the materials.

In levee construction on major waterways having flood peaks of long duration that cause maximum saturation of the levee, side slope ratios may range as high as 7:1. These flat slopes are necessary because levee materials are frequently of poor quality, are unstable, and frequently receive no compaction.

Most conservation levee construction is limited to those protecting farm lands on the upper tributaries. Since they are subjected to short-duration flood crests, minimum side slopes of 2:1 may be provided if they are properly protected at critical points. Levees that protect land from back flow after pump drainage, as in certain muck areas, tidal marshes, and swamp areas, should be designed and constructed by the same criteria established for dams.

10.8. Top Width. Top widths of dams vary with the height and purpose to which the dam is being put. The minimum top width for dams up to 5 m in height should be 2.4 m. If the top serves as a roadway, this minimum should be increased to 3.6 m to provide a 0.6-m shoulder to prevent raveling. Road water should be controlled to prevent erosion of the side slopes.

Top widths for dams exceeding 5 m in height may be designed by the empirical formula

$$W = 0.4H + 1 \tag{10.4}$$

where W = top width in meters,
H = maximum height of embankment in meters.

In levee construction top or crown width may vary from a minimum of 1 to 6 m or more. On levees subject to short-duration flood peaks, 1-m crowns are adequate. On levees subject to sustained peaks the width usually equals twice the square root of the height of the structure.

10.9. Freeboard. The distance between the maximum designed high water or flood peak level in the reservoir and the top of the settled dam or embankment constitutes the net freeboard. Reference is sometimes made to gross freeboard, or "surcharge," which is the distance between the crest of the mechanical spillway and the top of the dam. Net freeboard should be used in all design work. See Fig. 10.6.

The net freeboard should be sufficient to prevent waves or spray from overtopping the embankment or from reaching that portion of the fill that may have been weakened by frost action (that depth of soil loosened by frequent alternate freezing and thawing—this depth is seldom over 0.15 m).

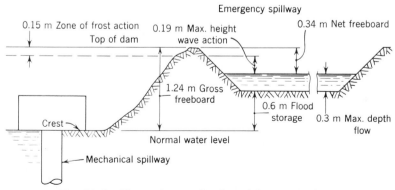

Fig. 10.6. Net and gross freeboard for reservoirs.

Wave height for moderate-size reservoir areas can be determined by Hawksley's formula,

$$h = 0.014(D_f)^{1/2} \tag{10.5}$$

where h = height of wave in meters from trough to crest under maximum wind velocity,
D_f = fetch or exposure in meters.

All freeboards should be based on the water level at the maximum flood heights for which the dam and spillways are designed. The maximum flood height for a given design storm is usually taken as the elevation of the emergency spillway crest for flood control reservoirs. In this case the emergency spillway is designed for a much larger storm. However, for farm ponds and other detention reservoirs with small mechanical spillways, the design flood height will be above the crest, but with a depth of flow in the spillway of not more than 0.3 m. The definition of net freeboard therefore depends on the definition of the maximum flood level. The flood storage depth, which is the difference in elevation between the mechanical and the emergency spillway crests, can be accurately determined by flood routing procedures given in Chapter 11. For small storage reservoirs with a 200- to 250-mm diameter mechanical spillway, this depth can be estimated from Fig. 10.7. The following example, as illustrated in Fig. 10.6, shows the procedure for estimating freeboard.

Example 10.2.   Determine the net and gross freeboard for a farm pond with a 0.6-ha (1.5-ac) water surface and an exposure length of 183 m (600 ft). Assume the frost depth is 0.15 m (0.5 ft), the 25-year design runoff peak is 4.00 m³/s (141 cfs), and the flow depth in the flood spillway is 0.3 m (1.0 ft).
Solution.   From Eq. 10.5

$$h = 0.014 (183)^{1/2} = 0.19 \text{ m } (0.6 \text{ ft})$$

From Fig. 10.7 read the flood storage depth of 0.6 m (2.0 ft).

Net freeboard = 0.15 + 0.19 = 0.34 m (1.1 ft)
Gross freeboard = 0.34 + 0.3 + 0.6 = 1.24 m (4.1 ft)

Additional freeboard should be added as a factor of safety where lives and high value property would be endangered by a dam failure. On large dams the freeboard is 50 percent more than the wave height and is increased for high winds. Levee freeboard is designed on the same basis as dams.

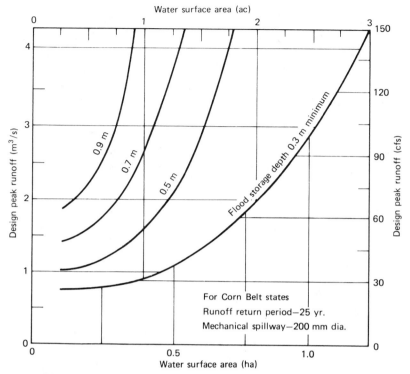

Fig. 10.7. Flood storage depth for small storage reservoirs.

10.10. Compaction and Settlement. The volume relationships of soil may be expressed by the formulas:

$$V = V_s + V_e \qquad (10.6)$$

where V = total in-place volume (L^3),
V_s = volume of solid particles (L^3),
V_e = volume of voids either air or water (L^3).

The void ratio is expressed by

$$e = V_e/V_s \qquad (10.7)$$

and the total porosity in percent by
$$n = (V_e/V) \times 100 \qquad (10.8)$$

The soil density is defined as the oven dry weight per unit volume of soil in place.

There is an optimum moisture content at which maximum density occurs for a given amount of energy applied during the compaction process. A standard test developed for disturbed soils is known as the Proctor density test. A typical Proctor density curve is shown in Fig. 10.8. In conducting a test the soil is placed in a container and compacted in layers with a weight dropped from a certain height for a definite number of times. The moisture content after each compaction is determined, and then water is added for the next test. After several of these tests a curve can be drawn as shown in Fig. 10.8. The density increases with moisture content up to a certain point, above which the density decreases. The maximum or Proctor density occurs at the optimum moisture content for compaction. This test is widely used in engineering construction.

A compacted earth fill is normally more dense than soil from the borrow area. If the volume of the borrow pit is desired, the above formulas apply as illustrated in the following example.

Example 10.3. In constructing an earth dam 1224 m³ (1600 yd³) of soil are required in the settled fill. If the desired void ratio of the dam is 0.6, what is the volume required from the borrow pit provided its void ratio is 1.1?

Solution. The volume of solids is the same in the fill as in the borrow pit, but the volume of voids will decrease in the fill. From Eq. 10.7 $V_e = 0.6V_s$ for the fill. Substituting in Eq. 10.6,

$$V = 0.6V_s + V_s = 1.6V_s = 1224 \text{ m}^3 \,(1600 \text{ yd}^3)$$
$$V_s = 765 \text{ m}^3 \,(1000 \text{ yd}^3)$$

For the borrow pit $V_e = 1.1V_s$. Substituting in Eq. 10.6,

$$V = 1.1V_s + V_s = 2.1V_s = 2.1 \times 765 = 1606 \text{ m}^3 \,(2100 \text{ yd}^3)$$

Rolled-filled embankments that have been placed in thin layers and compacted at optimum moisture and that are on an unyielding foundation will not settle more than about 1 percent of the total fill height. On deep plastic foundations settlement may reach 6 percent. Fills for small farm ponds and reservoirs usually do not receive as thorough compaction as the larger structures do; therefore, the fill is usually constructed to a height 5 to 10 percent higher than the designed settled fill height for all points along the embankment.

In levee construction a minimum allowance for settlement of 20 to 25 percent is usually made because dragline or conveyor placement without intensive compaction does not give a high degree of consolidation.

10.11. Wave Protection. If the water side of dams or levees is exposed to considerable wave action or current, it should be protected either by covering with nonerodible material called riprap or protected by energy dissipaters such as anchored floating logs.

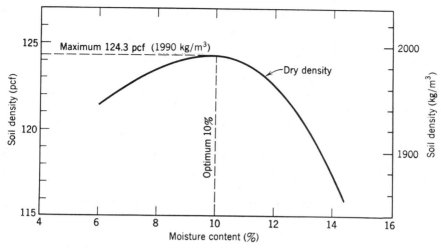

Fig. 10.8. A typical Proctor soil density curve.

Riprap may consist of hand-placed or dumped stone or various types of concrete block or slab construction. It should be carefully installed on a bedding or cushion layer of graded gravel or crushed stone from 0.2 to 0.5 m deep, depending on quality. This layer prevents the clay and silt from being washed out by wave action, thus causing the riprap stone to settle into the embankment.

A dense sod of a suitable grass will usually give slope protection on levees and small ponds that are not subjected to excessive wave action. In exposed areas riprapping or willow or cottonwood revetments must be used.

10.12. Mechanical Spillways. Most farm ponds and reservoirs in conservation programs are protected from overtopping by vegetated flood or emergency spillways. This vegetation will not survive if base flow entering the pond and frequent minor flood-flows keep the spillways unduly wet. Therefore, in order to protect these spillways and to carry low flows, various types of mechanical spillways are provided.

These consist of various types of structures (Chapter 9) having their entrance openings set at the elevation at which the pond is to be maintained and extending through the dam to a safe outlet point below the dam. The entrance elevation depends on the level of permanent storage desired.

Any durable material may be used for the mechanical spillway, provided it has a long life expectancy and is not subject to easy damage by settling, loads, or blows. The portion passing through the dam should be placed on a firm foundation of slowly permeable material, preferably on undisturbed soil, and should be located to the side of the original channel. When rigid conduits, such as concrete, cast-iron, or asbestos cement are placed on rock foundations, they must be cradled in concrete. Flexible conduits, such as steel or corrugated metal pipe, may be cradled in compacted impermeable soil material.

Where conduits are laid on foundations or are passing through fills that are subject to more than nominal settlement, consideration should be given to laying the conduit with a camber to compensate for this settlement. Detailed analytical procedure for determining the invert elevations for cambered conduits are given in Richey (1961). In such cases expansion joints of various types should be included in the conduit design so that the change in vertical alignment will not cause adverse stress with resulting cracking.

All such structures passing through the dam should be covered by thin, well-tamped layers of high-grade material. Where the fill surrounding the pipe is of low quality, concrete or metal seepage collars, extending out a minimum of 0.6 m in all directions, should be used. The number of collars should be sufficient to increase the length of the creep distance at least 10 percent. The creep distance is the length along the pipe within the dam, measured from the inlet to the filter drain or to the point of exit on the backslope of the dam. The increase in creep distance for an antiseep collar is measured from the pipe out to the edge of the collar (see Fig. 10.9a). For example, the increase in creep distance for a 1 × 1 m collar on a 300-mm diameter pipe would be twice the distance out, $2(0.5 - 0.15) = 0.7$ m. For small-diameter water pipe through the dam two 0.6 × 0.6 m collars are adequate because of better compaction around the small pipe and the long creep distance. All construction forms should be stripped from concrete structures before covering. All pipe joints should be carefully packed and sealed. Where concrete or mortar joints are used, it is advisable to complete the backfilling and tamping before it has set, thus avoiding danger of cracking the green concrete. A minimum of 1.2 m of protective fill should be hand-placed over the structure before heavy earth-moving equipment is allowed to pass over it.

The mechanical spillway may, in some instances, serve as a drain for cleaning, repairing, and restocking the pond. One of several types of installations is shown in Fig. 10.9b. All valve assemblies should be equipped to operate from above the surface of the water. All mechanical inlets should be protected with a screening device that prevents floating objects, turtles, and other foreign objects from entering and clogging the pipe. The device should be so designed that, after becoming loaded with debris, it will not retard the entrance of water. Outlets should be carried well below the toe of the dam and protected against scouring (see Chapter 9).

Mechanical spillways may be designed to carry a small percentage of the peak flow rate, such as the trickle pipe in a farm pond, or a much greater flow as in a flood control reservoir. For small watersheds and for flood spillways lined with concrete or other stable material, a pipe mechanical spillway may not be required. Where considerable flood storage volume is available, flood routing

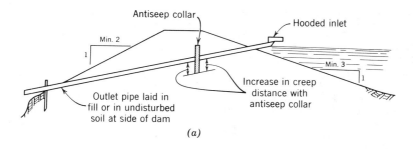

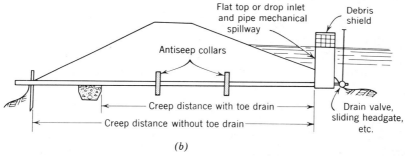

Fig. 10.9. Types of mechanical spillway designs. (a) Hooded inlet and pipe (b) Vertical riser and drain pipe combined.

procedures should be followed to determine the pipe size. The mechanical spillway should always be large enough to carry the seepage or base flow so that the flood spillway will stay dry and stable. For farm ponds with small watersheds, approximate pipe sizes are given in Table 10.1.

10.13. Flood or Emergency Spillways. All reservoirs must be equipped with one or more emergency spillways that will safely by-pass floods that exceed the temporary storage capacity of the reservoir. Where a natural spillway, such as a depression in the rim of the impounding area does not exist, it is necessary to construct one or more trapezoidal channels in undisturbed earth.

After careful estimation of the peak rates and amounts of runoff to be handled, application of the broadcrested weir formula and vegetated waterway design procedures will determine the cross-sectional area required (Chapters 4, 7, and 9). The following example illustrates the procedure for small structures.

Example 10.4. The design capacity of a trapezoidal-shaped grassed flood spillway for a dam is 3.4 m³/s (120 cfs). The flood spillway width is 7.6 m (25 ft), side slopes 3:1, and the design depth of flow is 0.3 m (1 ft). What

TABLE 10.1 Mechanical Spillway Pipe Diameters for Farm Ponds[a]

Peak Runoff Rate for 25-yr Return Period (m³/s)	*Water Surface Area at Normal Water Level in Hectares*				
	0.2	*0.4*	*0.8*	*1.2*	*2.0*
0.5 or less	200 mm	150 mm	150 mm	—	—
0.5 to 1.0	200	200	150	150 mm	—
1.0 to 1.5	250	250	200	200	150 mm
1.5 to 2.0	250	250	250	250	200
2.0 to 2.5	—	300	300	300	250

[a] Suggested sizes of corrugated metal pipe primarily for ground water or seepage flow. For smooth wall pipes, use the next smaller diameter.

width is required for the spillway so that it does not restrict the flow, that is, increase the design level in the reservoir?
Solution. From Chapter 9 the broad-crested weir formula with $C = 3.2$ will give a satisfactory estimate for the width of the control section,

$$L = 3.4/(0.552 \times 3.2 \times 0.3^{3/2}) = 11.7 \text{ m (38 ft)}$$

The grassed approach channel should be flared out on each side at least 0.2 m per meter of length as measured along the center line of the channel upstream from the control section. See Fig. 10.10.

The above example assumes that the flow passes through critical depth at the downstream end of the control section as shown in Fig. 10.10. For this condition to occur the slope in the exit section must be greater than the critical slope, otherwise the exit section could control the flow. The slope in the approach channel should not be less than 2 percent upstream to provide drainage and reduce inlet losses. The exit section should be adequate to carry the design flow, but at velocities not greater than those recommended for grassed waterways given in Chapter 7. Outlets should extend well below the dam or into an adjacent waterway and should be well-protected against scouring.

The design capacity for flood spillways on small reservoirs and ponds is normally based on a return period of 25 years from the contributing watershed. For larger structures where high value property may be damaged or where lives are endangered, a 100-year return period storm or a maximum probable storm may be considered. Flood spillway flow may be reduced if the capacity of the mechanical spillway is large compared to the flood flow. If the design flood storm occurs when the reservoir is full, temporary flood storage is not available to reduce the flow in the flood spillway.

10.14. Stability or Earth Embankments. The factors previously discussed in this chapter have a direct bearing on the stability and safety of low-height

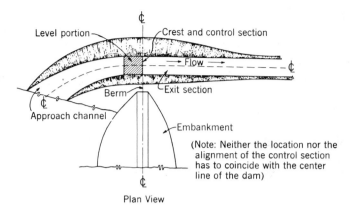

Plan View

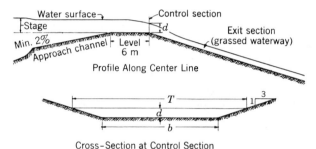

Profile Along Center Line

Cross–Section at Control Section

Fig. 10.10. Excavated flood spillway for reservoirs. (Redrawn from U.S. SCS, 1969.)

embankments, but if properly constructed and maintained, they should be permanently safe. For embankments in excess of 15 m in height, such factors as stability against headwater pressure, resistance to horizontal shear, and shear stress in foundations should be more fully investigated.

10.15. Construction. Construction according to plan and adjustment to soil conditions while the work is in progress are essential for rolled-fill dam success. All trees, stumps, and major roots should be removed from the site. Sod and topsoil should be removed and stock-piled for later spreading over the dam and exposed areas to secure good vegetation. Before the placement of any fill material, the original ground surface should be thoroughly plowed and disked parallel to the length of the dam. If necessary, it should be sprinkled to assure that it is at optimum moisture for compaction. Proper moisture conditions make it possible to force some of the fill material into the original surface, thus eliminating any dividing plane between the fill and the foundation.

After the site has been cleared, the cutoff trench is excavated. The compacted core should be constructed only of carefully selected materials laid down in thin blankets at optimum moisture. The fill material should be placed in thin layers, evenly, over the entire section of the dam. The thickness of layers for pervious soils should be limited to 25 cm in thickness; the more plastic and cohesive soils should not exceed 15 cm in thickness. In placing the fill material on the dam, the fill should be nearly horizontal with a slope of 20:1 to 40:1 away from the center of the dam. If, upon the approach of rain, the surface is left in a fairly smooth condition, the resultant surface drainage will keep the dam from becoming saturated.

Laboratory tests can be made to determine closely the optimum moisture content for each type of soil to achieve the necessary maximum consolidation. Generally, for small dams, a rule of thumb that may be used is: *the soil is at optimum moisture when it is too wet for good tilth but not wet enough to exude moisture under compaction.*

The degree of compaction to be achieved is specified as a ratio, in per cent, of the embankment density to a specified standard density for the soil. In earth dam construction, the degree of compaction should run 85 to 100 percent of the maximum Proctor density.

Moisture control, where the fill materials are too dry, may be obtained by sprinkling. Overwet soils can be dried more rapidly by disking and light working to give maximum exposure.

Soils with high clay contents should be carefully compacted with the soil slightly dryer than the lower plastic limit to prevent the formation of shear planes called *slickensides*. Pervious materials, such as sands and gravels, consolidate under the natural loading of the embankment. However, additional compaction does aid in passing the *critical density,* in increasing shear strength, and in limiting embankment settlement. Compaction of these noncohesive materials is best achieved when they are nearing saturation.

Selection of proper compaction equipment is important. The sheepsfoot roller is best suited for compacting fills. Its weight may be varied by adding water to the drum, and the compaction per unit area may be adjusted to various numbers and sizes of tamping feet.

Special precautions should be taken in compacting materials close to core walls, collars, pipes, conduits, and so on. All such mechanical structures should be constructed so that they are wider at the bottom than at the top; thus settlement of the soil will create a tighter contact between the two materials. Thin layers of soil, at moisture contents equal to the remainder of the fill, should be tamped into place next to all structures. This can best be done with hand-operated pneumatic or motor tampers. Heavy hauling equipment should be given varied routes over a fill to prevent overcompaction along the travel ways.

10.16. Protection and Maintenance. The entire reservoir area should be fenced to prevent damage to embankments, spillways, and banks. Where this is not practicable, the spillways and dam at least should be protected.

To minimize damage from sedimentation the entire watershed should be protected by adequate erosion control practices. Buffer or filter strips of dense close-growing vegetation should be maintained around the pond edge.

A properly designed and constructed earth dam, well-sodded and protected, should require a minimum amount of maintenance. However, as insurance on the investment, a regular inspection and maintenance program should be established. Particular attention should be given to surface erosion, the development of seepage areas on the downstream face or below the toe of the dam, the development of sand boils and other evidence of piping, evidence of wave action, and damage by animals or human beings. Early recognition and repair of such conditions will prevent development of dangerous conditons that will increase the cost of repair.

Excessive weed growth, development of insect breeding areas at the water edges, and in fish ponds depletion of the fertility level should all be corrected. Under no conditions should trees be permitted to grow on or near the embankment.

FARM PONDS

10.17. Types of Ponds and Reservoirs. The three types of reservoirs in common use are (1) dugout ponds fed by ground water, (2) on-stream ponds fed by continuous or intermittent flow of surface runoff, streams, or springs, and (3) off-stream ponds.

Dugout Ponds. Dugout ponds are limited to areas having slopes of less than 4 percent and a prevailing reliable water table within 1 m of the ground surface. Design is based on the storage capacity required, depth to the water table, and the stability of the side-slope materials.

On-stream Ponds. The on-stream type of reservoir depends on the runoff of surface water for replenishment. The designed storage capacity must be based both upon use requirements and upon the probability of a reliable supply of runoff. Where heavy usage is expected, the design capacity of the pond must be adequate to supply several years' needs in order to assure time for recharge in the event of a sequence of one or more years of low runoff. Spring- or creek-fed ponds consist of either a scooped-out basin below a spring or a reservoir formed by a dam across a stream valley or depression below a spring. The dam may be placed across a depression rather than a definite stream to catch diffuse surface water. A spring-fed pond should be designed to maintain the pond surface below the spring outlet. This eliminates the hazard of diverting the spring flow due to the increased head from the pond. When the spring flow is adequate to meet use requirements, surface waters should be diverted out of the pond to reduce sedimentation and to reduce spillway requirements.

Off-stream Storage Pond. The off-stream or by-pass pond is constructed adjacent to a continuously flowing stream, and an intake, through either a pipe or open channel, diverts water from the stream into the pond. Controls on the intake permit reduction in sedimentation, particularly if all flood water can be diverted from the pond. Proper location and diking are essential to protect against stream overflow damage.

10.18. Essential Requirements for Ponds. To assure an effective pond or reservoir, certain basic requirements must be met: (1) The topographic conditions at the pond site must allow economical construction; cost is a direct function of fill length or height, for these dimensions determine cubic content. (2) An adequate and reliable supply of water, free from mine, organic, or chemical pollution must be available. (3) Soil materials must be available to provide a stable, impervious fill. (4) All ponds must be equipped with adequate mechanical and flood spillway facilities to maintain a uniform water depth during normal conditions and to safely manage flood runoff. (5) Large ponds should be equipped with a drain pipe to facilitate maintenance and fish management. (6) Adequate safety equipment must be provided around drop-inlet structures and other hazardous portions of the dam. (7) All design specifications must be adhered to in construction, and a sound program of maintenance must be followed to protect against damage by wave action, erosion, burrowing animals, livestock, farm equipment, and careless recreational use. All of these must be carried out to assure safety of the structure and to prevent damage to property below.

10.19. Site Selection. Dams for livestock water, irrigation, or fire protection must be located where the impounded water may be used most effectively with a minimum of pumping and piping. Topographic features must be carefully studied to eliminate need for excessively large structures. As previously discussed, suitable soil for the embankment is important, as well as choice of a site requiring a minimum of fill. Channel slopes above the fill should range from 4 to 8 percent to provide adequate water surface area.

The site for water storage structures is also dependent upon a contributing watershed capable of supplying the necessary runoff. The probable rates and volumes of runoff should be determined by the methods outlined in Chapter 4. Approximate volumes of runoff may be determined from Fig. 10.11. These are general values that should be checked with local hydrologic data.

10.20. Water-Storage Requirements. The storage capacity of a pond will depend on the water needs, evaporation from the water surface, seepage into

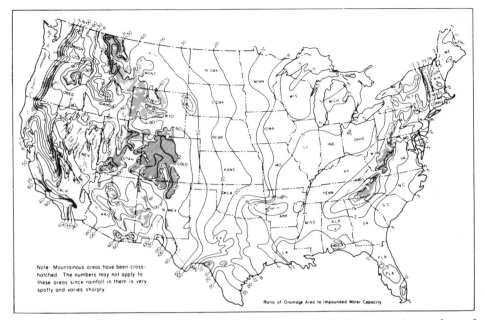

Note: Mountainous areas have been cross-
hatched. The numbers may not apply to
these areas since rainfall in them is very
spotty and varies sharply.

Ratio of Drainage Area to Impounded Water Capacity

Fig. 10.11. The approximate drainage area in acres required to store 1 acre-foot of water in farm ponds. (Redrawn from U.S. SCS, 1969.) 1 ac = 0.40 ha and 1 ac-ft = 0.1233 ha-m.

the soil or through the dam, storage allowed for sedimentation, and the amount of carry-over from one year to the next. Two to four meters of depth are needed for fish to survive where the water freezes over. Water needs for domestic uses, livestock, spraying, irrigation, and fire protection may be estimated from Table 10.2. Evaporation from a pond-water surface can be estimated by multi-plying local pan evaporation values published by the Weather Bureau by a factor of about 0.7. Evaporation can be reduced by selecting a site having a small surface area and deep depth. Seepage losses depend on soil properties and construction techniques. For this reason they are difficult to predict. Studies have shown that ponds with large ratios of storage volume to watershed area provide a reliable supply even in dry years. Thus, within reasonable limits, maximum capacity for a given site is desirable. Minimum storage is usually computed by estimating the total annual needs and allowing 40 to 60 percent of the total storage for seepage, evaporation, and other nonusable requirements.

Example 10.5. Design a pond in northeastern Missouri to provide water for 200 steers, for irrigating a 0.2-ha (0.5-ac) garden, and for a family of six persons. A suitable site for the dam has a drainage area of 27 ha (67 ac), and the estimated design peak runoff for a 25-year return period storm is 2.12 m³/s (75

TABLE 10.2   Water Requirements for Farm Uses

Type of Use	Average use in		
	Gallons[a] per day	Ac-ft per year	Ha-m per year
Household, all purposes, per person	50–100	0.08	0.01
Fire protection	—	0.23	0.03
Dry cow or steer, per 450 kg (1000 lb.) weight	9–18	0.015	0.002
Milk cow, 450 kg (1000 lb.) including milkhouse and barn sanitation	18–40	0.032	0.004
Horse or mule, 450 kg (1000 lb.)	8–12	0.011	0.0014
Turkeys per 100 head	10–15	0.014	0.0017
Chickens per 100 head	6–9	0.008	0.001
Swine per 45 kg (100 lb.)	1–1.5	0.002	0.0002
Sheep per 45 kg (100 lb.)	1–1.5	0.002	0.0002
Orchard spraying, per year of tree age per application	1		
Irrigation (humid regions) per season		1-1.5 ft	0.3-0.45 m
Irrigation (arid regions) per season		1-5 ft	0.3-1.5 m

[a] For air temperatures of 50 and 90°F., respectively.
Source: Midwest Plan Service (1968).

cfs). Seepage and evaporation losses from the pond are estimated as 60 percent of the storage capacity.

Solution. From Table 10.2 the annual water requirements are:

200 steers (200 × 0.002)	0.40	ha-m
Irrigation (0.2 × 0.45)	0.09	
Household use (6 × 0.01)	0.06	
Total	0.55	ha-m (4.46 ac-ft)

Storage requirements allowing 57 percent losses:

$$0.55 \times 100/(100 - 57) = 1.28 \text{ ha-m } (10.4 \text{ ac-ft})$$

From Fig. 10.11, read six acres of drainage area for each ac-ft of storage. Required watershed size is $10.2 \times 6/2.47 = 25.2$ ha (62.2 ac), which is adequate.

From a contour map of the reservoir area and a field investigation of the soils, a dam site was selected as shown in Fig. 10.12. By measuring the area

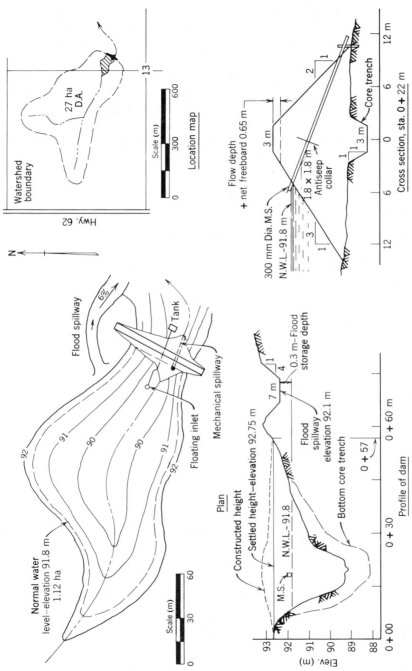

Fig. 10.12. Farm pond plan and layout for Example 10.5.

within each contour line with a planimeter, the water-level height was determined (see Table 10.3). The area was measured to the center line of the dam. This procedure is sufficiently accurate because most of the soil in the dam is usually taken from the reservoir area. (An approximate volume may be obtained by multiplying the water-surface area by 0.4 times the maximum water depth.) Storage at this site could be increased by raising the water level or by moving the dam further downstream. If by so doing, sufficient storage could not be obtained, another site would have to be selected.

By interpolation between contours at elevations of 91 and 92 m, 1.28 ha-m of storage is available at an elevation of 91.8 m, which is the normal water surface (see Table 10.3). The following specifications are determined:

- Crest elevation of mechanical spillway   91.8 m
- Flood storage depth from Fig. 10.7 for 1.12-ha surface   0.3 m
  area and 2.12-m³/s peak runoff rate (minimum)
                     Flood spillway elevation   92.1 m

  From Fig. 10.12 measure the water surface fetch, 200 m.
- Substituting in Eq. 10.5, $h = 0.014 (200)^{1/2} = 0.2$-m wave height.
- Assuming frost depth of 0.15 m, net freeboard = 0.2 + 0.15 =   0.35
- Flow depth in flood spillway (assumed)   0.3
         Elevation of top of dam (settled height)   92.75 m
- Allowance for settlement at sta. 0 + 22,

$$10\% \times (92.75 - 89.00) = 0.38 \text{ m } (1.2 \text{ ft})$$

- Top width of dam from Eq. 10.4,

$$W = 0.4 (92.75 - 89.00) + 1 = 2.5 \text{ m } (8.2 \text{ ft})$$

- Select side slopes of 3:1 upstream and 2:1 downstream.
- From Table 10.1, read 300-mm mechanical spillway diameter for 2.12 m³/s and 1.12 ha water surface.
- Antiseep collar for estimated 15-m pipe length and 10 percent increase in creep distance requires one 1.8×1.8 m collar or two 1×1 m collars.
- Livestock water pipe, select 32-mm (1¼-in.) diameter steel pipe approximately 30 m (100 ft) in length.
- Volume of fill in dam and in core trench is computed in Table 10.4 from fill height measured from profile in Fig. 10.12.

TABLE 10.3   Volume of Pond-Water Storage[a]

Contour Elevation (m)	Area Within Contour Line to Center Line of Dam (ha)	Average Area (ha)	Contour Interval (m)	Volume of Storage (ha-m)
89.5	0.05			
90	0.25	0.15	0.5	0.08
91	0.64	0.45	1.0	0.45
92	1.24	0.94	1.0	0.94
			Total	1.47 (11.9 ac-ft)

To elevation 91.8, storage is $1.47 - (0.94 \times 0.2/1) = 1.28$ ha-m and water surface area = $1.24 - 0.2 (1.24 - 0.64) = 1.12$ ha (2.8 ac)

[a] Computed for the pond in Fig. 10.12 and Example 10.5. The number of contour lines is reduced to simplify computations.

- Flood spillway width at control section (Eq. 9.6),

$$L = 2.12/(0.552 \times 3.2 \times 0.3^{3/2}) = 7.3 \text{ m (24 ft)}$$

- Flood spillway below dam with 6 percent slope, maximum velocity of 1.5 m/s (5 fps), and 4:1 side slopes for trapezoidal vegetated waterway will require a bottom width of about 10.7 m (35 ft) (see Chapter 7).

TABLE 10.4   Volume of an Earth Dam[a]

Station along Center Line	Height of Dam (m)	Cross-Sectional Area[b] (sq. m)	Average Cross-Sectional Area (sq. m)	Length of Section (m)	Volume of Section (m³)
0 + 00 m	0	0			
0 + 08	0.9	4.28	2.14	8	17.12
0 + 15	3.6	41.40	22.84	7	159.88
0 + 22	3.8	45.60	43.50	7	304.50
0 + 27	1.7	11.48	28.54	5	142.70
0 + 37	1.0	5.00	8.24	10	82.40
0 + 57	0	0	2.50	20	50.00
				Total fill	756.60

Core trench volume, avg. depth 1 m, avg. width 4 m, and length 40 m   160.00

Total fill and core trench   916.60 m³ (1199 cu. yd)

[a] Computed for dam in Fig. 10.12 and Example 10.5.
[b] Cross-sectional area = $2.5h^2 + 2.5h$, where h is the dam height.

REFERENCES

ASCE Task Committee on Cement and Clay Grouting (1958). "Symposium on Cement and Clay Grouting of Foundations." *Proc. ASCE* **84**, SM1, Papers 1544, 1552, February.

Casagrande, A. (1937). "Seepage through Dams." *New Eng. Waterworks Assoc. J.* **51**, 131–172.

Cedergren, H. R. (1977). *Seepage, drainage, and Flow Nets,* 2nd ed. John Wiley & Sons, New York.

Creager, W. P., et al. (1945). *Engineering for Dams,* Vols. I, II, and III. John Wiley, New York.

Edminster, T. W., and C. E. Staff (1961). "Plastics in Soil and Water Conservation." *Agr. Eng.* **42**, 182–185, 248–250.

Hamilton, C. L., and H. G. Jepson (1940). "Stock-water Developments; Wells, Springs, and Ponds." U.S. Dept. Agr. Farmers' Bull. 1959.

Holtan, H. N. (1950). "Sealing Farm Ponds." *Agr. Eng.* **31**, 125–130, 133, 134.

Midwest Plan Service (1968). "Private Water Systems." MPS Iowa State University, Ames, Iowa.

Richey, C. B. (ed.) (1961). *Agricultural Engineers Handbook.* McGraw-Hill, New York, pp. 475–491.

Sherard, J. L., et al. (1964). *Earth and Earth-Rock Dams.* John Wiley, New York.

Spangler, M. G., and R. L. Handy (1973). *Soil Engineering.* Intext Educational Publishers, New York.

Taylor, D. W. (1948). *Fundamentals of Soil Mechanics.* John Wiley, New York.

Terzaghi, K., and R. B. Peck (1948). *Soil Mechanics in Engineering Practice.* John Wiley, New York.

Tschebotarioff, G. P. (1951). *Soil Mechanics, Foundations and Earth Structures.* McGraw-Hill, New York.

U.S. Bureau Reclamation (1977). "Design of Small Dams," 2nd ed. U.S. GPO, Washington, D.C.

———— (1960). "Earth Manual." U.S. GPO, Washington, D.C.

U.S. Soil Conservation Service (1969). "Engineering Field Manual." (lithographed).

PROBLEMS

10.1. Determine the distance to set the slope stakes from the center line of a dam 9.1 m (30 ft) high with upstream slopes of 3:1 and downstream 2:1. Top width of dam is 4.3 m (14 ft) and the ground slope perpendicular to the center line is level. Compute these distances if the ground slope is 15 percent.

10.2. If the critical frost depth is 0.15 m (0.5 ft), maximum exposure of the water surface is 396 m (1300 ft), depth of flow in flood spillway is 0.3 m (1 ft), and flood storage depth is 1.2 m (4 ft), determine the net and the gross freeboard.

10.3. Determine the gross freeboard and top width for a farm pond. The elevation of the mechanical spillway crest is 24.4 m (80 ft) and the elevation of the base of the dam is 19.5 m (64 ft). Net freeboard is 0.46 m (1.5 ft) and elevation of the flood spillway crest is 25.6 m (84 ft). Flow depth in the flood spillway is 0.15 m (0.5 ft). Allowing 5 percent for settlement, what should be the elevation of the top of the dam before and after settlement?

10.4. Compute the volume of fill per unit length for an earth dam 9.1 m (30 ft) high, assuming a top width of 2.4 m (8 ft), side slopes of 3:1 upstream and 2:1 downstream, and a level base perpendicular to the center line of the dam.

10.5. A perforated drain pipe to control the seepage line in a dam has openings 14 mm (9/16 in.) in diameter. Using the criteria given in Appendix H and the following particle sizes obtained from distribution curves of available filter materials, will the filter meet these criteria? Show why.

Layers	D_{15} Size (mm)	D_{85} Size (mm)
Soil in dam	0.004	0.07
Graded sand	0.08	1.2
Graded gravel	3.0	36.0

NOTE: All size distribution curves are nearly parallel.

10.6. An earth dam is to be constructed having a volume after settlement of 1148 m³ (1500 yd³). The desired wet Proctor density of the dam is to be 1922 kg/m³ (120 pcf) with an optimum moisture content of 20 percent. If the void ratio of the dam is 0.75, how many cubic meters (cubic yards) of borrowed soil with a void ratio of 1.1 are required? What is the dry weight of the soil in the dam in metric tons (tons)?

10.7. A farm pond is to provide 0.42 ha-m (3.4 ac-ft) of water for 100 milk cows; seepage and evaporation losses are 45 percent of the stored volume; drainage area is 12.1 ha (30 ac); settlement allowance for fill is 5 percent; and flood spillway flow depth is 0.3 m (1.0 ft). Determine the elevation of (1) normal water level, (2) flood spillway, and (3) the top of the dam after settlement and (4) before settlement at the highest point where the base of the dam is at elevation 24.4 m (80 ft). Assume flood peak flow is 1.5 m³/s (53 cfs) and water surface area is 0.45 ha (1.1 ac). Frost depth is 0.15 m. Data shown on next page.

Contour Elevation		Length of Water Surface		Accumulated Storage	
m	(ft)	m	(ft)	ha-m	(ac-ft)
24.4	(80)	0	(0)	0	(0)
25.6	(84)	30	(100)	0.025	(0.2)
26.8	(88)	91	(300)	0.123	(1.0)
28.0	(92)	183	(600)	0.395	(3.2)
29.2	(96)	244	(800)	0.888	(7.2)

10.8. Immediately after construction of a dam, the embankment had a void ratio of 0.40. What is the settlement of the dam as a percentage of the height before consolidation if the void ratio after consolidation is 0.30?

10.9. Compute the seepage rate per unit length through a homogeneous earth dam resting on an impervious foundation if the top width is 4.3 m (14 ft), height of dam 12.8 m (42 ft), upstream side slope 3:1 and downstream 2:1, the hydraulic conductivity 0.00012 m/h (0.0004 fph), and the net freeboard 0.9 m (3 ft).

10.10. Assuming a storage loss of 50 percent by seepage and evaporation, determine the storage capacity of a farm pond to supply water for 100 steers, 50 milk cows, 50 hogs, and 1000 chickens, and irrigation of a 0.8-ha (2-ac) garden.

CHAPTER 11

Headwater Flood Control

Headwater flood control includes all measures that will reduce flood flows in watersheds of small rivers and their tributaries. The maximum size of a watershed for a headwater area is arbitrarily selected as 2590 square kilometers (1000 square miles). Floods may be classified as either small- or large-area floods. In this text a headwater flood is considered to be a small-area flood. All flood control activity downstream from headwater areas constitutes downstream engineering, which is beyond the scope of this text.

Headwater flood control measures are considerably different than downstream measures. In headwater areas the average rainfall intensity is much higher and the variation in those factors that affect runoff is much more extreme than in larger watersheds. The effects of crops, soils, tillage practices, and conservation measures are more important since the surface condition of the entire headwater area can often change completely from season to season and from year to year. Headwater floods are typically flash floods of short duration that occur rather frequently, that is, two or three times a year. The effectiveness of headwater measures decreases rapidly with distance downstream. Since downstream floods are more spectacular and damages more evident, the headwater flood is too often neglected.

Flood control is a relative term as it is not economical to provide protection for the largest flood that will occur. Since civilization began, floods have been one of the best recorded of all natural phenomena. After describing a series of historical floods, Hoyt and Langbein (1955) conclude the prologue to their book with the following concept of flood control that is too often overlooked:

> Somewhere in the United States, year 2000 plus or minus. Nature takes its inexorable toll. Thousand-year flood causes untold damage and staggering loss of life. Engineers and meteorologists believe that present storm and flood resulted from a combination of meteorologic and hydrologic conditions such as may occur only once in a millennium. Reservoirs, levees, and other control works which have proved effective for a century, and are still effective up to their design capacity, are unable to cope with enormous volumes of water involved. This catastrophe brings home the lesson that protection from floods is only a relative matter, and that eventually nature demands its toll from those who occupy flood plains.

Although the Federal Government develops and supervises most flood control programs through such agencies as the Corps of Engineers and the U.S. Soil Conservation Service and Forest Service, there are exceptions where local groups have organized and financed flood control programs. Outstanding among these is the Miami Conservancy District in Ohio, which is described by Morgan (1951). It was completed in 1923 at a cost of about $31 000 000. In general, state governments have not financed or individually administered flood control programs.

11.1. Economic Aspects of Flood Control. Flood control is an economic problem in which the protection of life and property, as well as the public benefits, must be evaluated in balancing the annual savings from flood control against construction and maintenance costs. According to Langbein and Hoyt (1959), more than 10 000 000 people in the United States reside and/or work on land subject to occasional overflow.

Damages. Flood damages may be classified as (1) *direct* losses to property, crops, and land, that can be determined in monetary values; (2) *indirect* losses, such as depreciated property, traffic delays, and loss of income; and (3) *intangible* losses not subject to monetary evaluation, including community insecurity, health hazards, and loss of life. Since the distinction between direct and indirect losses is one of degree, there has been a lack of uniformity in evaluating damages. Indirect losses are difficult to determine and are often estimated as a percentage of the direct losses. Although the damage to land frequently goes unnoticed, soil erosion from the uplands, sedimentation in reservoirs, stream channels, and flood plains, and pollution of water supplies greatly affect the economy of the entire watershed.

The flood losses from severe floods in the United States are shown in Fig. 11.1. These losses per square mile are high for small watersheds and in general decrease with an increase in drainage area. Damages from small areas often go unnoticed and do not receive the publicity that goes with the more spectacular large flood.

Annual flood losses for urban and agricultural land are shown in Table 11.1. Agricultural losses are much greater in headwater areas than downstream, while the reverse is true in urban areas. Without wise land use control in floodplains, the estimated national flood damage is expected to increase from 3.8 to 6 billion dollars by the year 2000. Goddard (1976) reported that 16 percent of urban lands were in floodplains (53 percent were developed) by 1973, while only 7 percent were subject to the 100-year flood.

Flood water causes damage by inundation and by high velocities. Though in some instances sediment deposits may be beneficial to farm lands, more frequently deposits of fine soil particles and sand have a damaging effect. Bridges, buildings, roads, farm lands, and stream channels are often destroyed by flood

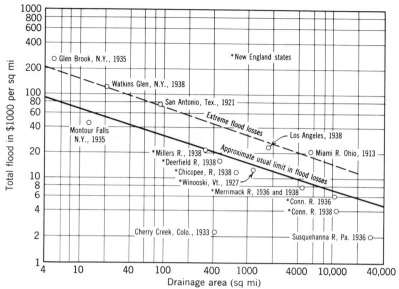

Fig. 11.1. Flood losses from severe floods in the United States. (Data from Barrows, 1948.)

waters having high velocities. Some of these damages may be classed as non-recurrent, depending on the nature of replacements and repairs. For example, a bridge replaced well above the high water level will probably not be in danger of subsequent damage. However, most damages, such as those from inundation and damage to land, are recurrent in nature.

Flood damages cannot always be prevented, but they may be reduced by flood forecasting and by using proper flood control measures. On large rivers flood warnings can be issued several days in advance of the flood. In this connection the U.S. Corps of Engineers employs large-scale models of the Mississippi River for predicting downstream river stages. Longer range forecasts can be made by using hydrologic data. About 15 percent of the annual

Table 11.1 U.S. Annual Flood Losses in Millions of Dollars

Land Use	Headwater	Downstream	Total
Urban and built-up	$ 330	$ 990	$1320
Agricultural	1130	682	1812
Other	340	378	718
TOTAL	$1800	$2050	$3850

Source: Goddard (1976).

flood loss could be prevented by accurate forecasting. Flood insurance is subsidized by the federal government for approved areas that have met suitable criteria. Insurance will only reduce the risk of economic loss, not the damage itself.

Benefits.   In general, benefits from flood control are based on reduction of losses. For example, when a reservoir reduces all floods by a given depth, the benefits are determined from the difference in damages at the two stages and by the number of floods during a specified period of time. Aside from benefits to individuals and industries, there are public benefits, such as economic stability and better social conditions.

11.2.   Downstream Flood Control.   Flood control activities on all navigable streams and on many tributaries that are not navigable would be classed as downstream. The Corps of Engineers from the beginning has been largely responsible for downstream measures.

Control on Lower Reaches.   Levees, channel improvement, and diversion floodways are the principal methods employed to reduce damage from floods. On the Mississippi River levees were the only means of control for many years. As levees were built further upstream, valley storage was reduced and flood heights were increased, resulting in a continuous process of raising and straightening of levees. In more recent years it has been necessary to include such procedures as stabilization and protection of channel banks, straightening of the channel by cutoffs, construction of floodways, and dredging operations in order to maintain a navigable stream and to provide flood protection.

Control on Upper Reaches.   On watersheds of about 13 000 square kilometers (5000 square miles) or less, reservoirs may provide adequate flood protection. As in the lower reaches, levees for local protection, channel improvement, channel straightening, and stream bank protection are applicable flood control measures. Such methods may be more economical than a system of reservoirs, but in general, levees speed the flow and accentuate the problem in unimproved areas below.

Reservoirs for flood control may be classified as natural or artificial. Lakes, swamps, and other low areas on the land surface have a tendency to reduce flood heights. Even a lake that is full at the beginning of a storm will have a regulating effect on stream flow. The extent of this effect depends upon the ratio of the surface area of the lake to the size of the watershed. A good example of a lake-regulated stream is the St. Lawrence River. The Great Lakes which drain into the river control the flow to such an extent that the maximum is not more than 20 percent greater than the minimum flow. This may be compared to the Missouri River which has a maximum stage 2900 percent greater than the minimum flow. If single-purpose flood control reservoirs are to be economically feasible, the protected area must be of high value. In Ohio the

Miami River and Muskingum Valley Conservancy Districts and in New England the Connecticut River and Merrimack River projects are examples of flood reservoir projects.

Where water is stored for two or more uses, as illustrated in Fig. 11.2, reservoirs are classified as multiple-purpose structures. Requirements for flood control conflict to some extent with other objectives. For maximum flood prevention the reservoir must be empty at the beginning of the storm period and the stored water must be discharged as rapidly as the capacity of the stream below will allow. Water for power, irrigation, navigation, and water supply is withdrawn gradually at times when it can be used to best advantage. For the most part, irrigation water is needed only during the growing season. Water requirements for generating electrical energy vary with daily and seasonal loads. Water released during the summer for navigation also aids in alleviating stream pollution and may provide a more uniform water supply for cities downstream. Where a reservoir can be constructed with a capacity greater than required for other purposes, the additional capacity may be available for flood control. The Hoover dam on the Colorado River, the Grand Coulee on the Columbia, Fort Peck dam on the Missouri River, and Norris dam in Tennessee are examples of multiple-purpose reservoirs although not all of them are for flood control. The conservation pool is permanent storage for recreation, wildlife, and sediment accumulation.

11.3. Types of Floods. A flood may be defined as an overflow or inundation from a river or other body of water. The distinction between normal discharge and flood flow is generally determined by the stage of the stream when bankful. As shown in Table 11.2, floods are a normal part of a stream's history and should not be looked upon as an unexpected occurrence. Most floods occur on the flood plains adjacent to rivers and streams and result from such natural causes as excessive rainfall and melting snow. Occasionally, tidal waves or hurricanes cause flooding. With respect to loss of life, reservoir failures have produced some of the worst floods, but fortunately these seldom occur.

Large-Area Floods. Large-area floods occur from storms of low intensity having a duration of a few days to several weeks. Since melting snow may also

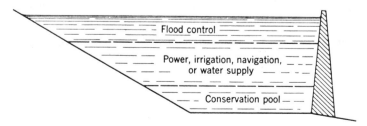

Fig. 11.2. Multiple-purpose reservoir.

Table 11.2 Flow Characteristics of a Stream

Relative Flow	*Frequency of Occurrence*
10 percent of channel capacity	50 percent of the time
50 percent of channel capacity	5 percent of the time
Bank full	Equalled or exceeded about twice a year
Moderate flooding	About once in 10 years
Severe flooding	About once in 50 years

contribute to or even cause large-area floods, seasons of maximum total precipitation do not necessarily coincide with the time of occurrence of large-area floods. In many parts of the country these floods come in the late winter or early spring. In many of the arid regions of the West these floods occur in late spring, while in Florida they come in the autumn. As illustrated in Fig. 11.3 for Ohio, the greatest number of large-area annual floods occurred in March and April, but the small-area floods were most prevalent in June, July, and August.

Small-Area Floods.   Small-area floods occur from storms of high intensity having a duration of 1 day or less. Harrold (1949) found that in Ohio small-area floods occurred in the summer between the months of May and September. Ninety percent of the annual amount of rain, falling at rates greater than 25 mm/h (1 iph) fell during this 5-month growing season. Since 85 percent of the annual soil loss occurred during this period, such floods cause great damage to agricultural land through soil erosion, which in turn results in sediment accumulation in rivers and reservoirs. These floods usually do not produce high runoff on large streams but often cause serious local damage.

11.4.   Flood Plain Zoning.   Any flood control program should be well coordinated among the many private and governmental interests. Regulation of land use in the flood plain by local governments can be more economical and effective than structural measures. Such action cannot be taken by private interests, but is a governmental responsibility.

Until recent years the federal government gave practically no attention to nonengineering methods of reducing flood damages, such as flood plain zoning. Occupancy of the flood plains in many areas has increased the potential damage. As an example, a study in Ohio showed that a recurrence of the rainfall which caused the 1913 flood would have caused more damage in 1960 than in 1913, despite the fact that millions of dollars had been spent for flood control.

11.5.   Flood Control Legislation.   Prior to 1936 there was no definite federal policy for the control of floods. With the exception of reclamation projects, federal undertakings were principally concerned with navigation. Laws enacted in several states created watershed districts primarily for watershed protection and flood control.

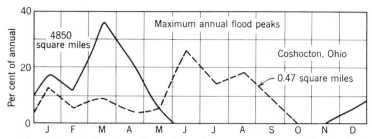

Fig. 11.3. Monthly distribution of small- and large-area annual floods in Ohio. (Redrawn from Harrold, 1961.)

Flood Control Act of 1936 As Amended. A federal flood control policy was established whereby the federal government authorized flood control activities on navigable streams and their tributaries. The act definitely specified that in planning for flood control the lands on which the flood waters originate must be considered. Federal investigations of watersheds and measures for runoff and waterflow retardation as well as erosion control were delegated to the U.S. Department of Agriculture. Investigations and improvements on rivers and waterways for flood control and allied purposes were delegated to the Defense Department. The program under the Department of Agriculture is carried out through the Soil Conservation Service for ranch and farm lands and through the Forest Service for ranch and range lands, both in cooperation with local, state, and other federal agencies.

Flood Control Act of 1944. This legislation supplemented the original act of 1936. The provisions of this act require that plans and investigations be referred to the states concerned for investigation and consultation before submission to Congress. This act provided the enabling legislation for the Pick-Sloan Plan in the Missouri River Basin.

Watershed Protection and Flood Prevention Act of 1954 as Amended. This act, commonly referred to as Public Law 566, along with subsequent amendments, authorized the Department of Agriculture to assist local organizations to prepare plans for engineering works and land treatment measures for the reduction of flood damages in headwater streams. The maximum watershed area was limited to about 100 000 ha (390 square miles). This act was the first major authorization by the federal government for engineering works in headwater areas. Several states have enacted laws to authorize special-purpose districts to carry out works of improvement under this law.

The National Flood Insurance Act of 1968, as Amended. Flood insurance is provided to private owners on the condition that the participating community adopts and enforces government floodplain management standards. Subsidized insurance covers only direct losses to buildings and their contents. Motor vehicles, growing crops, land, livestock, and so on are not insurable under the government program. Even though the insurance is subsidized by the government, acceptance has been slow. U.S. Dept. Housing and Urban Development (1978) reported that 14 500 communities had enrolled in the program.

11.6. Conservancy Districts. Conservancy districts are legal enterprises organized for the purpose of soil conservation and flood control. In some states flood control works constructed by the U.S. Corps of Engineers or the U.S. Department of Agriculture under legislation, such as Public Law 566, are sponsored by and function under a conservancy district established for that purpose. The district may arrange for the necessary local support and cooperation among landowners and provide the necessary maintenance after the project is completed. The Miami Conservancy District in Ohio is a major privately financed project organized under the conservancy laws. Conservancy districts are known by a variety of names in different states, such as watershed districts, water-management districts, and other special-purpose titles. Conservancy districts are authorized in a number of states, including Colorado, Ohio, Oklahoma, Indiana, Florida, Iowa, Minnesota, Kansas, and Texas.

METHODS OF HEADWATER FLOOD CONTROL

Two general classes of headwater flood control measures are (1) those that retard flow or reduce runoff by land treatment or reservoirs, and (2) those that accelerate the flow by channel improvement, channel straightening, and levees.

REDUCING FLOOD-FLOWS

Methods of reducing flood-flows include (1) watershed treatment in which the storage of water is increased on the surface and in the soil profile, (2) flood control reservoirs, and (3) underground storage. Underground storage is accomplished by spreading the flow over a considerable area. This method is applicable only in special situations, particularly in arid regions.

Measures that retard the flow or reduce runoff are economically and physically more desirable because (1) all visible evidence or danger of the flood is removed, (2) the flow in the stream is more uniform, thus providing greater recharge of the ground water and a more adequate water supply, (3) an important step toward the conservation of natural resources is achieved, (4) higher crop production results, especially in areas where conservation of moisture is important, and (5) reduction of sedimentation in lower tributaries is accomplished.

11.7. Watershed Treatment. Watershed treatment includes all practices applied to the land that are effective in reducing flood runoff and controlling erosion. Proper land use and conservation practices are necessary for adequate watershed control.

Land management may increase the (1) amount of surface storage, (2) rate of infiltration, and (3) capacity of the soil to store water. Runoff retardation by land management is largely dependent on vegetative cover and favorable soil surface conditions. For example, in Fig. 11.4a the results of soil differences show about 3 times as much runoff occurring from heavy soils in Mississippi as from permeable loess soils in Iowa. The effect of vegetation on runoff for varying amounts of rainfall is shown in Fig. 11.4b.

The effect of land management on the time distribution of flood-flows is shown in Fig. 11.5. The land before treatment was severely eroded and about two thirds of the watershed was in scrawny woodland that had been burned and grazed. After the land was retired from use and reforested, the peak flow from a comparable storm about 10 years later was reduced to only 15 percent of that before treatment. The volume of flow represented by the area under the hydrograph was about the same for the two storms. Land management is thus effective in reducing flows in small watersheds from short-duration storms, but its effectiveness diminishes rapidly with increased drainage area and magnitude of the storm.

Conservation practices, such as contouring, strip cropping, and terracing will reduce flood peaks as well as total runoff from small to medium storms during the growing season. For larger storms, the effect of contouring and strip cropping on runoff is usually much less than the reduction in soil losses discussed in Chapter 5. Level or absorptive-type terraces can have a considerable effect on flood peaks of rather large watersheds. For example, a 1953 flood from an 882-square mile watershed in northwestern Iowa could have been reduced from 1870 to 865 m^3/s (66 000 to 30 000 cfs) by installing level terraces on 37 percent of the watershed area. If only 17 percent of the area was terraced, the flow was estimated at 1274 m^3/s (45 000 cfs). These estimates were made by the Soil Conservation Service by flood routing procedures assuming that a terrace would hold 38 mm (1.5 in.) of water and that the infiltration rate was 8 mm/h (0.3 iph).

Graded terraces generally will reduce runoff volumes less than 25 percent. Land management and conservation practices have little, if any, effect on floods from winter storms and on catastrophic floods from large watersheds.

11.8. Reservoirs.   The two types of reservoirs are the flood regulated storage reservoir and the detention reservoir. The principal difference is that the detention reservoir operates automatically by discharging through one or more fixed openings in the dam, whereas the regulated reservoir discharges through adjustable gates. The chief advantage of the regulated reservoir is its flexibility of operation.

Detention reservoirs such as shown in Fig. 11.6 may have one or more discharge openings of fixed dimensions. These reservoirs have emergency spillways to handle runoff in excess of the design flood. Practically all headwa-

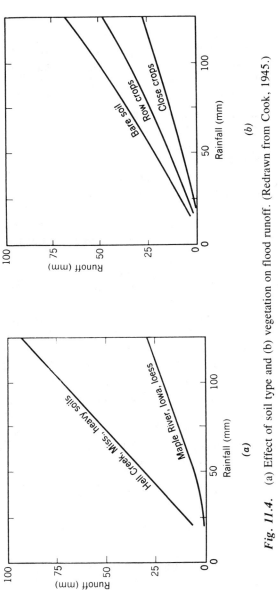

Fig. 11.4. (a) Effect of soil type and (b) vegetation on flood runoff. (Redrawn from Cook, 1945.)

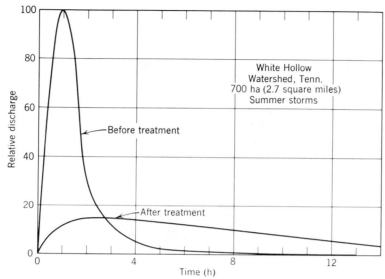

Fig. 11.5. Effect of land management on flood flow. (Data from Tennessee Valley Authority.)

ter flood control reservoirs are of the detention type. Reservoirs on the watersheds of the Trinity River in Texas, Sandstone Creek in Oklahoma, the Little Sioux River in Iowa, as well as in the Miami Conservancy District are examples of such structures. The principal advantages of the detention reservoir are its simplicity and automatic operation without personnel.

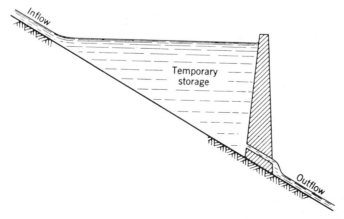

Fig. 11.6. Diagrammatic representation of detention reservoir operation.

Reservoirs for flood control reduce flood peaks, but not flood volumes. This reduction in peak flow diminishes rapidly with distance downstream since the main stream receives an increasing percentage of its runoff from other tributaries. The principal disadvantages of reservoirs are that some land must be flooded to protect other lower land and that the annual and initial costs are high. The annual cost may be reduced by controlling erosion, and thus sedimentation, which prolongs the useful life of the reservoir. Initial cost of storage per unit volume, in general, decreases with increasing reservoir storage.

The effect of a series of small headwater reservoirs on flood runoff will be illustrated by a hypothetical watershed shown in Fig. 11.7. The 600-square-mile watershed consists of ten 60-square-mile tributaries. From a storm that produced 25 mm (1 in.) of runoff in 4 hrs, the assumed hydrograph of flow at the mouth of each 10-square-mile watershed has a peak flow of 28.3 m³ (1000 cfs). By flood routing the runoff from watershed 1 to point B, the peak flow is reduced from 28 to 17 m³/s (1000 to 600 cfs) (curve 1 in Fig. 11.8). The area under the hydrograph or volume essentially remains unchanged. In Fig. 11.8 hydrographs 2 and 3 represent the flow from the 10-square-mile watersheds 2 and 3 each routed to point B, the outlet of the 60-square-mile area. The three hydrographs are then added together to obtain the composite hydrograph for

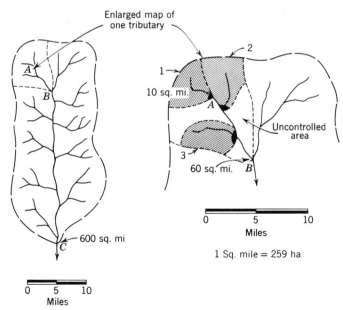

Fig. 11.7. Hypothetical watershed consisting of ten 60-square-mile tributaries and a series of reservoirs, each with a watershed of 10 square miles. (Redrawn from Leopold and Maddock, 1954.)

the three areas as shown by the dashed line in Fig. 11.8. If the flow from the remaining 30 square miles is added to the previous composite hydrograph, the resulting hydrograph at point *B* (top curve) is obtained.

Assume now that a reservoir with adequate capacity is constructed on each of the three 10-square-mile watersheds (1, 2 and 3), and that the outflow is controlled to 2.83 m³/s (100 cfs) at each dam. The flow from the remaining one half of the area or 30 square miles is not controlled because of lack of good reservoir sites or other reasons. The composite hydrograph for the three controlled watersheds, when routed to point *B*, is shown in Fig. 11.9. The flow from the uncontrolled 30 square miles is unchanged. When the uncontrolled area flow is added to the controlled flow, the resulting hydrograph with dams is obtained. The peak flow for this hydrograph is 56.6 m³/s (2000 cfs) compared to 96.3 m³/s (3400 cfs) without the dams, or a reduction of 41 percent. If only the three 10-square-mile watersheds with dams are considered, the peak flow at point *B* is reduced 82 percent.

The percentage reductions in peak flood flows resulting from reservoirs on these 10-square-mile watersheds for an 8-hour storm producing 25 mm (1 in.) of runoff are 88 percent for one reservoir on a 10-square-mile watershed, 41 per-

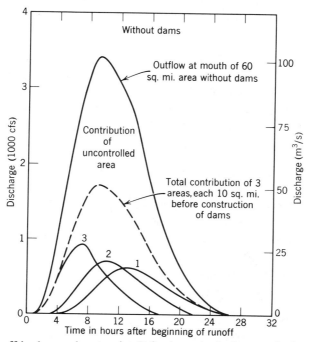

Fig. 11.8. Runoff hydrographs at point B for hypothetical watersheds without reservoirs. (Redrawn from Leopold and Maddock, 1954.)

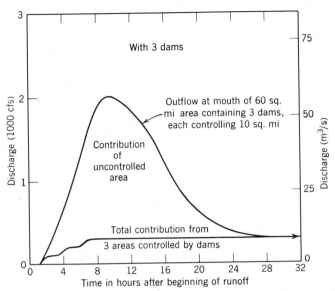

Fig. 11.9. Runoff hydrographs for a hypothetical 60-square-mile watershed showing the effect of reservoirs in reducing peak flows. (Redrawn from Leopold and Maddock, 1954.)

cent for three reservoirs on a 60-square-mile watershed, and 36 percent for 30 reservoirs on the 600-square-mile area. Each dam would provide 88 percent reduction at its outlet and give good protection for a designated reach of the channel downstream from the dam. The effect of the reservoir rapidly diminishes downstream and thus further works are needed. These other downstream works do no good upstream. Both headwater and downstream works are needed for complete protection.

The effect of size and location of flood control reservoirs on peak flow reduction at the outlet of a large watershed is shown in Table 11.3. Although the large reservoir has the greatest effect downstream, it does not reduce flows within the watershed as do the small and medium reservoirs. The 12 small and medium dams give only an additional 5 percent reduction downstream. Headwater reservoirs and downstream reservoirs serve different purposes and they are not generally interchangeable.

INCREASING CHANNEL CAPACITY

The purpose of increasing channel capacity is to decrease height and duration of floods and reduce flood damages. Increasing the capacity of a stream may be accomplished (1) by channel improvement, (2) by channel straightening, and (3) by levees.

Table 11.3   Flood Reduction by Reservoirs

Relative Size of Dam and Watershed			Reduction in Peak Flow of the Large Watershed (%)
Small	Medium	Large	
Number of Dams			
7			40
	5		38
7	5		45
		1	71
7	5	1	76

Source: Black (1972).

11.9.   Channel Improvement.   Channel improvement here includes those measures that increase the channel capacity, namely, enlarging the cross-sectional area and increasing the velocity. In narrow flood plains, channel improvement as well as channel straightening is usually more economical than levees.

Increasing Cross Section.   Increasing the cross section may be accomplished by deepening or widening the channel and by removing trees and sandbars from the watercourse. On small streams removal of trees and sandbars may increase the effectiveness as much as one third to one half, but in large streams the effect is considered negligible.

Increasing Velocity.   Removing debris and vegetation has a greater effect on the roughness coefficient in small streams than in large streams. The hydraulic radius and resulting velocity can be increased by widening or deepening the channel. For the same increase in cross-sectional area, deepening the channel is more effective than widening. The depth may be increased by using levees as well as by dredging or cleaning out the channel. The two ways of increasing the channel slope are (1) deepening the channel or lowering the water level at the outlet, and (2) straightening. Deepening the channel or lowering the water level at the outlet can be accomplished by increasing the cross-sectional area at the outlet, and removing debris or sandbars.

11.10.   Channel Straightening.   The principal method of straightening streams is to provide cutoffs. A cutoff is a natural or artificial channel that shortens a meandering stream, as shown in Fig. 11.10. The purpose of a cutoff is to increase the velocity, to shorten the channel length, and to decrease the length of levees. The length of the stream channel may be shortened as much as one-half the original length.

Cutoffs are desirable where (1) the stream capacity in the bend is less than the capacity in other parts of the channel, (2) the capacity of the entire channel

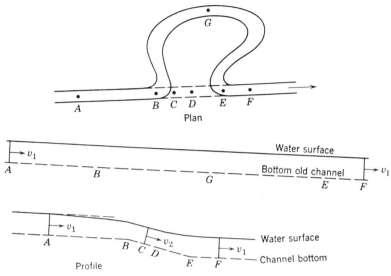

Fig. 11.10.   Effect of a cutoff on stream flow.

is to be increased with levees, (3) construction of the cutoff is more economical than increasing the capacity around the bend, and (4) the cutoff does not detrimentally affect the flow characteristics of the stream.

The effect of cutoffs is not always desirable. On alluvial streams cutoffs alone do not necessarily solve flood problems. They may cause serious streambank erosion above and below the cutoff and sediment deposits below. In some streams cutoffs cause extreme lowering of the channel and its tributaries. In others the increased gradient results in flow rates in excess of downstream channel capacity.

Cutoffs increase the velocity in the affected portion of the stream by increasing the hydraulic gradient. Because some water was formerly stored in the channel and in the flood plain along the bend *BGE* in Fig. 11.10, the cutoff also increases the stage downstream.

In a meandering stream cutoffs may occur naturally. Since erosive forces are a function of the velocity and depth of the stream as well as of the angle between the banks and the direction of flow, these forces are continually changing. Where there are bends in the stream, the soil is eroded from the concave bank, carried downstream on the same side, and deposited at the end of the bend, forming a bar that deflects the current to the other side of the channel. As the cutting continues, the stream becomes more crooked and a natural cutoff results.

The effect of a single cutoff, either natural or man-made, for steady flow conditions in the channel is shown in Fig. 11.10. Before the cutoff the stream flowed on a uniform slope around the bend *BGE*. After the stream is straightened, it flows from *B* directly to *E*, which is about one fourth of its previous length. Because of the increased slope from *B* to *E*, the velocity is increased from v_1 to v_2 with a corresponding decrease in the depth of flow. The decreased depth causes the water surface to be lowered to point *A* upstream. In section *CD* flow takes place at a uniform depth with velocity v_2. At *D* the velocity begins to decrease but is still greater at *E* than at *A*, and likewise the depth is less at *E* than at *A*. Below *E* there is a concave upward surface known as the backwater curve. At some point *F*, the lower end of the backwater curve, the velocity and depth is the same as at *A*. The cutoff is effective in lowering the stage through the cutoff section as well as above and below, to points *A* and *F*, respectively. The effect extends further upstream than downstream. Since cutoffs are usually short, the points *C* and *D* may overlap so that the velocity in the cutoff section does not correspond to the channel slope.

The effect of cutoffs becomes more complicated for unsteady flow and for a series of cutoffs. During a storm period the flow is constantly increasing and points, *A, C, D,* and *F* change with the stage of the stream. Unless the slope or cross section of a stream is changed between cutoffs, a series of cutoffs acts similarly to a single cutoff.

11.11. Levees. Levees are embankments along streams or on flood plains designed to confine the river flow to a definite width for the protection of surrounding land from overflow. Levees may be designed either to confine the river flow for a considerable distance or to provide local protection.

The effect of confining water between levees is (1) to increase the velocity through the leveed section, (2) to increase the water surface elevation during floods, (3) to increase the maximum discharge at all points downstream, (4) to increase the rate of travel of the floodwave, and (5) to decrease the surface slope upstream. Levees for protection of local areas have less effect on flood-flow; however, the end result of any levee system is a reduction in valley storage.

The location, spacing, and height of levees must be adjusted to provide adequate capacity between the levees, to provide protection to the flood plain area, and to be economical in cost. The design and construction of levees is discussed in Chapter 10.

PREVENTIVE MAINTENANCE

Preventive measures for maintaining the capacity of the stream channel include those that affect erosion in the channel itself and those that reduce sediment from upper tributaries. Maintenance in the channel is required to prevent the collection of debris and to reduce sediment from caving banks.

11.12. Bank Protection. The two classes of bank protection are (1) those that retard the flow along banks and cause deposition and (2) those that cover the banks and prevent erosion.

Retarding the flow along stream banks is desirable to control meandering, to protect the bank, thereby reducing deposition below, and to protect highways, railroads, and other structures near the channel. A common method of control is to build retards extending into the stream from the banks. Materials to construct these retards include piles, trees, rocks, and steel framing. Such retards, sometimes referred to as jetties, serve to decrease the velocity along the concave bank and, hence, increase deposition of sediment.

A method of locating retards is shown in Fig. 11.11. The first major retard at *A* is located by the intersection of the projected center line of flow with the concave bank. In locating the second major retard *C* a line *HB* is drawn parallel to the above projected center line and through the end of the retard *A*. The intersection of this line with the concave bank locates point *B*. *AC* is then made equal to twice *AB*. Additional retards are located by the intersection of a line connecting the end points of the two previous retards with the concave bank (see *D*). An auxiliary retard at *K* is located a distance *AB* upstream from *A* and is extended into the stream about one-half the length of the other retards. The retards should extend into the stream at an angle of 45 degrees for a distance of about 30 percent of the channel width. On small streams the spacing of the retards may be made equal to the stream width, and the length, 0.25 times the spacing. On 30-degree curves or larger, continuous bank protection should be provided rather than retards.

Vegetative or mechanical control measures are methods of preventing stream-bank erosion. Plants suitable for vegetative control are grass, shrubs, and trees. Mechanical measures to cover the stream bank include such devices as wood and concrete mattresses, rock or stone, asphalt, and sacked or monolithic concrete.

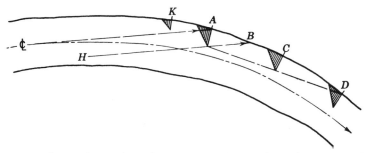

Fig. 11.11. Design and location of retards. (Redrawn from Saveson and Overholt, 1937.)

11.13. Reduction of Sediment and Debris. Sediment and debris in stream channels can be reduced by deposition in suitable settling basins or by land treatment.

Sediment from high-velocity streams in cultivated watersheds is deposited on flood plain areas and in the stream channels. Such sediment reduces the effectiveness of drainage ditches and the productivity of agricultural land. Although settling basins are often satisfactory, good land treatment accompanied by channel cleanout may be more practical.

Sedimentation and debris basins have three essential features: an inlet, a settling basin, and an outlet. Sediment-laden water from a stream may be diverted into a large settling basin where a portion of the sediment is deposited as a result of greatly reduced velocities. At the lower end of the basin the flow is then returned to the stream channel. Such settling basins are eventually filled with sediment, thus necessitating cleanout or the use of a new area. In western Iowa settling basins have been used to reduce sedimentation in channels across a wide flood plain.

The barrier system of removing debris and sediment from mountain streams was developed in Utah. Large debris is deposited as the flood spreads out at the mouth of the canyon and the finer material settles out in a settling basin. Additional features of the system consist of (1) a barrier or cross dike, (2) lateral dikes, and (3) temporary drift dams.

FLOOD ROUTING

Flood routing is the process of determining the stage height, storage volume, and outflow rate from a reservoir or a stream reach for a particular hydrograph of inflow.

11.14. Principles of Flood Routing. Flood routing involves inflow into the reservoir or stream reach, outflow through the dam or stream reach, and storage in the reservoir or stream valley and channel. In routing a flood through a reservoir, the objective is to determine the height of the dam for a given size outlet structure. The problem in flood routing is to determine the relationship among inflow, outflow, and storage as a function of time. This problem can be solved by the following continuity equation for unsteady flow:

$$idt = odt + sdt \qquad (11.1)$$

where i = inflow rate for a small increment of time,
$\quad\quad o$ = outflow rate for a small increment of time,
$\quad\quad s$ = rate of storage for a small increment of time,
$\quad\quad t$ = time.

This equation must be satisfied at any and all times during the period from the beginning of inflow until outflow has stopped. Hence, it must also be satisfied for any given time interval between the above limits. Equation 11.1 may also be written

$$t\frac{i_1 + i_2}{2} = t\frac{o_1 + o_2}{2} + S_2 - S_1 \qquad (11.2)$$

where S = volume of storage,
              1 and 2 = subscripts denoting beginning and end of the time interval, respectively.

The assumption here is that evaporation and seepage losses are negligible. In applying the equation to large reservoirs the backwater effect, which increases reservoir storage, should be considered, but this effect is negligible in small headwater reservoirs.

11.15. Methods of Flood Routing. Of the two general methods of flood routing, the first method involves the division of the inflow hydrograph into time intervals of short duration so that during each period inflow and outflow rates may be assumed to be constant. Storage- and outflow-stage relationships must be known. Either observed or developed inflow hydrographs are necessary in the solution. The second method involves analytical integration procedures in which a given flood hydrograph is replaced by an equivalent flood of uniform intensity. The available storage curve and outflow rates are represented by empirical formulas. The first flood routing method gives more accurate results, although analytical integration methods provide a more direct solution to the continuity equation (Eq. 11.1).

The many procedures for solving Eq. 11.1 include numerical, graphical, mechanical, and electrical methods. Mechanical procedures may involve slide rules, flood routing machines, and electrical procedures, such as electronic flood routing machines. Numerical and graphical procedures are basically trial-and-error solutions. Analog and digital computers are ideally suited for flood routing.

11.16. Elements of Flood Routing. In any flood routing procedure the factors that must be considered are (1) inflow hydrograph, (2) outflow or spillway discharge, (3) available storage, and (4) outflow hydrograph.
Inflow Hydrograph. Inflow hydrographs are the same as runoff hydrographs, discussed in Chapter 4.
Outflow. The outflow curve represents the depth-discharge relation of the reservoir spillway structure or the lower end of a stream reach. It may also include the flow through the emergency spillway. The rate of discharge from a reservoir is influenced by the hydraulic head and by the type and size of

spillway. The most common types of structures for detention reservoirs are drop inlet pipe spillways with box or orifice inlets (see Chapter 9). In the case of stream routing the outflow for the reach above is the same as the inflow to the next reach below.

Available Storage.   The available storage curve represents the depth-capacity relation of a reservoir above the elevation of the mechanical spillway crest, or for a stream above the elevation of some arbitrarily selected stage. The volume of storage for various stages is determined from the topography of the storage basin or stream valley. Where considerable accuracy is desired, the volume in a reservoir can be computed from a contour map by either the average end area method or by the prismoidal formula given in Appendix G. On streams several valley cross sections are often determined to obtain storage.

Outflow Hydrograph.   The outflow hydrograph shows the rate of outflow (spillway discharge) as a function of time. This hydrograph must be determined by flood routing.

11.17.   Reservoir Flood Routing Procedure.   Many graphical and numerical procedures have been developed for solving Eq. 11.2, the continuity equation. The procedure to be described is partly numerical and partly graphical. It was selected because it can be adapted for programming with the digital computer. The procedure will apply to either reservoir or stream flood routing.

In most flood routing procedures the size of the mechanical spillway cannot be determined directly. A size is selected and the flood is routed through the reservoir. If the peak outflow rate or the maximum water stage in the reservoir does not meet site or cost limitations, another size spillway is selected and the flood routing procedure is repeated. With most drop inlet pipe spillway structures, the discharge rate when the pipe first flows full will be only slightly less than the maximum outflow rate. With peak runoff, total runoff, drainage area, and available storage known, the rate of outflow and consequently the approximate size of the pipe may be determined from an equation developed by Culp (1948),

$$\frac{q_0}{q} = 1.25 - \left(\frac{1500V}{RA} + 0.06\right)^{1/2} \tag{11.3}$$

where q_0 = rate of outflow when the pipe first flows full in m³/s,
   q = peak inflow in m³/s,
   V = available storage in ha-m,
   R = runoff in mm,
   A = drainage area in ha.

The procedure for reservoir flood routing will be illustrated in Example 11.1.

Example 11.1. Design a combination flood control reservoir and farm pond for a site that has a drainage area of 48.58 ha (120 ac). The total runoff for a 50-year return period is 88.9 mm (3.5 in.) and the peak runoff rate is 5.38 m³/s (190 cfs). A depth of 2.44 m (8 ft) for the pond is available below an elevation of 29.26 m (96 ft). The storage capacity of the reservoir above 29.26 m (0 stage) is shown in Fig. 11.12. A box-inlet spillway and circular concrete outlet pipe are to be used in the outlet structure. The maximum allowable stage in the reservoir is 1.62 m (5.3 ft) at elevation 30.88 m (101.3 ft). By flood routing procedure, determine the size of the outlet structure, the actual water stage, elevation of the flood spillway crest, and the maximum height of the dam, allowing a net freeboard of 0.61 m (2 ft) and a flow depth of 0.30 m (1 ft) in the flood spillway.

Solution.   With a stage of 1.62 m (5.3 ft) above the crest of the mechanical spillway, the storage from Fig. 11.12 is 5.86 m³/s-h (207 cfs-hours) or 2.109 ha-m (17.1 ac-ft). From Eq. 11.3 the rate of outflow when the pipe first flows full is

$$q_0/ 5.38 = 1.25 - \left[(1500 \times 2.109)/(88.9 \times 48.58) + 0.06\right]^{1/2}$$
$$q_0 = 5.38 \times 0.36 = 1.94 \text{ m}^3\text{/s (68.4 cfs)}$$

Assume a 0.9×1.1 m (3×3.5 ft) box inlet (crest length 2.9 m) and a 762-mm (30-in.) outlet pipe. (Box-inlet area should be about twice the area of the pipe.) Using the weir formula with $C = 3.0$ and the pipe flow formula with $K_e = 1.0$, $n = 0.014$, and $L = 33.5$ m (110 ft), compute the spillway discharge curve shown in Fig. 11.12. q_0 is 1.81 m³/s (64 cfs) and is close enough to the estimated value to be satisfactory. Assume an inflow hydrograph as shown in Fig. 11.13. Such a hydrograph may be developed as described in Chapter 4 or otherwise. The volume of runoff, which is the area under the hydrograph, is 4.32 ha-m ($88.9 \times 10^{-3} \times 48.58$) or (35 ac-ft).

Compute the routing curve shown in Fig. 11.14 for 12- min ($\Delta t = 0.2$ h) time increments. The procedure is given in Table 11.4. The expression $(S/Dt + o/2)$ is to be evaluated to simplify the solution of Eq. 11.4 as will become clear later in the example. The time interval should be short enough to define the inflow with the desired accuracy (see U.S. SCS, 1957). It may be estimated as 10 to 15 percent of the time of peak. Select sufficient number of outflow rates to adequately define the spillway discharge curve. From Fig. 11.12 read the storage (Column 2) for the corresponding outflow rate. For example, an outflow rate of 0.28 m³/s (10 cfs) in line (3) corresponds to a stage height of 0.15 m (0.5 ft), and the storage to this height is read as 0.17 m³/s-h (6 cfs-hours). In column 3 and line (3) dividing the storage by the time interval gives $0.17/0.2 = 0.85$ m³/s (30 cfs). By adding one-half the outflow rate 0.14 m³/s (5 cfs) to this flow, the value, 0.99 m³/s (35 cfs) in Column (5), is obtained. The remaining values are computed and then plotted against the outflow rate as shown in Fig. 11.14.

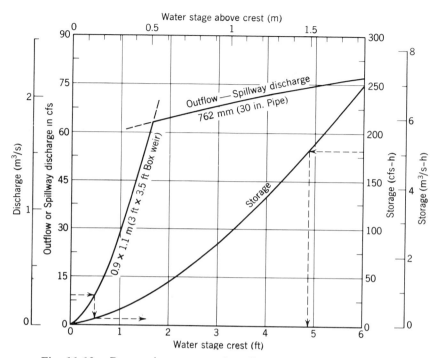

Fig. 11.12. Reservoir storage and outflow curves for Example 11.1.

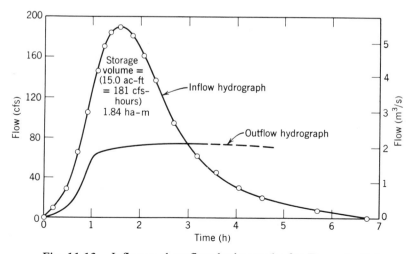

Fig. 11.13. Inflow and outflow hydrographs for Example 11.1.

Table 11.4 Computations for Routing Curve in Fig. 11.14

Outflow or Spillway Discharge (m^3/s)	Storage, S (m^3/s-h)	$S/\Delta t^a$ (m^3/s)	$o/2$ (m^3/s)	$S/\Delta t + o/2$ (m^3/s)
(1)	(2)	(3)	(4)	(5)
0	0	0	0	0
0.17	0.10	0.48	0.09	0.57
0.28	0.17	0.85	0.14	0.99
0.57	0.34	1.70	0.28	1.98
0.85	0.48	2.41	0.42	2.83
1.13	0.62	3.11	0.57	3.68
1.42	0.76	3.82	0.71	4.53
1.70	0.93	4.67	0.85	5.52
1.81	1.13	5.66	0.91	6.57
1.93	2.27	11.33	0.96	12.29
2.04	3.88	19.40	1.02	20.42
2.15	6.17	30.87	1.08	31.95

^{a} $\Delta t = 0.2$ h.

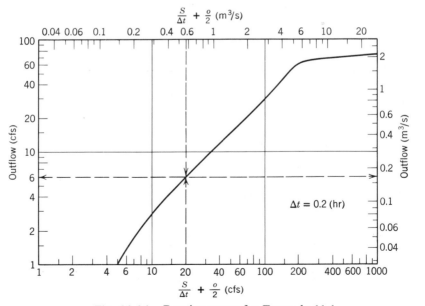

Fig. 11.14. Routing curve for Example 11.1.

For plotting the routing curve select log-log paper or arithmetic scales which will give the desired accuracy throughout the range of work. This curve greatly simplifies the computations.

Determine the outflow hydrograph as illustrated in Table 11.5. Values in Column (2) are obtained from the inflow hydrograph in Fig. 11.13. The average inflow for each time interval is computed in Column (3). The computations are simplified by writing Eq. 11.2 in the following form

$$\frac{S_2}{\Delta t} + \frac{o_2}{2} = \frac{i_1 + i_2}{2} + \left(\frac{S_1}{\Delta t} + \frac{o_1}{2}\right) - o_1 \qquad (11.4)$$

The expression on the right-hand side of Eq. 11.4 at 0.2 hr in Table 11.5 is 0.14 + 0 − 0 = 0.14, and the corresponding outflow rate for this expression in Column (4) as read from Fig. 11.14 is 0.03 m³/s (1 cfs). The next value in Column (4) is 0.45 + 0.14 − 0.03 = 0.56, and o = 0.17 m³/s (6 cfs). The routing is continued until the outflow rate exceeds the inflow rate that occurs at about 3.0 hr. If desired, the routing may be continued until the outflow is zero;

Table 11.5 Computations for Flood Routing in Example 11.1

Time (h) $\Delta t = 0.2$	i	$(i_1 + i_2)/2$	$S/\Delta t + o/2$	o
		All units in m³/s		
(1)	(2)	(3)	(4)	(5)
0	0		0	0
0.2	0.28	0.14	0.14	0.03
0.4	0.62	0.45	0.56	0.17
0.6	1.36	0.99	1.38	0.40
0.8	2.55	1.95	2.93	0.88
1.0	3.74	3.14	5.19	1.61
1.2	4.76	4.25	7.83	1.87
1.4	5.21	4.98	10.94	1.90
1.6	5.38[a]	5.30	14.34	1.95
1.8	5.15	5.27	17.66	1.98
2.0	4.67	4.93	20.61	2.01
2.2	4.11	4.39	22.99	2.07
2.4	3.54	3.82	24.74	2.10
2.6	2.97	3.26	25.90	2.10
2.8	2.46	2.72	26.52	2.12
3.0	2.01	2.24	26.64	2.12[b]
3.2	1.78	1.90	26.42	2.12
3.4	1.50	1.64	25.94	2.10

[a] Maximum inflow.
[b] Maximum outflow is about equal to inflow rate. Time of maximum water level in reservoir and maximum storage.

however, in this case it is not necessary to do so. Maximum storage may be computed from Column (4) where

$$S = (26.64 - 2.12/2)0.2$$
$$= 5.12 \text{ m}^3/\text{s-h or } 1.84 \text{ ha-m } (181 \text{ cfs-hours or } 15.0 \text{ ac-ft})$$

After plotting the outflow hydrograph from Columns (1) and (5) from Table 11.5 in Fig. 11.13, storage volume may be checked by measuring the area between the inflow and the outflow hydrograph.

From Fig. 11.12 the maximum water level height corresponding to a storage of 5.12 m³/s-h (181 cfs-h) is 1.49 m (4.9 ft). The crest elevation of the flood spillway is 29.26 + 1.49 = 30.75 m (100.9 ft). Maximum settled height of the dam is 2.44 + 1.49 + 0.61 + 0.30 = 4.84 m (15.9 ft). The maximum water level height and the desired maximum outflow rate are close enough to the design requirements to give a satisfactory solution to the problem. If the flood is to be routed above the crest of the emergency spillway in the above example, the depth-discharge curve must be changed to include both the outflow rate through the mechanical spillway and the emergency spillway.

11.18. Streamflow Routing Procedure. The basic principles of routing a flood in a stream channel are the same as for a reservoir. Reservoir storage is replaced by valley storage that increases rapidly with depth, especially when the water level exceeds the bankful stage. In streamflow routing the channel is divided into reaches of convenient length. Such factors as changes in cross sections, channel gradient, and location of tributaries affect the selection of reaches. Several cross sections are necessary within the reach to adequately determine the stage-storage relationship. The routing interval may be estimated as the time of travel though the reach at bankful stage. The outflow of a reach becomes the inflow of the reach below if routing is continued. Local inflow into the stream within the reach being routed, if large enough to be important, should be added to the inflow hydrograph or to the outflow hydrograph. The choice will depend on the nearness of the local flow to the upper or lower end of the reach. Detailed procedures are described in many references, such as U.S. Soil Conservation Service (1957) and Linsley et al. (1975).

REFERENCES

Barrows, H. K. (1948). *Floods: Their Hydrology and Control.* McGraw-Hill, New York.

Black, P. E. (1972). "Flood Peaks as Modified by Dam Size and Location." Water Resources Bulletin 8 (No. 4).

Cook, H. L. (1945). "Flood Abatement by Headwater Measures." *Civil. Eng.* **15**, 127–130.

Culp, M. M. (1948). "The Effect of Spillway Storage on the Design of Upstream Reservoirs." *Agr. Eng.* **29**, 344–346.

Goddard, J. E. (1976). "The Nation's Increasing Vulnerability to Flood Catastrophe." *J. Soil and Water Conservation* **31**, 48–52.

Harrold, L. L. (1949). "Has the Small-Area Flood Been Neglected?" *Civil Eng.* **19** (10), 38–39.

Harrold, L. L. (1961). "Hydrologic Relationships on Watersheds in Ohio." *Soil Cons.* **26**, 208–210.

Hoyt, W. G., and W. B. Langbein (1955). *Floods,* Princeton University Press, New Jersey.

Langbein, W. B., and W. G. Hoyt (1959). *Water Facts for the Nation's Future.* Ronald Press, New York.

Leopold, L. B., and T. Maddock (1954). *The Flood Control Controversy.* Ronald Press, New York.

Linsley, R. K., et al. (1975). *Hydrology for Engineers.* McGraw-Hill, New York.

Morgan, C. E. (1951). *The Miami Conservancy District.* McGraw-Hill, New York.

Muckleston, K. W. (1976). The Evolution of Approaches to Flood Damage Reduction." *J. Soil and Water Conservation* **31**, 53–59.

Saveson, I. L., and V. Overholt (1937). "Stream Bank Protection." *Agr. Eng.* **13**, 489–491.

U.S. Congress (1954). "Watershed Protection and Flood Prevention Act." Public Law 566, 83d Congress, H. R. 6788, August.

U.S. Department Housing and Urban Dev. (1978). "Questions and Answers on National Flood Insurance Program," Pamphlet HUD-471-FIA(2), May. Washington, D.C.

U.S. Soil Conservation Service (1957). *Hydrology, National Engineering Handbook,* Section 4 and Supplement A (lithograph). Washington, D.C.

Yevdjevich, V. M. (1960). *Flood Routing Methods, Discussion and Bibliography.* U.S. Geological Survey (lithographed).

PROBLEMS

11.1. Before a cutoff was made on a meandering stream the length of the channel around the bend (BGE in Fig. 11.10) was 1280 m (4200 ft), and the stream gradient was 0.08 percent. After the cutoff was made, the distance *BE* was 640 m (2100 ft). If the velocity in the old channel was 0.6 m/s (2 fps), how much has the cutoff reduced the time of flow from *B* to *E*, assuming the same hydraulic radius and roughness coefficient for the old and the new channel?

11.2. Design a system of retards for a stream 15 m (50 ft) wide where the channel makes a 35-degree turn on an 8-degree curve. Determine the length and spacing of retards by making a scale drawing of the stream.

11.3. By flood routing, determine the maximum water level for the reservoir in Example 11.1, using all available storage for flood control. Elevation of the crest of the box inlet is 26.8 m (88 ft), and the elevation of the center of the pipe at the outlet is 25.6 m (84 ft). The accumulated storage available at each 0.6-m (2-ft) stage above the crest is 0, 0.06, 0.17, 0.41, 0.76, 1.32, 2.10, 3.22 ha-m (0, 0.5, 1.4, 3.3, 6.2, 10.7, 17.0, and 26.1 ac-ft). Use a 0.9 × 0.9-m (3 × 3 ft) box inlet and 610-mm (24-in.) diameter pipe having the same coefficients and length of pipe as in Example 11.1.

11.4. Design the outlet structure (inlet and pipe size) for Problem 11.3, using the hydrograph described below and the runoff volume and peak of 80.8 mm (3.18 in.) and 6.12 m³/s (216 cfs), respectively. Maximum elevation of the flood spillway (maximum water level in the reservoir) should not exceed 30.2 m (99 ft).

Time (min)	q [m³/s (cfs)]	Time (min)	q [m³/s (cfs)]
0	0 (0)	87	5.86 (207)
10	0.31 (11)	97	5.21 (184)
23	0.91 (32)	110	4.42 (156)
32	2.15 (76)	129	3.06 (108)
42	3.43 (121)	151	2.01 (71)
52	4.70 (166)	171	1.47 (52)
58	5.52 (195)	193	0.99 (35)
64	5.95 (210)	217	0.68 (24)
74	6.12 (216)	270	0.25 (9)
		322	0 (0)

11.5. Route a flood through a 3048-m (10 000-ft) reach of a stream, whose bankful capacity is 22.66 m³/s (800 cfs). From a unit hydrograph the inflow hydrograph was obtained, and the outflow-storage data were developed from weighted averages of four cross sections in the reach. Route the flood until the outflow is less than 2.83 m³/s (100 cfs). How much was the flood peak reduced within the reach?

Inflow Hydrograph		Outflow Storage	
Time (min)	q (avg) [m³/s(cfs)]	Outflow [m³s (cfs)]	Storage [m³/s-h (cfs-h)]
0	0 (0)	0 (0)	0 (0)
0.75	21.24 (750)	1.42 (50)	1.98 (70)
1.50	63.72 (2250)	4.25 (150)	4.64 (164)
2.25	106.20 (3750)	8.50 (300)	7.02 (248)
3.00	148.68 (5250)	22.66 (800)	18.44 (651)
3.75	157.18 (5550)	42.48 (1500)	36.87 (1302)
4.50	131.69 (4650)	99.12 (3500)	93.46 (3300)
5.25	106.20 (3750)	141.60 (5000)	128.58 (4540)
6.00	80.71 (2850)	198.24 (7000)	159.16 (5620)
6.75	55.23 (1950)		
7.50	29.74 (1050)		
8.25	8.50 (300)		
9.00	0 (0)		

CHAPTER 12

Surface Drainage and Land Forming

Land forming is defined as the process of changing the natural topography so as to control the movement of water onto or from the land surface. It includes one or a combination of practices, such as land leveling for irrigation; land grading or shaping for irrigation, drainage, and moisture conservation; and shallow field ditches, which can be crossed with farm machinery. Land forming also includes grading work for erosion control, for example, contour benching or earthwork for parallel terracing. Land smoothing is generally referred to as the final operation of removing the minor differences in elevations that result from the operation of scrapers or other large earth-moving equipment. Land grading, shaping, and leveling are synonymous. Terminology varies in different sections of the country.

Land grading is essential to the development of surface irrigation systems. This practice has been adopted in more humid regions as a method for improving surface drainage on flat land. Grading land for both irrigation and drainage is entirely practical.

SURFACE FIELD DITCHES

The selection of surface drainage facilities for individual field areas depends largely on the topography, soil characteristics, crops, and availability of suitable outlets. Surface drainage systems must be suitable for mechanized operations on various types of topography, such as pot hole areas, flat fields, and gently sloping land. In irrigated areas such drains are needed to collect waste water from surface irrigation. Pothole areas are frequently found in glaciated regions where the topography is relatively flat and geologic erosion has not had time to develop natural outlets. Flat or level land having impermeable subsoils with shallow topsoil frequently requires surface drainage because subsurface drains are not practicable or economical. Claypan, hardpan, or tight alluvial soils are examples. On these flat fields water may accumulate because of excess rainfall, flooding from uplands, or overflow from streams. Flat land is defined as land with slopes less than 2 percent, the major portion of which is less than 1 percent. The two primary methods of surface drainage are land grading and field ditches. Field ditches include (1) bedding, in which the plow deadfurrows serve as drains; (2) field ditches with wider spacings than deadfurrows; and (3) parallel open ditches that cannot be crossed with farm machinery.

The importance of surface drainage is indicated by Gain (1964) who estimated that over 40 500 000 ha (100 000 000 ac) in the eastern United States and about 3 200 000 ha (8 000 000 ac) in the eastern provinces of Canada would benefit from surface drainage. The extent of these drainage areas is shown in Fig. 12.1.

Excessive surface water can be removed by one or more of the following processes: drainage by natural or constructed channels, by infiltration, or by evaporation and transpiration. Evaporation is usually inadequate, and, if the soil is impervious, surface drainage is the only remaining method. Shallow surface drains cannot remove ground water and give the benefits incident to good pipe drainage. However, surface drainage may be required even though pipe drains are installed.

12.1. Random Field Ditches. These drains are best suited to the drainage of scattered depressions or potholes (Fig. 12.2) where the depth of cut is not over 1 m. In cross section they are a flat "V" or parabolic in shape.

The design of field ditches is similar to the design of grass waterways, as discussed in Chapter 7. Where farming operations cross the channel, the side slopes should be flat; that is, 8:1 or greater for depths of 30 cm or less and 10:1 or greater for depths over 60 cm. Minimum side slopes of 4:1 are possible if the field is farmed parallel to the ditch. The depth is determined primarily by the topography of the area, outlet conditions, and the capacity of the channel. The grade in the channel should be such that the velocity does not cause erosion or sedimentation. Maximum allowable velocities for various soil conditions are given in Chapter 13. Minimum velocities vary with the depth of flow; however, these range from about 0.3 to 0.6 m/s for depths of flow less than 100 cm. The maximum grade for sandy soil is about 0.2 percent and for clay soils 0.5 percent. The roughness coefficient in the Manning equation may be taken as 0.04 if more reliable coefficients are not available. The capacity of the drain is usually not considered for areas less than 2 ha provided the minimum design specifications are met. However, where the area is larger than 2 ha, the capacity should be based on a 10-year return period storm, making allowances for minimum infiltration and interception losses. Since most field crops are able to withstand inundation for only a short period of time without damage, it is desirable to remove surface water within 12 to 24 hours.

The layout of a typical random field drain system is shown in Fig. 12.2. Normally, the channel should follow a route that provides minimum cut and least interference with farming operations. Where possible several potholes should be drained with a single ditch. The outlet for such a system may be a natural stream, constructed drainage ditch, or protected slope if no suitable ditch is available. Where the outlet is a broad, flat slope, the water is permitted

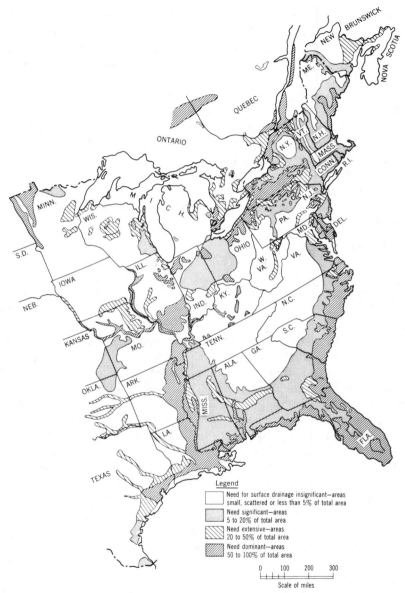

Fig. 12.1. Extent of lands needing surface drainage in eastern United States and Canada. (From Gain, 1964.)

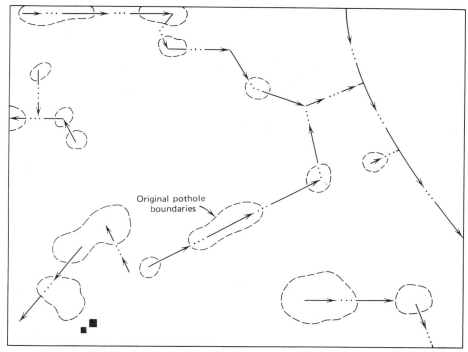

Fig. 12.2. A random field ditch system in central Iowa.

to spread out on the land below. This type of outlet is practical only if the drainage area is small.

12.2. Bedding. Bedding (also called humps and hollows) is a method of surface drainage consisting of narrow-width plow lands in which the deadfurrows run parallel to the prevailing land slope (see Fig. 12.3). The area between two adjacent deadfurrows is known as a bed. Bedding is most practicable on flat slopes of less than 1.5 percent where the soils are slowly permeable and pipe drainage is not economical. Studies in southern Iowa showed that level land gave slightly better yields than bedded land.

The design and layout of a bedding system involves the proper spacing of deadfurrows, depth of bed, and grade in the channel. The depth and width of bed depends on the land slope, drainage characteristics of the soil, and cropping system. Bed widths recommended for the Corn Belt region of the U.S. vary from 7 to 11 m for very slow internal drainage, from 13 to 16 m for slow internal drainage, and from 18 to 28 m for fair internal drainage. The length of the beds may vary from 90 to 300 m. In the bedded area the direction of farming operations may be parallel or normal to the deadfurrows. Tillage practices

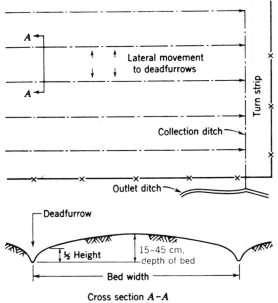

Fig. 12.3. Bedding system of surface drainage.

parallel to the beds have a tendency to retard water movement to the dead-furrows. Plowing is always parallel to the deadfurrows.

12.3. Parallel Field Ditch System. Parallel field ditches are similar to bedding except that the channels are spaced farther apart and may have a greater capacity than the deadfurrows. This system is well adapted to flat, poorly drained soils with numerous small depressions that must be filled by land grading.

The design and layout is similar to that for bedding except that ditches need not be equally spaced and the water may move in only one direction. The layout of such a field system is shown in Fig. 12.4. As in bedding, the turn strip is provided where ditches border a fence line. The size of the ditch may be varied, depending on grade, soil, and drainage area. The depth of the ditch should be a minimum of 20 cm and have a minimum cross-sectional area of 0.5 m². The side slopes should be 8:1 or flatter to facilitate crossing with farm machinery. As in bedding, plowing operations must be parallel to the channels, but planting, cultivating, and harvesting are normally perpendicular to them. The rows should have a continuous slope to ditches. The maximum length for rows having a continuous slope in one direction is 180 m, allowing a maximum spacing of 360 m where the rows drain in both directions. In very flat land with little or no slope, some of the excavated soil may be used to provide the

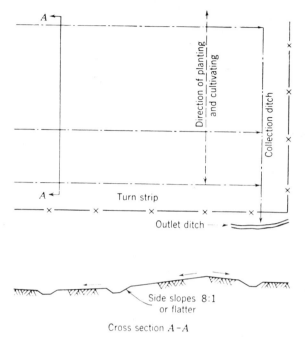

Side slopes 8:1
or flatter

Cross section *A-A*

Fig. 12.4. Parallel field ditch system of surface drainage.

necessary grade. However, the length and grade of the rows should be limited to prevent damage by erosion. On highly erosive soils that are slowly permeable, the slope length should be reduced to 90 m or less.

The cross section for field ditches may be V-shaped, trapezoidal, or parabolic. The W-ditch shown in Fig. 12.5 is essentially two parallel single ditches with a narrow spacing. All of the spoil is placed between the channels, making the cross section similar to that of a road. The advantages of the W-ditch are (1) allows better row drainage because spoil does not have to be spread, (2) may be used as a turn row, (3) may serve as a field road, (4) can be constructed and maintained with ordinary farm equipment, and (5) may be seeded to grass or row crops. The disadvantages of the W-ditch are (1) the spoil is not available for filling depressions, (2) a greater quantity of soil must be moved, and (3) a larger area is occupied by drains. The minimum spacing for W-ditches varies from about 5 to 15 m, depending on the size. The W-ditch is best adapted to relatively flat land where the rows drain from both directions.

12.4. Parallel Lateral Ditch System. The parallel lateral ditch system is similar to the field ditch system except that the ditches are deeper. These drains, illustrated in Fig. 12.6, cannot be crossed with farm machinery. For clarity the minimum size for open ditches is here given as 60 cm deep and side slopes less

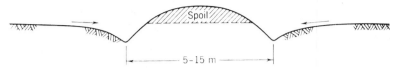

Fig. 12.5. W-ditch or double field ditch for surface drainage.

than 4:1. The purpose of lateral open ditches is to control the ground water table and to provide surface drainage. These ditches are applicable for draining peat and muck soils to obtain initial subsidence prior to subsurface drainage. For the same depth to water table, ditches provide the same degree of subsurface drainage as pipe drains.

The design specifications for lateral ditches are given in Table 12.1. Since lateral ditches are considerably deeper than collection ditches, overfall protection must be provided at outlets 1 and 2 indicated in Fig. 12.6. This protection may be obtained with a suitable permanent structure, by providing a gradual slope near the outlet, or by establishing a grassed channel.

Since these ditches are too deep to cross with farm machinery, farming operations must be parallel to the ditches. A collection ditch or quarter ditch should be provided for row drainage. As in other methods of drainage on flat land, the surface must be graded and smoothed and large depressions filled or drained by random field ditches.

For water table control during dry seasons, dams with removable crest boards are placed at various points in the open ditches to maintain the water surface at the required level. During wet seasons the crest boards are removed, and the system provides surface drainage. In highly permeable soils, such as sand, peat, and muck, crop yields may be increased by controlled drainage.

Table 12.1 Ditch Specifications for Water Table Control

	Sandy Soil	*Other Mineral Soils*	*Organic Soils, Peat and Muck*
Maximum spacing	200 m (660 ft)	100 m (330 ft)	60 m (200 ft)
Minimum side slopes	1:1	1½:1	Vertical to 1:1[a]
Minimum bottom width	1.2 m (4 ft)	0.3 m (1 ft)	0.3 m (1 ft)
Minimum depth	1.2 m (4 ft)	0.8 m (2.5 ft)	0.9 m (3 ft)

Source: Beauchamp (1952)

[a] Vertical for raw peat to 1:1 for decomposed peat and muck.

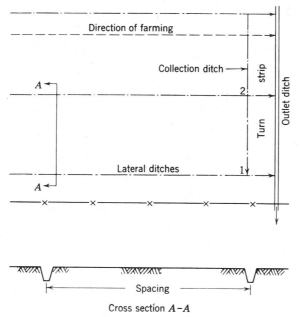

Fig. 12.6. Parallel lateral ditch system for water table control and surface drainage.

Deep permeable soils underlain with an impervious material provide the best conditions for successful water table control. The water level may be regulated by gravity, pumping, or a combination of gravity drainage and pumping. The depth at which the water table is to be maintained depends largely on the crop to be grown, soil, seasonal conditions including the quantity of water available, topography, and climatic conditions. In organic soils a high water table is desirable to provide water for plant growth, to control subsidence, and to reduce fire and wind erosion hazards. In these soils the water level should be maintained from 45 to 120 cm below the surface, depending on the crop.

12.5. Cross-Slope Ditch System. The drainage of sloping land may be feasible with cross-slope ditches. Such channels usually function both for surface drainage and for erosion control. Where designed specifically for the control of erosion, these drains are called terraces. Diversion ditches (see Chapter 8) are sometimes utilized to divert runoff from low-lying areas, thus reducing the drainage problem.

The cross-slope ditch system was developed in Wisconsin. It is primarily adapted to soils with poor internal drainage where subsurface drainage is not practicable and for land with slopes of 4 percent or less having numerous shallow depressions. This land is generally too steep for bedding or field ditches since farming up and down the slope results in excessive erosion.

12.6. Construction. The selection of equipment and procedure for construction of field ditches varies with the depth of cut and quantity and distribution of excavated soil. For depths of cut up to 75 cm, blade graders, scrapers, and heavier terracing machines are suitable. Construction of field ditches by the above methods may require spreading the spoil from both sides of the drain to prevent ponding back of the spoil. For deep cuts over 75 cm, bulldozers equipped with push or pull back blades and carryall scrapers may be used to fill the pothole area or other depressions near the point of excavation.

The field layout of a random surface ditch is shown in Fig. 12.7. Stakes are usually set every 15 m (50 ft) along the center line. Slope stakes are placed about 2 m further out from the center line than the computed distance so as not to be removed by construction equipment.

Example 12.1. Determine the volume of cut for the random field ditch shown in Fig. 12.7 (formulas given in Appendix G).
Solution. The cuts below were determined by instrument survey and were computed for a channel grade of 0.15 percent.

Sta (m)	Cut (m)	1/2 Top Width (m)	Cross-sectional Area (m²)	Avg. Cross-sectional Area (m²)	Distance (m)	Volume of Cut (m³)
0 + 00	0[a]	0	0			
				0.27	15	4.05
0 + 15	0.23	2.3	0.53			
				1.07	15	16.05
0 + 30	0.40	4.0	1.60			
				3.62	15	54.30
0 + 45	0.75	7.5	5.63			
				3.74	15	56.10
0 + 60	0.43	4.3	1.85			
				0.93	15	13.95
0 + 75	0	0	0			
				Total Volume		144.45 m³

[a] If the spoil is not placed in the pothole area, this cut should be increased to 0.15 m.

Several methods of establishing the grade in the channel are shown in Fig. 12.8. The laser system described in Chapter 15 is generally used by contractors. Final grade should be checked with a surveying instrument. The method shown in Fig. 12.8a is convenient for shallow ditches constructed with blade equipment. The hub stake in Fig. 12.8b may be left in place for later reference. This procedure is also suitable for terrace construction. The procedure in Fig. 12.8c is convenient for the operator, but considerable labor is required for layout.

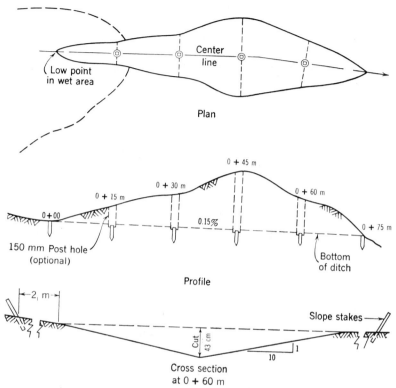

Fig. 12.7. Construction layout for a random field ditch.

12.7. Maintenance. Field surface ditches can usually be maintained by normal tillage operations. Such maintenance is particularly important on flat land since a very small obstruction in the channel may cause flooding of a sizable area. Tillage implements should be lifted when crossing ditches to avoid blocking the channel. If this procedure is not followed, the channel should be kept open by dragging or shaping a smooth channel in the bottom of the drain. When the soil is soggy and wet, equipment should not cross deadfurrows, field ditches, or grass waterways. Livestock may also damage such channels during rainy seasons. Pasturing at other times, however, is desirable. Plowing parallel to shallow surface ditches usually is adequate for maintenance. Minor depressions between ditches should be filled by land grading.

LAND GRADING

Although land leveling is the term generally associated with surface irrigation, land grading is synonymous but somewhat more descriptive. For most conditions, a sloping plane surface rather than a level surface is desired.

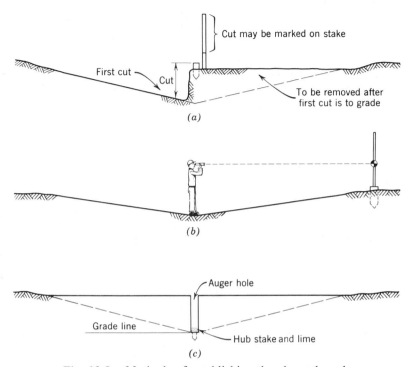

Fig. 12.8. Methods of establishing the channel grade.

12.8. Factors Influencing Design. Slopes, cuts, and fills are influenced by the soil, topography, climate, crops to be grown, and method of irrigation or drainage. The major problem with land grading is the effect of removing the topsoil and its influence on plant growth. Reduced growth may occur on the fill areas, although the exposure of subsoil in the cuts is usually a more serious problem. Establishment of a uniform design slope is more important for surface irrigation than for drainage. In Fig. 12.9 the graded surface for drainage is variable with less cut and fill than for the uniform grade for irrigation. Having a variable slope for drainage is not usually objectionable, provided flow velocities are not erosive. The topography thus places a severe limitation on the length and degree of slope as well as the location of slope changes. Row lengths for irrigation will be discussed in Chapter 19. The required accuracy of leveling will depend largely on its effect on crop production. For crops sensitive to excesses or shortages of water, greater precision is required. For flood irrigation the land slope in both directions may be restrictive, whereas for furrow irrigation the length-of-run and furrow grade are the most critical. In semihumid to humid climates land grading can be made compatible both for drainage and irrigation. For irrigation purposes, the largest flow occurs at the upper end of

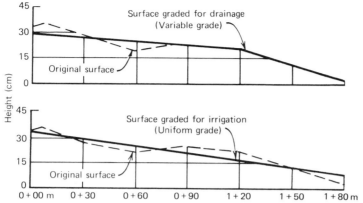

Fig. 12.9. Comparison of a surface graded for drainage or for irrigation.

the slope, while for drainage, rainfall enters along the entire slope length with the highest runoff at the lower end. These factors should be carefully considered in design.

Several methods of computing cuts and fills for land grading will be described. Field data are normally obtained from an instrument survey with ground elevations taken to the nearest 1 cm or 0.1 ft on a 30-m (100-ft) square grid for horizontal control. Elevations are taken at other critical points, such as highs and lows between grid stakes, and the water surface in the supply ditch or in the drainage outlet. After obtaining the desired balance between cuts and fills, the cut volume and fill volume are computed.

12.9. Plane Method. This method assumes that the area is to be graded to a true plane. The average elevation of the field is determined, and this elevation is assigned to the centroid of the area. The centroid is located by taking moments about two perpendicular reference lines, as shown in Fig. 12.10 for an irregularly shaped field. This procedure has been simplified by locating the grid system so that each grid point is at the center of the grid square and represents nearly equal areas (100 by 100 ft). In Fig. 12.10 the centroid is located at $X_c = 3.75$, $Y_c = 2.84$, and the average elevation is 8.37 ft. Any plane passing through the centroid will produce equal volumes of cut and fill. The general equation for a plane surface is

$$E = a + S_xX + S_yY \tag{12.1}$$

where E = elevation at any point,
a = elevation at the origin,
S_x and S_y = slope in the x and y directions, respectively.

The slope of any line in the X or Y direction is determined by the statistical

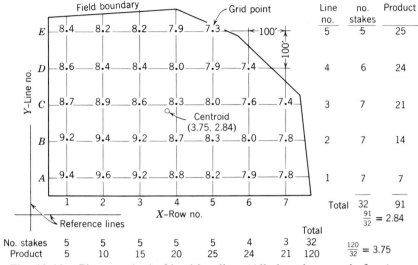

Fig. 12.10. Plane method of land leveling. (All elevations are in feet.)

least-squares procedure presented by Chugg (1947). The least-squares plane by definition is that which gives the smallest sum of all the squared differences in elevation between the grid points and the plane. It is called the plane of best fit. This method does not necessarily provide the best slope for irrigation. These slopes can be computed from two simultaneous equations, which are

$$(\textstyle\sum X^2 - nX_c{}^2)S_x + \left[\sum (XY) - nX_cY_c \right]S_y = \sum (XE) - nX_cE_c \quad (12.2)$$

$$(\textstyle\sum Y^2 - nY_c{}^2)S_y + \left[\sum (XY) - nX_cY_c \right]S_x = \sum (YE) - nY_cE_c \quad (12.3)$$

where n = total number of grid points,
       X_c = X distance to the centroid,
       Y_c = Y distance to the centroid,
       E_c = elevation of the centroid (avg elevation of all points).

For rectangular fields the terms involving XY become zero, and

$$S_x = \frac{\sum (XE) - nX_cE_c}{\sum X^2 - nX_c{}^2} \quad (12.4)$$

The slope S_y can be obtained from Eq. 12.4 by substituting Y for X. Although Eq. 12.4 is valid only for rectangular fields, a satisfactory solution can often be obtained by taking one or more arbitrarily selected rectangular areas within the field and extending the slopes of the plane to the remaining areas. The following example will illustrate computational procedure for an irregularly shaped field.

Example 12.2. Determine the equation for the plane of best fit for the field shown in Fig. 12.10. All elevations are in ft.

Solution. As shown in Fig. 12.10, $n = 32$, $X_c = 3.75$, and $Y_c = 2.84$.

$$E_c = (9.4 + 9.2 + 8.7 + \ldots + 7.8 + 7.4)/32 = 8.37 \text{ ft}$$

$$\sum X^2 = 1^2 + 1^2 + 1^2 + 1^2 + 1^2 + 2^2 + \ldots + 6^2 + 7^2 + 7^2 + 7^2 = 566$$

$$nX_c^2 = 32 \times 3.75^2 = 450$$

$$\sum (XY) = (1 \times 1) + (1 \times 2) + (1 \times 3) + \ldots + (7 \times 1) + (7 \times 2) + (7 \times 3)$$
$$= 327$$

$$nX_cY_c = 32 \times 3.75 \times 2.84 = 340.8$$

$$\sum (XE) = (1 \times 9.4) + (1 \times 9.2) + (1 \times 8.7) + \ldots + (7 \times 7.8) + (7 \times 7.4)$$
$$= 975.80$$

$$nX_cE_c = 32 \times 3.75 \times 8.37 = 1004.40$$

Substituting in Eq. 12.2,

$$(566 - 450)S_x + (327 - 340.8)S_y = 975.80 - 1003.87$$

From similar calculations and substituting in Eq. 12.3,

$$(327 - 340.8)S_x + (319 - 258.1)S_y = 749.40 - 760.27$$

These two simultaneous equations result in

$$S_x = -0.27 \text{ ft}/100 \text{ ft}$$

$$S_y = -0.24 \text{ ft}/100 \text{ ft}$$

Since the plane of best fit must pass through the centroid, substituting the above values in Eq. 12.1 gives the elevation of the origin as 10.07 ft. Thus, the plane of best fit is

$$E = 10.07 - 0.27X - 0.24Y$$

Elevations at each grid point are then computed to the nearest 0.01 ft. After adding the settlement allowance, to be discussed later, cuts and fills are computed to the nearest 0.1 ft.

Digital computer programs were developed by Smerdon et al. (1966) and Benedict et al. (1964) to solve land grading problems. In addition to minimizing cuts and fills the print-out gave both the least total volume of earth moved and the least distance transported. Where such facilities are available, the large number of arithmetic operations is of little consequence. Other methods to be described are also simpler than the plane method. These are also easily adapted to frequent changes in slope and different slope lengths.

12.10. Profile Method. With this method, ground profiles are plotted and a grade is established that will provide an approximate balance between cuts and fills as well as reduce haul distances to reasonable limits. The field in Fig. 12.10 is again shown in Fig. 12.11 to illustrate the profile procedure. The profiles are drawn along the X direction with a datum elevation assigned to each line A, B, etc. Cuts and fills are estimated by trial and error and determined graphically from the original and design profiles. If needed, profiles can be plotted at right angles to the original lines. In this way proposed grade lines can be adjusted to meet cross-slope and downgrade criteria.

12.11. Plan-Inspection Method. With this method the grid point elevations are recorded on the plan, and the design grade elevations are determined by inspection after a careful study of the topography. This method is illustrated in Fig. 12.12 using the same field as in Fig. 12.10. It is largely a trial and error procedure keeping in mind downgrade and cross-slope limitations. As will be explained later, the desired cut-fill ratio and volumes of earthwork are estimated from the summation of all cuts and fills as shown in Fig. 12.12. Although this method does not assure minimum cuts and fills or the shortest length of haul, it is a rapid method.

12.12. Contour-Adjustment Method. To apply this method a contour map is drawn and the proposed ground surface is shown on the same map by drawing in new contour lines. The uniformity of slope is controlled by properly spacing the new contours. As with the plan-inspection method, it is largely a trial and error procedure. The proper balance between cuts and fills is estimated graphically at the grid points by interpolating between contour lines and by taking the difference in elevation between the original and the new surface.

12.13. Earthwork Volumes. The average end area or the prismoidal formulas given in Appendix G are suitable for making earthwork calculations, but these are time consuming. A more common procedure, called the four-point method by the U.S. SCS (1959), is sufficiently accurate for land grading. Volume of cuts for each grid square is

$$V_c = \frac{L^2(\sum C)^2}{4(\sum C + \sum F)} \tag{12.5}$$

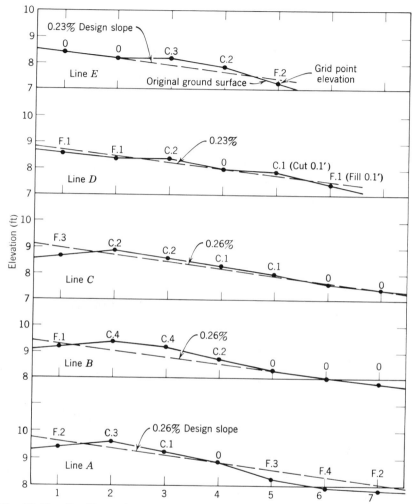

Fig. 12.11. Profile method of land leveling. (All cuts and fills are in feet.)

where V_c = volume of cut (L^3),
L = grid spacing (L),
C = cut on the grid corners (L),
F = fill on the grid corners (L).

For computing the volume of fills V_f, $(\sum C)^2$ in the numerator of Eq. 12.5 is replaced by $(\sum F)^2$. If the volume for only a portion of a grid square is desired, the full grid volume is reduced in proportion to the reduced area. For example, if V_c is 34 m³ on a 30 × 30-m grid, the volume for a 15 × 30-m area would be 17 m³. Another more approximate method is to assume that the cut or fill at a grid

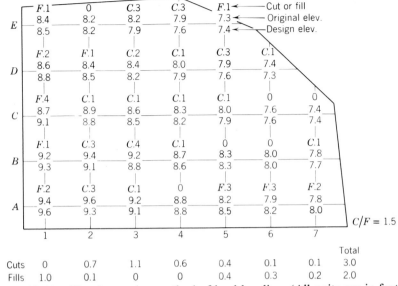

Fig. 12.12.　Plan-inspection method of land leveling. (All units are in feet.)

point represents the average for 15 m from the point in four directions. This method is only suitable for preliminary estimates.

Experience has shown that, in leveling, the cut-fill ratio should be greater than 1. Compaction from equipment in the cut area which reduces the volume and also compaction in the fill area which increases the fill volume needed are believed to be the principal reasons for this effect. Marr (1957) stated that on level ground between stakes the operator has an optical illusion of a dip in the middle, and therefore in filling, crowning often occurs. The ratio of cut to fill volume is usually 1.3 to 1.6, but may range from 1.1 to 2.0. With the plane method of computing cuts and fills, a settlement correction for the whole field is more convenient to apply. The settlement allowance or the amount of lowering of the elevation may range from 0.3 to 1 cm for compact soils and from 1.5 to 5.0 cm for loose soils. A small change in elevation will cause a considerable change in the cut-fill ratio. If extra quantities of soil are needed outside the area to be leveled, such as for a roadway or depression, the plane surface can be lowered by the amount of earthwork required. In computing costs normally the volume of cut is the basis for computation. The following example illustrates earthwork computations.

Example 12.3.　Determine the volume of cuts and fills and the cut-fill ratio for the field in Fig. 12.12 using the four-point method. All elevations and distances are in feet. Grid square A1, B1, A2, B2 is designated (AB12), and so on.

Solution. For all the full size 100×100-ft grid squares, substituting the cuts and fills in Eq. 12.5 from the lower left $(AB12)$, $(AB23)$, and so on, to upper left $(ED12)$ grid squares in Fig. 12.12 gives

$$\sum V_c = \frac{100 \times 100}{4 \times 27} \left[\frac{(0.3 + 0.3)^2}{(0.3 + 0.3 + 0.1 + 0.2)} + (0.3 + 0.4 + 0.1 + 0.3) + \ldots + 0 \right]$$

$$= 766.9 \text{ yd}^3$$

$$\sum V_f = \frac{100 \times 100}{4 \times 27} \left[\frac{(0.1 + 0.2)^2}{(0.1 + 0.2 + 0.3 + 0.3)} + 0 + \cdots + (0.1 + 0 + 0.1 + 0.2) \right]$$

$$= 248.3 \text{ yd}^3$$

For the partial grid squares assume that the nearest cut or fill is the same at the field boundary. Summation of these volumes starting at $A1$ and ending at the rectangle between $A1$ and $B1$ gives,

$$\sum V_c = \frac{100 \times 100}{4 \times 27} \left[0 + \tfrac{1}{2} \frac{(0.3 + 0.3)^2}{(0.3 + 0.3 + 0.2 + 0.2)} + \cdots + 0 \right] = 217.4$$

$$\sum V_f = \frac{100 \times 100}{4 \times 27} \left[\tfrac{1}{4}(0.2 \times 4) + \tfrac{1}{2} \frac{(0.2 + 0.2)^2}{(0.2 + 0.2 + 0.3 + 0.3)} + \cdots \right.$$

$$\left. + \tfrac{1}{2}(0.1 + 0.1 + 0.2 + 0.2) \right] = 403.2$$

$$\text{Total } V_c = 766.9 + 217.4 = 984.3 \text{ yd}^3$$

$$\text{Total } V_f = 248.3 + 403.2 = 651.5 \text{ yd}^3$$

$$C/F \text{ ratio} = 1.51$$

12.14. Construction and Maintenance. Prior to making the survey, heavy vegetative growth and residue should be removed. The surface of the soil should be fairly compact and as smooth as possible. Heavy carrier-type scrapers or pans especially designed for accurate depth control at shallow cuts and with laser grade control systems are most satisfactory. These are powered with crawler or rubber-tired units. Grades should be maintained as accurately as possible, with no reverse grade permitted. Cuts and fills are made between grid stakes as shown in Fig. 12.13. After the grade is checked, the ridges or islands are removed and the land smoothed. In land smoothing, leveling equipment should go over the area several times. For example, levelers should be

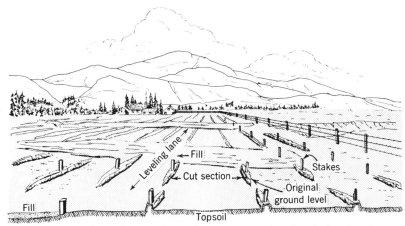

Fig. 12.13.   Field surface after heavy equipment operation, but before land smoothing. (With laserplane grade control no stakes are required.)

operated in both directions across the field and then diagonally both ways. Smoothing operations may be required for several years since fill material has a tendency to settle.

REFERENCES

Beauchamp, K. H. (1952). "Surface Drainage of Tight Soils in the Midwest." *Agr. Eng.* **33**, 208–212.

Benedict, R. et al. (1964). "Land Grading and Leveling Programs for Digital Computers." *Arkansas Agr. Expt. Sta. Bull.* **691**.

Chugg, G. E. (1947). "Calculations for Land Gradation." *Agr. Eng.* **28**, 461–463.

Coote, D. R. and P. J. Zwerman (1970). "Surface Drainage of Flat Lands in the Eastern U.S." Cornell University Ext. Bull. 1224.

Gain, E. W. (1964). "Nature and Scope of Surface Drainage in Eastern United States and Canada." *Am. Soc. Agr. Eng. Trans.* **7**(2), 167–169.

Marr, J. C. (1957). "Grading Land for Surface Irrigation." *Calif. Agr. Expt. Sta. Cir.* **438** (rev.).

Phelan, J. T. (1960). Bench Leveling for Surface Irrigation and Erosion Control, *Am. Soc. Agr. Eng. Trans.,* **3**(1), 14–17.

Phelan, J. T. (1961). "Land Leveling and Grading." in C. B. Richey et al., *Agricultural Engineers' Handbook,* McGraw-Hill, New York.

Phillips, R. L. (1958). "Land Leveling for Drainage and Irrigation." *Agr. Eng.* **39**, 463–465, 470.

Smerdon, E. T., et al. (1966). "Electronic Computers for Least-Cost Land-Forming Calculations." *Am. Soc. Agr. Eng. Trans.* **9**, 190–193.

U.S. Soil Conservation Service (1959). Irrigation, *Natl. Eng. Hbk.*, Section 15, Chapter 12, Land Leveling (lithograph), Washington, D.C.

Wood, I. D. (1951). "Land Preparation for Irrigation and Drainage," *Agr. Eng.* **32**, 597–599.

PROBLEMS

12.1. Compute the volume of soil to be excavated in cubic meters for a random field drain with 8:1 side slopes if the cuts at consecutive 15-m (50-ft) stations are 15 (0.5), 30 (1.0), 73 (2.4), 55 (1.8), 24 (0.8), and 0 cm (0 ft).

12.2. Determine the equation for the plane of best fit by the least-squares procedure for the field in Fig. 12.10, using only the 25 grid point elevations on the first five lines in the X and in the Y directions. Compute the cut or fill for points $A2$ and $C7$. How does your solution compare to that in Example 12.2?

12.3. Determine the cut and fill volumes for the data shown in Fig. 12.11, by assuming that the cut or fill at each grid point represents the average for the 100-ft grid square. What is the cut-fill ratio?

12.4. Determine the cut and fill volumes for grid squares bounded by lines $AB12$ and by lines $DE56$ shown in Figs. 12.10 and 12.11 using the four-point method.

12.5. Determine the cuts and fills by the plane method for a field assigned by the instructor, assuming a settlement allowance of 1.2 cm (0.04 ft.) Compute the volume of cuts and fills.

$$A = \frac{1}{2} \left[\sum x_i \left(y_{i-1} - y_{i+1} \right) \right]$$

CHAPTER 13

Open Channels

In this chapter open channels refer to open ditches for drainage and to canals for carrying irrigation water. Broad, shallow open ditches, which can be crossed with field machinery, are called surface drains and are discussed in Chapter 12. Open ditches provide outlets for tile and surface drains and remove surface water directly. Open ditch systems generally drain large areas and often involve several property owners. The design of lined or earth canals which convey water from storage reservoirs or wells is basically the same as for open ditches. However, for irrigation, flows can be regulated and canals can be lined with stabilizing materials giving greater flexibility in selecting a more efficient cross section.

The United States Census of 1960 shows that more than 235 000 km (146 000 mi) of open ditches, mostly in organized drainage districts, have been constructed in 38 states. During the period 1950–1960 over 53 000 km (33 000 mi) of such ditches were enlarged or installed. These figures would be much greater if all privately owned drainage ditches were included. Several million hectares of fertile land have been drained and total crop production greatly increased. According to a survey by USDA (1962) 22 percent of the total cropland in the United States has excess water as a dominant problem. The length of irrigation canals would be quite large considering that about 20 000 000 ha (50 000 000 ac) are irrigated in the West.

An earth-lined open channel properly designed should provide (1) velocity of flow such that neither serious scouring nor sedimentation will result, (2) sufficient capacity to carry the design flow, (3) hydraulic grade at the proper depth for good water management, (4) side slopes that are stable, and (5) minimum initial cost and maintenance. For carrying irrigation water additional requirements, such as low seepage loss, must be met. The engineer should have a knowledge of the capabilities and limitations of the various types of construction equipment and should consider these factors in the design of the system.

13.1. Channel Discharge Capacity. The discharge capacity of an open channel is computed from the Manning formula, which was discussed in Chapter 7. The depth of the constructed channel should provide a freeboard allowance ($D - d$ in Fig. 13.1) of 20 percent of the total depth, D. The freeboard

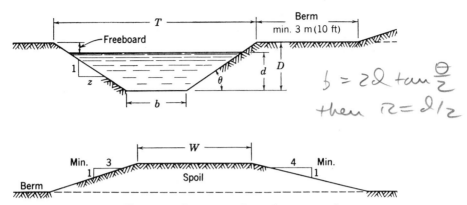

Fig. 13.1. Elements of an open channel cross section.

provides a safety factor for overtopping and allows a reasonable depth for sediment accumulation. On large earth irrigation canals in the West the freeboard commonly varies from 30 to 35 percent, and for lined canals it ranges from 15 to 20 percent as reported by Houk (1956). The freeboard for lined canals refers only to the top of the lining. The total freeboard is the same for lined and unlined channels. The freeboard should be increased on the outside edge of curves.

Wherever practical, the channel should be designed for a high hydraulic efficiency. In earth channels the stability of the soil places limitations on channel grade and side slopes. The topography and desired water levels may limit the design grade and the velocity of flow.

The roughness coefficient for open channels varies with the height, density, and type of vegetation; the physical roughness of the bottom and sides of the channel; variation in the size and shape of the cross section; alignment; and hydraulic radius. Primarily because of differences in vegetation, the roughness coefficient varies from season to season. In general, conditions which increase turbulence increase the roughness coefficient. Values of n can be obtained from Appendix B.

Experience and knowledge of local conditions are helpful in the selection of the roughness coefficient, but it is one of the most difficult problems in channel design. In general, a value of 0.035 is satisfactory for medium-sized earth ditches with bottom widths of 1.5 to 3 m, and 0.04 is suitable for smaller ditches with bottom widths of 1.2 m.

The design procedure for open channels is illustrated in the following example.

Example 13.1. Design an open channel to carry 4.42 m³/s (156 cfs) on a slope of 0.09 percent, assuming ditch side slopes of 2:1 and a bottom width of 1.2 m (4 ft).

Solution.   From Table B.1, select a roughness coefficient of 0.04 for small drainage ditch. Make a trial solution, using a flow depth of 1.5 m (5 ft) using symbols in Fig. 13.1.

$$R = \frac{bd + zd^2}{b + 2d(z^2 + 1)^{1/2}} = \frac{1.8 + 4.5}{1.2 + 3(5)^{1/2}} = 0.80 \text{ m (2.6 ft)}$$

From Fig. B.1, read $v = 0.65$ m/s (2.1 fps) or substitute in the Manning formula, $v = R^{2/3}s^{1/2}/n$, calculate:

$$q = av = 6.3 \times 0.65 = 4.10 \text{ (m}^3\text{/s) (0.32 m}^3\text{/s too low)}$$

Next try $d = 1.56$ m (5.1 ft), from which $R = 0.82$ m (2.7 ft), and determine $v = 0.66$ m/s (2.16 fps):

$$q = 6.74 \times 0.66 = 4.45 \text{ m}^3\text{/s (157 cfs)}$$

which is close to the desired capacity. If the freeboard is 20 percent, the total depth is $(1.56/100 - 20)100 = 1.95$ m (6.4 ft).

13.2.   Cross Section.   The design dimensions of an open channel cross section are shown in Fig. 13.1. Earth channels and lined canals are normally designed with trapezoidal cross sections. The size of the cross section will vary with the velocity and quantity of water to be removed.

Side Slopes.   Channel side slopes are determined principally by soil texture and stability. The most critical condition for caving occurs after a rapid drop in the flow level that leaves the banks saturated. For the same side slopes, the deeper the ditch the more likely it is to cave. Side slopes should be designed to suit soil conditions and not the limitations of construction equipment. Suggested side slopes are shown in Table 13.1. Whenever possible, these slopes should be verified by experience and local practices. Very narrow ditches should have slightly flatter side slopes than wide ditches because of greater reduction in capacity in the narrow ditches should caving occur.

Bottom Width.   After the channel grade, depth, and side slopes are selected, the bottom width can be computed for a given discharge. The bottom width for the most efficient cross section and minimum volume of excavation is determined by the formula

MINIMUM $b = 1.2$ m (4')

$$b = 2d \tan \theta/2 \qquad\qquad (13.1)$$

where b = bottom width,
         d = design depth,
         θ = side slope angle.

FOR ANY SIDE SLOPE
$R = d/2$

Table 13.1 Maximum Side Slopes for Open Channels

Soil	Side Slopes—Horizontal:Vertical	
	Shallow Channels up to 1.2 m (4 ft)	*Deep Channels—1.2 m* (4 ft) and Over
Peat and muck	Vertical	¼:1
Stiff (heavy) clay	½:1	1:1
Clay or silt loam	1:1	1½:1
Sandy loam	1½:1	2:1
Loose sandy	2:1	3:1

Source: By permission, from *Land Drainage and Flood Protection,* by Etcheverry, copyright, 1931, McGraw-Hill Book Co.

(See Fig. 13.1.) For any side slope it can be shown mathematically that, for a bottom width computed from Eq. 13.1, the hydraulic radius is equal to one half the depth. The minimum bottom width should be 1.2 m except in small laterals. It is not always possible to design for the most efficient cross section because of construction equipment limitations, allowable velocities, and increased maintenance.

Spoil Banks. The excavated soil may be placed on one or both sides of the channel, depending on the type of equipment, use of the channel, and the size of the cross section. If the spoil bank is to serve as a levee, the spoil is normally placed on only one side. For drainage ditches, the spoil bank should be spread until it blends into the adjoining field, thus permitting cultivation near the edge of the ditch. Where levees or access roads are desired, the spoil should be spread to conform to the cross section shown in Fig. 13.1. The berm width or the distance from the edge of the ditch to the edge of the spoil provides a degree of protection against caving and may be used as a place for the ditching machine to operate. The steeper the side slopes, the greater should be the berm width. For ditches with side slopes of 1:1 the berm width should be twice the depth, and for side slopes of 2:1 the berm width should be equal to the depth. The minimum berm width should be 3 m. The depth and top width of the spoil bank vary with the size of the ditch. Unless the spoil bank is to be used as a levee, breaks or gaps that allow surface water to drain to the channel should be provided, or tube inlets may be installed through the spoil bank (Chapter 9).

For irrigation canals spoil banks are often compacted and placed so as to raise the water level above the original ground surface. In this case the spoil side slope should be the same as for the channel. If the soil is sufficiently stable or if the canal is lined, the berm is sometimes omitted. In deep cuts berms are provided for stability as well as for roadways. The land side slope of canal spoil banks may be made steeper than that for drainage ditches, unless flat slopes are required for maintenance, control of seepage, or if the land is to be farmed.

13.3. Velocities. In earth channels optimum velocities of flow are based on (1) the selection of limiting velocities or on (2) the computed values of the critical tractive force. Tractive force is defined as the shearing force per unit area on the periphery of the channel.

No definite optimum velocity can be prescribed, but minimum and maximum limits can be approximated. Velocities should be low enough to prevent scour but high enough to prevent sedimentation. Usually an average velocity of 0.5 to 1.0 m/s for shallow channels is sufficient to prevent sedimentation. In channels that flow intermittently, vegetation may retard the flow to such an extent that adequate velocities at low discharges are difficult, if not impossible, to maintain. For this reason maximum or even higher velocities should be obtained, since scouring for short periods of time at high flows may aid in maintaining the required cross section.

Typical velocity distribution in an open channel is shown in Fig. 13.2. The maximum velocity occurs near the center of the stream and slightly below the surface. The average velocity in this stream, having a rather straight and relatively smooth channel, was 1 m/s with a maximum of 1.3. Around curves the maximum velocity and the velocity contours move toward the outer bank. Since the Reynolds' number is about 500 for turbulent flow in open channels, velocities are almost always high enough for turbulent flow.

Limiting Velocity Method. From a survey of irrigation canals in western United States, Fortier and Scobey (1926) recommended the limiting velocities shown in Table 13.2. Their data have been widely accepted. These velocities may be exceeded for some soils where the streamflow contains sediment because deposition may produce a well-graded channel bed resistant to erosion. Where a powerful abrasive is carried in the water, these velocities should be reduced by 0.15 m/s. For depths over 0.9 m Fortier and Scobey (1926) permitted velocities 0.15 m/s higher than shown. Where the channel is winding or curved, the limiting velocity should be reduced about 25 percent.

Tractive Force Method. In 1950 the U.S. Bureau of Reclamation began studies based on the tractive force theory for improving the design of earth-lined irrigation canals. As shown in Fig. 13.3, the tractive force (shear) is equal to and in the opposite direction to the force the bed exerts on the flowing water.

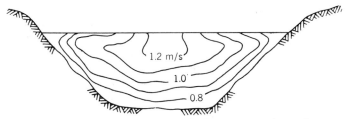

Fig. 13.2. Velocity distribution in an open channel.

TO PREVENT SEDIMENTATION

0.5 – 1.0 m/s USUALLY GOOD

Table 13.2 Fortier and Scobey's Limiting Velocities with Corresponding Tractive Force Values (Straight Channels After Aging)

| Material | n | For Clear Water | | Water Transporting Colloidal Silts | |
		Velocity (fps)[a]	Tractive Force (psf)	Velocity (fps)	Tractive Force (psf)
Fine sand colloidal	0.020	1.50	0.027	2.50	0.075
Sandly loam noncolloidal	0.020	1.75	0.037	2.50	0.075
Silt loam noncolloidal	0.020	2.00	0.048	3.00	0.11
Alluvial silts noncolloidal	0.020	2.00	0.048	3.50	0.15
Ordinary firm loam	0.020	2.50	0.075	3.50	0.15
Volcanic ash	0.020	2.50	0.075	3.50	0.15
Stiff clay very colloidal	0.025	3.75	0.26	5.00	0.46
Alluvial silts colloidal	0.025	3.75	0.26	5.00	0.46
Shales and hardpans	0.025	6.00	0.67	6.00	0.67
Fine gravel	0.020	2.50	0.075	5.00	0.32
Graded loam to cobbles when noncolloidal	0.030	3.75	0.38	5.00	0.66
Graded silts to cobbles when colloidal	0.030	4.00	0.43	5.50	0.80
Coarse gravel noncolloidal	0.025	4.00	0.30	6.00	0.67
Cobbles and shingles	0.035	5.00	0.91	5.50	1.10

[a] SI conversions: 1 psf = 4.88 kg/m^2 = 47.88 N/m^2; 1 fps = 0.305 m/s.
Source: From Lane (1955).

– reduce if a powerful abrasive in water
– if channel is windy or curved
reduce limits by 25%

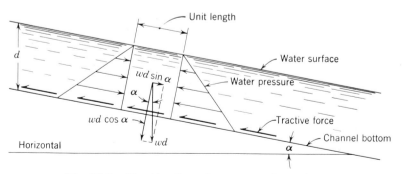

Fig. 13.3. Tractive force in an open channel.

In a uniform channel of constant slope and constant flow, the water is moving in a state of steady, uniform flow without acceleration because the force tending to prevent motion is equal to the force causing motion. For small channel slopes, sin α is equal to the tan α or the slope, and the tractive force,

$$T = wdsK \qquad (13.2)$$

where T = tractive force (F/L²),
        w = unit weight of water (F/L³),
        d = depth of flow (L),
        s = slope (hydraulic gradient) (L/L),
        K = ratio of the tractive force for noncohesive material necessary to start motion on the sloping side of a channel to that required to start motion for the same material on a level surface. For an infinitely wide channel of uniform depth this ratio is 1. The ratio is less than 1 on the sloping sides of a channel because the force of gravity is also acting on the particle tending to roll or slide it down the slope.

For noncohesive material the above ratio, K, is further defined by Lane (1955) in terms of the angle of repose of the material with the horizontal and the side slope of the channel. For cohesive and fine single-grain materials, the cohesive forces are so great in proportion to the gravity forces causing particles to roll down the slope, that the gravity component may be neglected, in which case K equals 1. A tractive force equation has been derived by Smerdon and Beasley (1961) for nonuniform flow, such as would occur in a terrace channel or in an irrigation canal where water is removed along its length. They also found that the critical tractive force for cohesive soils was correlated with the plasticity index, dispersion ratio, particle size, and clay content. Laflen and Beasley (1960) concluded that the critical tractive force increased linearly as the void ratio decreased, but the values varied with different soil types. Similarly, Russian data published in 1936 for cohesive soils showed that the critical tractive force increased as the soil compaction increased.

Critical tractive force values were computed by Lane (1955) for the corresponding velocity of flow and roughness coefficient indicated in Table 13.2. In making the computations a depth of 0.9 m, bottom width of 3.1 m, and a side slope ratio of 1.5:1 were assumed. Thus, these tractive force values were not independently derived, but design may be based on these values if more reliable data are not available. The application of tractive force theory to design is in the developmental stage, but it does provide a theoretical rather than a purely empirical basis for design. Design values for cohesive soils are still being developed, but for noncohesive soils critical tractive forces have been well established. The USBR (1962) has developed equations for cohesive soils which involve more elaborate soil tests than indicated above. The design procedure using critical tractive force is illustrated in the following example.

Example 13.2. Design a stable channel with a side slope ratio of 1:1 to carry a flow of 4.42 m³/s (156 cfs) on a slope of 0.25 percent. The soil is stiff clay, very colloidal, n = 0.025, and the water is transporting colloidal silts.

Solution. From Table 13.2 the critical tractive force at the bottom of the channel when the particles are in a state of impending motion is 2.24 kg/m² (0.46 psf). Since the soil is cohesive, tractive force on the side slopes does not limit the design. If the soil was noncohesive, the tractive force on the side slope would control in the analysis. Substituting in Eq. 13.2, the maximum depth with $K = 1$ is

$$d = \frac{2.24}{1000 \times 0.0025} = 0.9 \text{ m (2.95 ft)}$$

Assuming $b = 2.44$ m (8 ft), compute $a = 3.01$ m² (32.3 ft²) and $p = 4.99$ m (16.36 ft). Substituting in the Manning equation, $v = 1.43$ m/s (4.7 fps) and $q = 4.30$ m³/s (152 cfs), which is close enough to the required capacity of 4.42 m³/s (156 cfs).

An alternate solution can be obtained by using the maximum permissible velocity of 1.52 m/s (5.0 fps) in Table 13.2 rather than the critical tractive force. In this case, the 1.52 m/s (5.0 fps) velocity would result in a greater depth of flow and a smaller bottom width than shown above. For practical purposes either criterion is satisfactory. Selection of design criteria should be based on the best available data. For more detailed design procedure see Chow (pp. 164–179, 1959).

13.4. Water Surface Profiles. For simple design problems the depth of flow and the velocity can be assumed constant. For this uniform flow condition the energy grade line, hydraulic grade line, and the channel bottom are all parallel. However, if an obstruction, such as a culvert, is placed in the channel, a concave water surface called a "backwater curve" develops upstream of the structure. A change from a gentle to a flat grade could produce the same effect. If the channel slope changed from a flat to a steep grade, a convex water surface would develop. These examples illustrate two common types of gradually varied flow conditions. Such a water surface profile for uniform-shaped channels can be computed from the continuity equation

$$\frac{v_1^2}{2g} + d_1 + S_0 x = \frac{v_2^2}{2g} + d_2 + Sx \tag{13.3}$$

where v is the velocity of flow and all other symbols are defined in Fig. 13.4. Such an equation can be easily solved with digital computers.

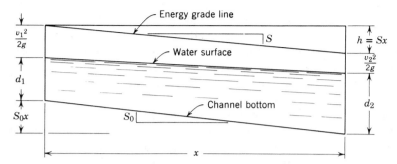

Fig. 13.4. Gradually varied flow in open channels.

Where the water surface profile is to be computed, the general procedure is to (1) determine the depth of flow at the control section, (2) assume a new depth at a short distance (reach) from the control section, (3) evaluate the slope of the energy grade line, S, and the velocity head, (4) substitute these values in Eq. 13.3, (5) compute the length of the reach, and (6) continue the trial and error procedure for each reach repeating Steps (1), (2), (3), and (4). The process is continued until the desired profile is obtained. Variations in depth should not result in velocity changes of more than 20 percent. Where the depth of flow is greater than the critical depth, profile computations should proceed upstream. Where the depth is less than critical, they should proceed downstream. The following example illustrates the procedure for a backwater curve.

Example 13.3. Determine the water surface profile in a drainage ditch in which a culvert restricts the normal flow. The channel has a 1.22 m (4 ft) bottom width, side slopes 1:1, n = 0.04, a normal depth of flow of 0.91 m (3.0 ft), and a channel slope of 0.2 percent. The culvert at Station 30 + 00 (ft) causes the backwater to rise 0.21 m (0.7 ft) at the culvert. Elevation of the bottom of the ditch at the culvert is 16.15 m (53.0 ft).

Solution. From the Manning equation the velocity is 0.73 m/s (2.38 fps) at a depth of 0.91 m (3.0 ft) and the discharge is 1.415 m³/s (50 cfs). Computation of the length of successive reaches between selected depths of flow is illustrated in the following table (all units in ft).

Thirteen types of water surface profiles have been classified according to the nature of the channel slope and the relative depths of flow. Chow (1959) has developed a graphical-integration method for determining the shape of the flow profile which has broad application. Because the subject is beyond the scope of this text, the reader should refer to Chow (1959), King and Brater (1963), and many other excellent hydraulics books for further details.

13.5. Alignment. Where changes in direction are necessary, gradual curves should be provided to prevent excessive bank erosion. The radius of curvature

(1)	(2)	(3)	(4)	(5)	(6)	(7)	(8)	(9)	(10)
	Elevation								
Station	Bottom Channel	Water Surface	d	$d + \dfrac{v^2}{2g}$	$\dfrac{d_1 + d_2}{2b}$	$(1/K')^2$	S	$S_0 - S$	x
30 + 00	53.00	56.70	3.7	3.7476	0.900	0.409	0.00100	0.00100	192
28 + 08	53.38	56.88	3.5	3.5563	0.850	0.457	0.00126	0.00074	256
25 + 52	53.89	57.19	3.3	3.3669	0.812	0.596	0.00151	0.00049	189
23 + 63	54.27	57.47	3.2	3.2731	0.788	0.690	0.00170	0.00030	310
20 + 53	54.89	57.99	3.1	3.1801	0.763	0.782	0.00193	0.000074	1254
7 + 99	57.40	60.40	3.0	3.0880					

Col. (1): Computed from distances determined in Col. (10).
Col. (2): Computed for slope of 0.2 percent to correspond to station numbers.
Col. (3): Computed from slopes given in Col. (8) or Col. (2) plus Col. (4).
Col. (4): Depths arbitrarily assumed depending on the accuracy desired.
Col. (5): Specific energy head computed for each depth in Col. (4).
Col. (6): Average depth between stations divided by b; needed to obtain values in Col. (7).
Col. (7): Obtain from tables in King and Brater (1963).
Col. (8): Compute from Manning's equation or by equation, $S = (Qn/K'b^{8/3})^2 = 0.00243(1/K')^2$. Use the average depth between successive stations with Manning's equation.
Col. (10): Solve for x in Eq. 13.3, which can be obtained by dividing the change in specific energy in Col. (5) at successive stations by Col. (9).

depends on the velocity of flow and the stability of the side slopes. If gradual curves will not eliminate erosion in the channel, it may occasionally be necessary to decrease the velocity by increasing the width or side slopes or to provide bank protection. It is convenient to express the radius of a curve by the degree of curve, which is defined as the angle subtended at the center of a circle by a 30.48-m (100-ft) chord. The relationship between the degree of curvature and radius of curvature shown in Fig. 13.5 is expressed by the equation

$$R = \frac{15.24}{\sin D/2} \tag{13.4}$$

where　R = the radius of curvature in m,
　　　　D = the degree of curvature.

The maximum degree of curvature for unprotected earth channels usually ranges from 4 to 10 degrees for large channels and up to 20 degrees for small channels. The degree of curvature should be reduced as side slopes and the velocity of flow are increased.

Circular curves are generally satisfactory in open channel design. The procedure for laying out such curves is given in Appendix G.

13.6. Junctions. The junction of one channel with another should be such that serious bank erosion, scour holes, or sedimentation will not occur. Where the general direction of the main is perpendicular to that of the lateral, the lateral may be curved near the junction. For small channels having low velocities, the angle at which the two channels join is less important than for larger ditches having higher velocities.

The bottom of the main and the lateral should join at the same elevation. If the lateral is shallower than the main, the overfall at the junction may be eliminated by increasing the grade of the lateral in the first 60 to 90 m or by

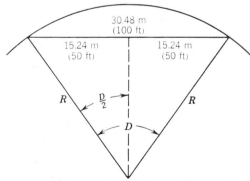

Fig. 13.5. Degree of curve.

increasing the slope on the entire lateral. Some engineers recommend excavating the first 15 to 90 m of the lateral at a level grade to provide a recessed area for sediment storage until the channel stabilizes. Since maximum velocities occur at the higher stages, the increased slope near the outlet may not be serious because the water in the main will back up into the lateral. The most serious condition exists when the main is at low flow and the lateral has a large discharge.

DRAINAGE DITCHES

For a given watershed area the required capacity of an open ditch is considerably different than the design capacity of a grassed waterway. Open ditches generally have flatter slopes, lower velocities, steeper side slopes, and a greater depth of channel flow than do grassed waterways. While grassed waterways are designed to carry peak runoff, open ditches are designed to remove water much more slowly but still rapidly enough to prevent serious damage to the crop.

13.7. Design Capacity. The rate of conveyance of water by open ditches is influenced by (1) rainfall rate, (2) size of the drainage area, (3) runoff characteristics including slope, soil, and vegetation, (4) potential productivity of the soil, (5) crops, (6) degree of protection warranted, and (7) frequency and height of flood waters from rivers and creeks. Although the degree of protection is very important in design, it is one of the most difficult factors to evaluate because costs must be compared to anticipated flood damages. More frequent flooding is permissible for agricultural land than for homes and building sites. Likewise, timberland requires less intensive drainage than does cultivated land.

The runoff for open ditch design may be expressed as a drainage coefficient or rate of flow per acre, hectare, or square mile. The drainage coefficient is defined as the depth of water that is to be removed in a 24-hr period from the entire drainage area.

A method of determining the required rate of water removal is by an empirical formula,

$$q = 0.0283C(A/259)^x \tag{13.5}$$

where q = runoff in m^3/s,
C = a constant,
A = watershed area in hectares,
x = an exponent.

The constants C and x vary with the degree of drainage desired and the location (see Table 13.3). The U.S. SCS (1973) presented a general procedure for computing the constant C from rainfall excess. Runoff volume is determined for a 2-

FOR DIFFT WS

Q_1 ① $4 \approx B$ $Q_{MAX} = Q_A + Q_B$ $Q_{DESIGN} = Q_2 + \Delta Q$

Q_2 ② $A >> B$ $Q_c = k(A_A + A_B)^x$

③ $20-40$ (%SMALL ONE IS OF TOTAL -20)($Q_1 - Q_2$) = ΔQ

CULTIVATED $Q = 37 A^{0.85}$ good drainage

$= 72 A^{0.7}$ excellent drainage

$Q = k A^x$

PASTURE $= 24 A^{0.85}$ (low value)

HIGH VALUE CROPS $= 150 A^{0.6}$

Table 13.3   Runoff Constants for Drainage Ditch Design

Land Resource Region No.[a]	Drainage Area 1000 ha (sq mi)	Constants		Remarks
		C	x	
Northern Humid Area				
	1–52 (4–200)[b]	150	0.6	Not max runoff, but good overflow protection, A curve.
K, L, M, R, S, N (north 37° Lat)	0.5–26 (2–100)[b]	70	0.7	B curve, excellent drainage.
	0.5–52 (2–200)[b]	37	0.83	For grain crops, C curve, good drainage.
	0.5–52 (2–200)[b]	24	0.85	For grasses (improved pasture), D curve, fair drainage.
Southern Humid Area				
H and N (south 37° Lat), I, J, O, P, T, U	0.3–26 (1–100)	43	0.83	Approx average for delta and coastal areas.
N (south 37° Lat), O, P, T, U	0.3–26 (1–100)	30	0.83	For grasses (improved pasture).
O, P, T, U	0.3–26 (1–100)	10	0.83	For forest drainage in Gulf and Atlantic areas.
F, G, and H (north 37° Lat)	0.5–52 (2–200)	24	0.85	Normal conditions.
H and N (south 37° Lat), I, J, P	0–4 (0–17)	131	0.7	Max for hill areas.
	0–41 (0–160)	15	0.83	Range land.

[a] Location of land resource regions is given in Chapter 1.
[b] Data from curves for flat watersheds having average slopes less than 5 m per km (25 ft per mi) (approximately 0.5 percent).

to 5-year return period storm for a 48-hr period. Rainfall excess is taken as one half this value for the 24-hr depth.

Quite frequently watersheds with upper tributaries having rather steep slopes and high runoff-producing characteristics drain into an alluvial valley. Open ditches are then often necessary to carry the water across the flood plain from the hill area to the natural outlet. Relying on experience and judgment the engineer should modify Eq. 13.5 to meet such local conditions.

13.8. Grades. The engineer frequently has little choice in the selection of grades for open ditches since the grade is determined largely by the outlet elevation, elevation and distance to the lowest point to be drained, and depth of ditches. Where open ditches drain flat land, the grade should be as steep as possible, provided maximum permissible velocities are not exceeded.

The depth at all points along the channel should be sufficient to adequately drain the area. Where subsurface drains outlet into the ditch, a minimum depth of from 1 to 2 m is required. In peat and muck soils the ditch should be made deeper to allow for subsidence. Because of reduced velocities, sediment accumulates more readily and vegetation grows more abundantly in shallow than in deep ditches. In some instances an allowance is made for accumulation of sediment, depending on channel velocities and soil conditions.

13.9. Controlled Drainage Structures. Controlled drainage with open ditches requires structures in the channel to maintain the water at the required level.

Several types of control structures are the burlap bag dam, timber or sheet piling cutoff, reinforced concrete structure, and a combination culvert and control gate. The elevation views of some of these structures are illustrated in Fig. 13.6. Since these structures are designed for organic soils, the cutoff wall is larger than required for more impermeable soils. All control dams should be provided with an erosion-controlling apron constructed below the dam as for a drop spillway. Burlap bag dams with a facing of timber (Fig. 13.6a) are suitable for low heads and small drainage areas. These bags are filled with a weak cement mixture and tamped together before wetting. Timber or sheet piling (Fig. 13.6b) may be placed to the side and under the ditch. This type of dam provides a greater barrier against seepage and is most suitable in deep organic soils or other highly permeable materials. Reinforced concrete dams (Fig. 13.6c) are better suited for handling large quantities of drainage flow. Where seepage is a problem, piling may be required both under and to the sides of the structure. Combination culvert and control gate structures (not illustrated) with crestboards at the upper end of the conduit may provide an economical and practical dam. Crestboards are placed in the openings when the water level is to be increased or removed when drainage is required.

LOCATION AND LAYOUT

Open ditches are arbitrarily designated as mains, submains, laterals, and field ditches. The drainage plan should incorporate as needed levees, pump installations, field and open ditch systems, and pipe drains into a coordinated system.

13.10. Preliminary Survey. The preliminary survey is an initial investigation of the proposed project. The preliminary survey should include an estimate of the potential productivity of the soil, and physical features of the watershed, as well as rainfall and runoff data, carried out in sufficient detail to provide a tentative design and to make possible a valid estimate of the economic feasibility of the plan.

Since the location of open ditches requires experience and good judgment combined with a careful study of local conditions, only a few general rules can

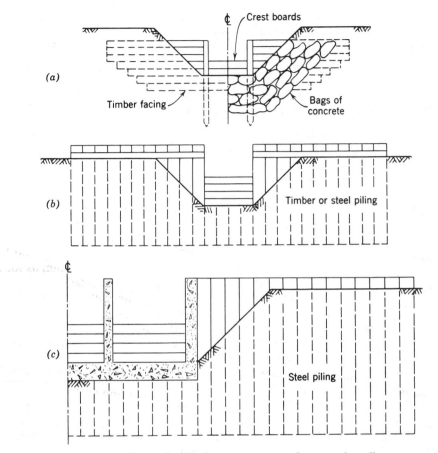

Fig. 13.6. Controlled drainage structures for organic soils.

be given: (1) follow the general direction of natural drainageways, particularly with mains and submains; (2) provide straight channels with gradual curves, especially for large ditches; (3) locate drains along property lines if practicable; (4) make use of natural or existing ditches are much as possible; (5) use the available grade to best advantage, particularly on flat land; and (6) avoid unstable soils and other natural conditions that increase construction and maintenance costs.

13.11. Location Survey. After the ditch system has been designed and approved, and construction has been authorized, the engineer proceeds with the field layout. The general location and alignment of the channel has normally been determined by the preliminary survey. If the system is small, the preliminary and location surveys may be made at the same time. In staking the field layout, minor changes in location or design are sometimes desirable.

13.12. Legal Aspects. *Mutual Drainage Enterprises.* Many states have laws that provide for the organization of mutual drainage enterprises, also called drainage associations. To establish such an enterprise the landowners involved must be fully in accord with the plan of operation and with the apportionment of the cost. After the agreement has been drawn up and signed, it must be properly recorded in the drainage record of the county or other political subdivision. The local court may be asked to name the district officials, sometimes called commissioners, who are responsible for the functioning of the district, or they may be named in the agreement. The principal advantage of the mutual district is that less time is required to establish an organization and the costs are held to a minimum. Because it may be difficult for several landowners to come to an agreement, particularly on the division of the costs, such districts are difficult to organize where the number of landowners is large or where considerable area is involved. Although a mutual enterprise cannot assess taxes, it may petition to become an organized district in some states to overcome this disadvantage. A large number of mutual enterprises are in existence and much drainage has been accomplished in this manner.

Organized Drainage Enterprises. Such an enterprise, often called a county ditch or a drainage district, is a local unit of government established under state laws for the purpose of constructing and maintaining satisfactory outlets for the removal of excess surface and subsurface water. It is different from a mutual enterprise in that minority landowners can be compelled to go along with the project. Levee districts formed for the purpose of keeping out excess flood water are similar to drainage districts in their organization. For further details on drainage enterprises consult the laws or publications of the state involved. The method of organization, powers of district officials, and methods of assessing benefits, damages, and financing vary greatly from state to state. A uniform law for drainage, irrigation, or flood control districts has been proposed by Harman (1922).

IRRIGATION CANALS

The location and layout of laterals and farm ditches for distributing irrigation water is normally a part of a surface irrigation system, which is described in Chapter 19. The discussion to follow applies mainly to larger canals beyond the farm boundaries but may apply to farm ditches as well.

13.13. Design Flow. The design flow for irrigation canals should be adequate to supply the maximum rate of water use by plants. Peak use rates according to Houk (1956) are often estimated as 10 to 15 percent higher than maximum monthly rates. This maximum rate depends on the acreage irrigated, climatic conditions, and the type and stage of growth of different crops within the area. Allowances should be made for additional flow from storm rainfall, for conveyance losses, and for unavoidable delivery and application losses. Canal seepage losses will be highest during the early part of the growing season. For large projects with diversified crops, maximum demand usually occurs during the middle of the growing season or slightly later.

Houk (1956) reported that seepage losses on large irrigation projects, where only small proportions of the canals are lined, may vary from 15 to 45 percent of the water diverted. He recommended that the design flow of laterals per unit area should be 10 to 15 percent greater than for the main canals and for sublaterals and farm ditches, 25 to 50 percent greater.

13.14. Location and Layout. Canals are located so that the water may flow by gravity from the canal to the point of use. However, under some conditions the topography or soil formation may make another location more economically feasible. Most problems in location arise because of varying surface and subsurface conditions. The water surface in the canal should be kept above the natural ground surface at the point of delivery and should permit the measurement of flow where required. Gradual curves are more important for canals than for drainage ditches since capacity flow must be carried for a longer period of time. Layout surveys are similar to those for drainage ditches (see previous section of this chapter and Appendix G).

13.15. Canal Linings. Irrigation canals are lined for several purposes, namely: (1) to reduce seepage losses, thus increasing conveyance efficiency and decreasing drainage problems; (2) to insure against uninterrupted operation due to breaks; (3) to provide a more efficient cross section by increasing side slopes, by reducing the roughness coefficient, by eliminating weed and moss growth, and by increasing channel slope without danger of erosion; and (4) to reduce maintenance. In areas where water is in short supply one of the most important benefits from lining canals is the saving of water, which then becomes available for other beneficial uses. As discussed previously, losses in earth canals are sometimes a high percentage of the water diverted. One of the reasons for emphasizing linings is that canal losses are usually somewhat easier to control than losses due to poor application and water distribution on a field.

Canal linings may be constructed with a large number of materials, such as concrete, rock masonry, brick, colloid clay-soil mixtures, soil cement, asphalt, rubber, and plastic. Materials which are most satisfactory meet all the purposes described above. In laterals and farm ditches, Portland cement concrete and asphalt are the two most common materials. The selection of a lining material will depend largely on cost and availability of materials, soil conditions, cross section and length of the canal, and comparative annual costs. Average annual cost, including maintenance and value of the water saved, is the best basis for making a decision.

Concrete more nearly meets all of the requirements for a lining than any other material. Its principal disadvantages are high initial cost and possible damage by soil chemicals and freezing and thawing. Concrete linings vary in thickness from 25 to 150 mm and may or may not be reinforced. Concrete may be placed by pouring in alternate panels, by shooting it on pneumatically (called shotcrete or Gunite), by plastering, by fitting precast slabs, or by continuous pouring with slip-form equipment. Because of its favorable cost, slip-form placement is being widely accepted for surface channels. Similar equipment has also been developed for placing underground distribution pipe. Precast slabs with tongue and groove joints filled with asphalt provide a slightly flexible lining that will adjust to minor movements of the subgrade. Stone and brick linings can be placed in a similar manner; however, the labor cost would be higher.

Asphalt linings may be applied as an asphaltic concrete that includes sand and gravel, hot-sprayed asphalt membranes, and prefabricated liners. Membrane-type linings may or may not be covered with soil or other material. Although asphalt is considerably lower in initial cost, it has a shorter life and is more subject to physical damage than concrete. Vegetation also grows through asphalt membranes, unless the soil is first sterilized. Lauritzen (1961) estimates that the life of exposed asphalt liners is probably less than 10 years. See USBR (1963) and Houk (1956) for further details on linings.

13.16. Conveyance and Control Structures. Most of the conservation structures discussed in Chapter 9 have application to irrigation. On-farm structures are discussed in Chapter 19. Inverted siphons, tunnels, flumes, and flow regulating structures are often necessary for proper water control. In crossing natural depressions or canyons, flumes or inverted siphons may be constructed. Inlet and outlet transition sections between the structure and the canal must be carefully designed. In any lined channel where the construction cost is high, considerable effort should be made to design structures and facilities which have a high hydraulic efficiency. Model studies may be required in some cases.

CONSTRUCTION

Prior to construction open ditches or irrigation canals must be designed and the location marked in the field with center line and slope stakes. Where hub stakes must be preserved, a baseline offset from the channel at a suitable distance should be established. Slope stakes must be properly set in order to establish the desired side slope ratio. Where the land is sloping or irregular normal to the center line, slope stakes must be located in the field by a trial and error procedure, such as illustrated in Appendix G.

13.17. Types of Equipment. As shown in Table 13.4, ditch construction equipment may be classified according to function and type of earth-moving

Table 13.4 Classification of Earth-moving Equipment for Open Channel Construction

	Action	
Function	*Continuous*	*Intermittent*
Excavation	Wheel excavator (trench type) Plow-type ditcher[a] Template excavator Blade grader[a] Elevating grader[a] Hydraulic dredge	Dragline (scraper-bucket excavator) Clamshell Hoe (also template) Shovel Scraper Bulldozer Pull-back blade
Spoil spreading	Blade grader[a,b] Tillage machines[a] Terracing machines[a]	Bulldozer[b] Scraper Pull-back blade[b]

[a] Continuous except for turning at the ends.
[b] Either continuous or intermittent, depending on the method of operation.

action involved. In regard to function, construction equipment may either cut and carry the soil or cut, spread, and push the spoil. In regard to type of earth-moving action, it may be either continuous or intermittent. Several types of machines are illustrated in Fig. 13.7.

Continuous-action machines generally have a higher output than intermittent-motion equipment. Examples of continuous-action machines are the wheel and template excavators. The wheel excavator is similar to the wheel-type trenching machine (Fig. 15.1), except that the resulting side slopes are less than vertical. The template excavator digs a somewhat larger channel than the wheel-type machine, but both types excavate channels whose cross sections cannot be varied except by modifying the equipment. Both machines construct the channel as they move along the ditch line. Plow-type ditchers are similar in construction and operation to an ordinary lister, but the ditcher is a much larger machine. Blade and elevating graders, though not precisely continuous in action, are suitable only in soils where sufficient traction can be obtained in the bottom of the ditch. Continuous-action machines for spreading the spoil include the bulldozer and the blade grader.

Intermittent-action machines are quite generally used for open channel construction (see Table 13.4.). For cleanout work, draglines are sometimes equipped with trapezoidal-shaped buckets, with wide tracks for straddling the channel, and with a special gear to provide continuous movement of the machine. The hoe is known by a variety of names, such as the back hoe, trench hoe, and drag hoe. In wet areas these machines may be mounted on floats. As land excavators, they may be equipped with rubber tires or with crawler-type

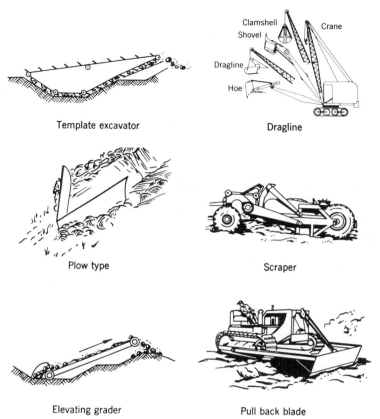

Fig. 13.7.   Open-channel excavation equipment.

tread. Since scrapers, shovels, and bulldozers require channel bottoms that are firm, they may not be suitable for drainage work. However, tractors with pull-back blades are satisfactory. The bulldozer is perhaps the most widely used machine for spreading soil banks, but scrapers may be satisfactory for small channels. Such machines as the dragline and clamshell, however, may deposit the spoil so that very little, if any, spreading is required.

13.18.   Factors Influencing the Selection of Equipment.   The shape and design dimensions of the channel are influenced by the type of equipment available. For maximum efficiency, machinery should be selected to fit the requirements of the individual job. In many instances the engineer is called upon to evaluate bids of contractors. Such bids may be rejected if the contractor does not have suitable equipment to do the work.

The three types of operations in earth-work construction are digging, hauling, and placing. A machine that can perform all three operations is desirable since such a machine usually provides the most economical construction. For example, a dragline can dig the soil, move it from the channel to the spoil bank, and place the spoil in the desired position.

The selection of equipment depends on such factors as moisture conditions, type of soil, degree of accuracy required, shape and dimensions of the channel and the spoil bank, moving requirements, volume of work, and financial considerations. For drainage ditch construction the selection depends largely on moisture conditions and to a degree on the nature of the soil.

13.19. Explosives. Blasting with dynamite may be satisfactory for constructing new ditches or cleaning out old ones in such areas as swamps and wet natural channels that are not readily accessible with earth-moving machinery. It is generally more economical and practical to use machinery than to use explosives. At best, blasting is hazardous work and should be attempted only by experienced and qualified persons.

13.20. Estimating Costs. In preparing open channel designs, the engineer should determine the economic feasibility of the proposed project. Such an analysis includes a comparsion of the costs with the expected benefits. Costs vary widely from section to section and from year to year. Costs for a drainage or irrigation district project include (1) construction costs, frequently determined by contractor bids; (2) right-of-way costs; (3) damage to land, roads, bridges, fences, and railroads; (4) engineering expenses; (5) attorneys' fees and other legal expenses; and (6) fees of commissioners, assessors, and other district officials. Construction expenses include charges for excavation, bridges, clearing the land of timber and brush, and for auxiliary structures. On an individual farm the expenses are primarily those of construction and engineering. Although it is difficult to estimate actual operating time on a percentage basis, many authorities on heavy earth-moving equipment estimate that the actual production time is as low as 30 to 50 percent of the available working time.

MAINTENANCE

Maintenance is a continuing problem and it may be required soon after construction is completed. Maintenance may be divided into two phases: preventive maintenance before failure, and corrective maintenance after partial or complete failure.

13.21. Causes for Deterioration of Open Channels. The failure of open channels usually results from one or more of three conditions: poor design, improper construction, and lack of adequate maintenance.

The major causes for deterioration of open channels are (1) sedimentation in the channel, (2) excessive growth of vegetation, (3) channel and bank erosion,

(4) high sediment load in the water, (5) poor location and alignment, (6) improper depth or width, (7) inadequate culvert and bridge capacity, (8) failure to provide the necessary legal arrangements to fix responsibility and to collect maintenance expenses for channels involving more than one landowner, and (9) general lack of interest in maintenance by the public, landowners, and district officials.

13.22. Preventive Maintenance. Although the best possible design may have been developed and the channel may have been constructed according to plan, maintenance measures are always necessary for proper functioning. The control of excessive tree, brush, grass, and weed growth along the channels is an important preventive maintenance measure. The effect of weed growth on drainage ditch capacity is shown from a survey by Ramser (1947) in Missouri, Arkansas, Mississippi, and Illinois. The study showed that ditch capacities were reduced as much as 75 percent of the original capacity. The primary methods of control consist of spraying, mowing, grubbing and clearing, grazing, underwater cutting, and burning. A number of chemical sprays have largely replaced more costly hand-grubbing and clearing methods. The rate of application, method of application, and effectiveness of chemical sprays are adequately covered in many other publications. Burning vegetation along channels must be accomplished when the material is dry; flame burners are suitable when the vegetation is green. Burning and spraying of ditches may be required several times during the growing season. Since grazing may be detrimental, particularly in channels with steep side slopes and in membrane-lined irrigation canals, this practice should be avoided.

Bank erosion resulting in channel sedimentation may be reduced by spreading and seeding the spoil bank. If the spoil banks are sufficiently flat, they may be farmed with the adjoining field. In organic soils where wind erosion causes serious filling of the ditches, permanent vegetation (grass) on the spoil banks is satisfactory. Proper land use combined with good conservation practices in the upper tributaries will greatly reduce maintenance of drainage ditches.

13.23. Corrective Maintenance. After the original installation, changes in cross section, grade, or alignment of the channel may be necessary for proper functioning of the system. An improper outlet or inadequate grade may cause sedimentation in the channel. Side slopes that cave and fill the ditch need to be reshaped to a more stable slope. Sediment bars or sharp curves causing meandering may necessitate straightening the channel or flattening the curves. Widening of the channel and enlarging of culverts and other obstructions may also be necessary. Where serious scouring occurs, the grade may be reduced and drop spillways constructed if more economical measures are not adequate.

Even though all possible steps have been taken to reduce sedimentation, cleanout work is often eventually necessary. In general, equipment used for constructing the original ditch is suitable.

13.24. Maintenance Costs. Because the amount of maintenance varies so widely with different conditions, it is difficult to estimate these costs. Preventive maintenance measures are generally cheaper than corrective maintenance measures, and timely maintenance is more economical than delayed maintenance. Because of the small volume of soil to be moved, cleanout work per unit volume may be three to five times the original excavation cost. Where maintenance has been badly neglected, it may cost more to reclaim an old channel than to construct a new one. Where several landowners are involved, responsibility for maintenance must be clearly established and funds provided or assessed against the land to assure that the work will be done.

REFERENCES

Chow, V. T. (1959). *Open-Channel Hydraulics.* McGraw-Hill, New York.

Fortier, S., and F. C. Scobey (1926). Permissible Canal Velocities." *Am. Soc. Civil Engrs. Trans.* **89**, 940–984.

Houk, L. E. (1956). *Irrigation Engineering,* vol. II. John Wiley, New York.

King, H. W., and E. F. Brater (1963). *Handbook of Hydraulics* (5th ed.). McGraw-Hill, New York.

Kraatz, D. B. (1971). *Irrigation Canal Lining.* Food and Agr. Org., United Nations, Rome.

Laflen, J. M., and R. P. Beasley (1960). "Effects of Compaction on Critical Tractive Forces in Cohesive Soils." *Missouri Agr. Expt. Sta. Res. Bull. 749.*

Lane, E. W. (1955). "Design of Stable Channels." *Am. Soc. Civil Eng. Trans.* **120**, 1234–1260.

Lauritzen, C. W. (1961). "Lining Irrigation Laterals and Farm Ditches." *U.S. Dept. Agr., Agr. Inform. Bull. 242.*

Linsley, R. K., and J. B. Franzini (1979). *Water Resources Engineering,* 3rd ed. McGraw-Hill, New York.

Smerdon, E. T., and R. P. Beasley (1961). "Critical Tractive Forces in Cohesive Soils." *Agr. Eng.* **42**, 26–29.

U.S. Bureau Reclamation (1963). *Linings for Irrigation Canals.* U.S. GPO, Washington, D.C.

——— (1962). "Studies of Tractive Forces of Cohesive Soils in Earth Canals." *Hydraulics Br. Report,* Hyd-504, Denver, Colorado.

U.S. Census: 1960 (1962). *Irrigation of Agricultural Lands,* vol. III. Bureau of the Census, U.S. Government Printing Office, Washington, D.C.

U.S. Department of Agriculture (1962). "Basic Statistics of the National Inventory of Soil and Water Conservation Needs." *Statist. Bull. 317,* August.

U.S. Soil Conservation Service (1973). "Drainage of Agricultural Lands." Water Information Center, Inc., Port Washington, New York.

PROBLEMS

13.1. Determine the design runoff for an open ditch to provide good drainage in your area for a flat watersheds of 40.5 ha (100 ac), and 2590 ha (10 mi^2).

13.2. Compute the most efficient bottom width for an open channel with a flow depth of 2.44 m (8 ft) in silt loam soil. What are the velocity and the channel capacity if the hydraulic gradient is 0.09 percent?

13.3. A farmer in your area desires to drain 259 ha (640 ac) of alluvial sandy loam land. A topographic survey indicates that the maximum slope is 0.19 percent (10 ft/mi). Assuming very good drainage, design the ditch cross section including the spoil bank.

13.4. Design an open ditch to carry the runoff from a 3626-ha (14-mi^2) watershed in your area. The slope of the land along the route of the ditch is 0.27 percent, the average slope of the watershed is 0.5 percent, and the soil is heavy clay. Assume that an excellent degree of drainage is desired. The ditch should provide sufficient depth for pipe drains.

13.5. Determine the deflection angles for the layout of an 8-degree curve if the ditch makes a 45-degree change in direction (see Appendix G). The point of curvature is at Station 1+07 m (3+50 ft).

13.6. What is the design capacity of an irrigation canal with a total depth of 1.52 m (5.0 ft) (including freeboard), bottom width 1.22 m (4.0 ft), 2:1 side slopes, and a hydraulic gradient of 0.09 percent? Assume $n = 0.04$ and the freeboard is 20 percent.

13.7. If the critical tractive force on the bottom of a channel in cohesive soil is 0.98 kg/m^2 (0.2 psf), what is the maximum average velocity of flow at maximum slope for a trapezoidal channel where the depth of flow is 1.22 m (4 ft), bottom width 1.52 m (5 ft), $n = 0.03$, and side slopes 1:1?

CHAPTER 14

Subsurface Drainage Design

The design of a subsurface drainage system includes the layout and arrangement of the drain lines, selection of a suitable outlet, proper depth and spacing of laterals, determination of the length and size of drains, selection of good quality materials of adequate strength, and the design of such accessories as surface inlets and outlet structures.

14.1. Benefits of Subsurface Drainage. Subsurface drainage is an important conservation practice. Wet lands are usually topographically situated so that when drained they may be farmed with little or no erosion hazard. Sloping land may thus be used less intensively to achieve the same over-all production of intertilled crops. Many soils having poor natural drainage are, when properly drained, rated among the most productive soils in the world.

Specific benefits of subsurface drainage are (1) aeration of the soil for maximum development of plant roots and desirable soil microorganisms, (2) increased length of growing season because of earlier possible planting dates, (3) decreased possibility of adversely affecting soil tilth through tillage at excessive moisture levels, (4) improvement of soil moisture conditions in relation to the operation of tillage, planting, and harvesting machines, (5) removal of toxic substances, such as salts, that in some soils retard plant growth, and (6) greater storage capacity for water, resulting in less runoff and a lower initial water table following rains. Through these benefits drainage enhances farm productivity by (1) adding productive acres without extending farm boundaries, (2) increasing yield and quality of crops, (3) permitting good soil management, (4) assuring that crops may be planted and harvested at optimum dates, and (5) eliminating inefficient machine operation caused by wet areas in fields. In arid regions irrigation and drainage are complementary practices. In some areas leaching of soluble salts through a drainage system is essential before the land can be developed. Drainage is often a necessity as a result of excess water that accumulates from low efficiencies in the conveyance and application of water for irrigation. Although these losses can be reduced, they cannot be entirely eliminated. The benefits of drainage can be realized only when the soil is potentially productive if drained. In many areas it may be desirable to leave soil undrained and utilize the land as range land or as a recreation and wildlife area.

The benefits and costs from drainage in a humid area are estimated from measured corn yields in Table 14.1. Subsurface drainage gave more than twice the increase in yields compared to surface drainage. However, surface drainage resulted in a higher benefit-cost ratio because of much lower investment cost than for subsurface drainage. Costs were based on 1975 prices, and they will vary greatly from field to field, especially for surface drainage. Costs also vary from region to region.

14.2. Drainage Requirements and Plant Growth. Excess water affects plant growth by reducing aeration, for plants are not adversely affected even in total water culture if air is provided. Although for design purposes the water table is a convenient term for reference, it is not satisfactory for explaining the water relationships with respect to plant root environment. Because of the capillary fringe above the water table, saturation may occur for some distance above it, especially in a clay soil. A better criterion for drainage depth would be the "upper surface of saturation," or a depth based on the oxygen diffusion rate of the soil. Several investigators have found that roots do not generally penetrate deeper than approximately 30 cm above a static water table. Soils that have a relatively small volume of noncapillary pore space may provide an unfavorable environment for root growth even though not saturated.

Drainage requirements for optimum plant growth are determined largely by the volume and content of the soil air. These needs depend on the type of crop, the soil, availability of plant nutrients, climatic conditions, biological activity, and soil and crop management practices. The rooting depth and the tolerance to excess water are the most important crop characteristics. For example, rice is tolerant because it has special internal mechanisms for obtaining oxygen from the air; whereas, tobacco is sensitive to excess water. Plants have widely varying oxygen requirements, sometimes even greater than the concentration in the atmosphere.

As shown in Fig. 14.1, a sharp reduction in yield with an increase or decrease in water table depth from an optimum of 0.3 m may be expected for a peat soil. As the soil texture becomes heavier, maximum yields are obtained at greater water table depths.

The relationship of drainage and nitrogen availability is illustrated in Fig. 14.2. A major effect of drainage is the increased nitrogen supply that can be obtained from the soil. These amounts, which are estimated by projecting the dashed curves to the abscissa, are about 56 and 157 kg/ha of N at water table depths of 0.4 and 1.5 m, respectively. The practical implication is that N applications increase crop yields more at high than at low water tables. With a shallow water table nitrogen gives a good response because the crop cannot obtain it from the soil. The data in Fig. 14.2 were obtained in The Netherlands where rainfall rates are low and a constant water table can be maintained. Be-

Table 14.1   Drainage Benefits and Costs for Corn in a
Silty Clay Soil in Northern Ohio

	Drainage System		
	Surface only	*Subsurface only*	*Surface plus Subsurface*
Corn Yields (kg/ha) (11-yr avg.)	5580	7090	8030
Increased Yields (undrained 4400 kg/ha)	1180	2690	3630
Benefits at $0.10/kg ($2.50/bu)	$118	$269	$363
Costs/ha[a]	$25	$74	$99
Benefit—Cost Ratio	4.7	3.6	3.7
Benefits less Costs	$93	$195	$264

[a] Estimates do not include fertilizer and other production costs as these would be similar for all drainage systems. Costs were computed assuming 2 percent depreciation (50-yr life) per year, 8 percent interest on average investment, and $1.00/ha per year for maintenance for subsurface drains; and 8 percent annual interest and $5.00/ha per year for maintenance of surface drainage, only. Initial investment was $247/ha for surface drains and $1210/ha for subsurface drains (12 m spacing, 1 m deep).

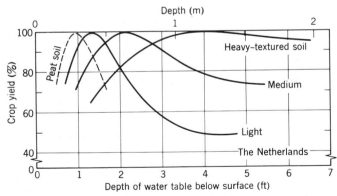

Fig. 14.1. General relationship between crop yield and constant water table depth during the growing season in the Netherlands. (Redrawn from Visser, 1959.)

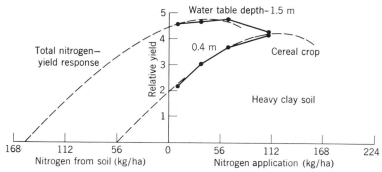

Fig. 14.2. Influence of water table depth on nitrogen supplied by the soil. (Redrawn from Van Hoorn, 1958.)

cause a constant water table prevents root development below it, these results may not be applicable where widely fluctuating water table conditions prevail, as in the humid area of the United States.

The effect of water table depth on relative rye yields is given in Fig. 14.3. The curves show lines of equal yield for conditions where the crop uses water from the water table. Yields for a constant water table down to a depth of about 75 cm increase and then decrease rapidly below 100 cm. This decrease in yield at the lower depths is not typical of heavy soils. At point *B* the average water table depth is 85 cm and the deviation 15 cm, thus resulting in a winter depth of 70 cm and a summer depth of 100 cm. With the average water table depth less than 125 cm, yields decreased as fluctuations in the water table increased; however, when the average water table was greater than 125 cm fluctuations generally caused an increase in yield.

Soil and air temperatures affect oxygen diffusion rates as well as soil organisms and the biological processes in the plant. Increasing the temperature usually causes a decrease in oxygen and an increase in carbon dioxide concentration because of the increase in respiration rates of roots and organisms. Kramer and Jackson (1954) found that tobacco plants grown at a water temperature of 28°C (68°F) recovered from flooding if drained within one day, while plants grown at 34°C (93°F) died under the same conditions. Normally, during the dormant season plants are not affected by flooding because at low temperatures the biological processes are slow. Drainage influences soil temperature because of the differences between wet and dry soil in specific heat, thermal conductivity, and evaporation. The specific heat of soil particles is about 0.2 for most dry soils as compared to 1.0 for water. The thermal conductivity of dry soil is one third to one half that of water. Evaporation from the surface requires solar energy that would otherwise be available to warm the soil. However, in the field, only differences of a few degrees in surface soil temperatures have

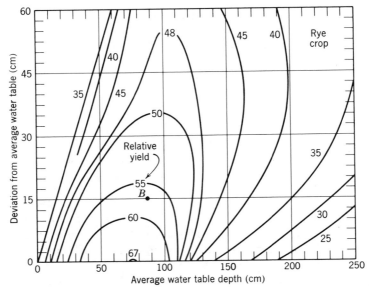

Fig. 14.3. Relative rye crop yields at varying water depths for a sandy soil in the Netherlands. (Redrawn from Wesseling et al., 1957; original by Bloemen.)

been measured between points directly over a tile line and midway between lines where the poorest drainage occurs. This small difference is probably due to the large mass of the earth and equalization of temperatures by conduction.

Flooding of plant roots causes a rapid reduction in transpiration, reduced absorption of oxygen and other plant nutrients with a corresponding increase in the carbon dioxide content of the soil, a disturbance of microbiological activity, and a reduced mass of soil from which nutrients can be obtained. Kramer (1949) found that excess carbon dioxide reduced absorption much sooner than a deficiency of oxygen. Death of the plant may be due to toxic substances moving up from the roots, but the effect is complex and not fully understood. When aeration is reduced, iron and manganese may become high enough in concentration to be toxic to the roots. Decomposition of organic matter may produce hydrogen sulfide which is also toxic. Changes in the oxygen and carbon dioxide balance may affect the growth of disease organisms. With prolonged flooding and reduced biological activity, soil structure may be destroyed, which in turn reduces aeration. In heavier textured soils wetting and drying can have a beneficial effect. In sandy and organic soils, such as peat and muck, a high water table is less critical than in heavy-textured soils.

The relationship of subsurface drainage to root development is illustrated in Fig. 14.4. A high water table in the spring inhibits root development, leaving a plant with an inadequate root system during the summer months, but a low water table in the spring allows maximum root development.

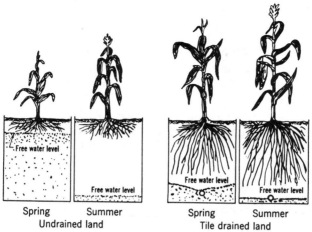

Fig. 14.4. Root development of crops grown on drained and undrained land. (Redrawn from Manson and Rost, 1951.)

14.3. Pipe Drains. These drains include concrete and burned clay tile, corrugated plastic tubing, or other perforated conduit. Corrugated steel pipe with a high structural strength is suitable to withstand high soil loads, to cross unstable soils that require the rigidity of a long pipe, and to provide a stable outlet into open ditches.

Tile for lateral drains (drains intended to receive water directly from the soil as contrasted with main lines which received most of their flow from laterals) are usually 0.3-m lengths, 75 to 150 mm inside diameter, and have a wall thickness of about 1/12 their diameter. In peat and muck soils and in irrigated regions 0.6- to 1.0-m lengths and tile with fitted ends similar to sewer pipe are common. The tile are laid end to end in the bottom of a trench that is then backfilled. Some tile are also perforated. Water enters the tile line through perforations and cracks between the ends of adjoining tile.

Corrugated plastic tubing (CPT) in the U.S. was first commercially produced in 1967, and by 1973 it had largely replaced concrete and clay tile for the smaller-size drain laterals. High density polyethylene (HDPE) tubing is widely used in the U.S., while polyvinylchloride (PVC) is more prevalent in Europe. Corrugated tubing is light in weight (about 1/25 of concrete or clay tile), durable, resistant to soil chemicals, extruded in long lengths, and easy to join and handle in the field. It is especially suitable for installation with a mole plow, and less labor is required than for tile. CPT is subject to damage by rodents and its hydraulic roughness is higher than tile. It will float in water and tend to stay curved as in the shipping coil. CPT is perforated or slotted with three or more rows of openings and may be made with a fabricated porous covering to prevent inflow of fine particles in sandy soils. One type is arch shaped with a perforated, flexible flat bottom permitting it to be rolled in a coil partially collapsed.

14.4. Mole Drains. Mole drains as shown in Fig. 14.5 are cylindrical channels artificially produced in the subsoil without digging a trench from the surface. They are similar to pipe drains except that they are not lined with tile or other stabilizing material.

Moling is a temporary method of drainage. Where soil conditions are suitable, moles function efficiently for the first few years and then gradually deteriorate. Mole drains have been successful in England, New Zealand, and several European countries. Their maximum life is 10 to 30 years. Except in some soils in Louisiana, Florida, and California, mole drainage is generally not practiced in the U.S.

Mole drains fail principally because the soil is not sufficiently stable to maintain a channel. Although high clay soils are generally the most suitable, the clay content is not necessarily a good index for moling. In Iowa a study of three soils showed that the greater the clay content the more rapid the failure of the channel. The moisture content at the time of moling should be as high as possible, provided the soil can support the tractor. Mole channels usually range in diameter from 60 to 100 mm. The small sizes are more stable than larger ones.

Where mole drains are suitable, they are generally pulled across and over pipe drains, using the pipe drains as outlets. Moles may be directly connected to the pipe, or gravel may be placed in the pipe trench. The depth varies from 0.5 to 1.2 m depending on the depth of a stable soil layer. Spacings will range from 1 to 10 m. Length of drain is usually less than 500 m depending on the grade, which may range from nearly level to 5 percent.

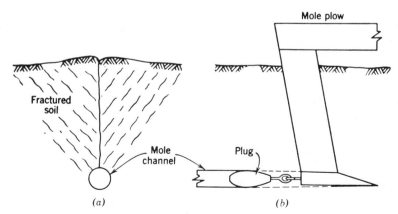

Fig. 14.5.   Mole drainage: (a) Cross section of a mole channel. (b) Method of forming a mole drain.

TYPES OF SYSTEMS

Tile drainage systems in general use are the natural, herringbone, gridiron, and interceptor types shown in Fig. 14.6. Combinations of two or more of these types are frequently required for the complete drainage of an area.

14.5. Natural or Random. The natural or random system is widely adopted in fields that do not require complete drainage with equally spaced laterals. This system is quite flexible as well as economical since the drain lines follow natural draws or other low depressions. Such a system is particularly adapted to the drainage of small or isolated wet areas.

14.6. Herringbone. The herringbone system is adapted to areas that have a concave surface or a narrow draw with the land sloping to it from either direction. The main line is laid out nearly normal to the slope and follows the low area. Despite the large amount of double drainage (land drained both by the

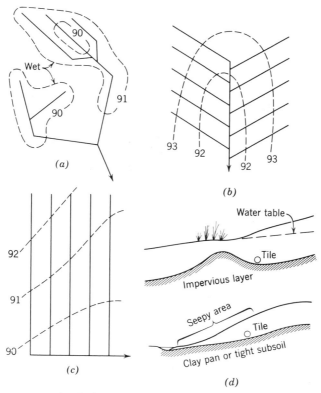

Fig. 14.6. Common types of pipe drainage systems. (a) Natural or random. (b) Herringbone. (c) Gridiron. (d) Cutoff or interceptor.

laterals and the main or submain), the herringbone system is particularly suitable where the laterals are long and the waterway requires thorough drainage.

14.7. Gridiron. The gridiron system is similar to the herringbone system except that the laterals enter the main from only one side. The gridiron system is the most common. The gridiron pattern is more economical than the herringbone system because the number of junctions and the double-drained area are reduced. Where the waterway is of considerable width, a main is placed on both sides of the waterway. This system, known as the double main system, is essentially two separate gridiron patterns. Since the mains need not cross the waterway at critical points, serious erosion problems may be prevented. Although right-angle junctions are shown in Fig. 14.6c, main and laterals may intersect at angles less than 90 degrees.

14.8. Cutoff or Interceptor. The cutoff or interceptor drain is normally placed near the upper edge of a wet area as shown in Fig. 14.6d. The usual cause for such wet conditions is the outcropping of impermeable strata on or near the surface. Frequently, this condition exists along waterways. To drain such areas the interceptor drain is frequently installed on both sides of the waterway.

OUTLETS

14.9. Types of Outlets. The two principal types of outlets for pipe drains are gravity and pump. Pump outlets (see Chapter 16) may be considered where the water level at the outlet is higher than the bottom of the pipe outlet for any extended period of time.

Gravity outlets, by far the most common, include other pipe drains, constructed waterways, natural channels, or wells. Outlet ditches should have sufficient capacity to carry surface runoff and drain flow. Where the drainage system is connected to other pipe drains, the outlet should have sufficient capacity to carry the additional discharge. As shown in Fig. 14.7, corrugated metal or other rigid pipe is recommended. Some type of grill or flap gate over the end is desirable to prevent entry of rodents. If there is danger of flood water backing up into the drain, an automatic flood or tide gate may be installed in place of the flap gate. The end of the outlet pipe should be 0.3 m or more above the normal water level in the ditch. To prevent damage due to high velocities in the ditch or failure from snow loads, the exposed end of the pipe should not extend beyond the bank more than one third its total length. The minimum total length should be 5 m, and the diameter should be the same or larger than the drain pipe size. In connecting the metal pipe to the drain, a concrete collar may be installed. Where available, drop inlets and other permanent structures described in Chapter 9 are suitable for stabilizing the outlet.

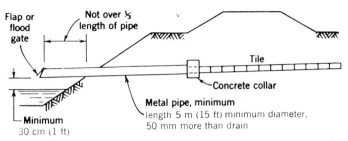

Fig. 14.7. A suitable gravity outlet for pipe drains.

Vertical drainage outlets (wells) are essentially wells extending into a porous soil layer or open rock formation in the lower horizons. The existence of a substratum that can continually take in large quantities of water and that can be reached without prohibitive cost is the exception rather than the rule. Since there is no positive method for locating or for predicting the permanent capacity of such outlets, their use involves considerable risk. Also, if drain outflow is contaminated with sediment, sewage, or dissolved salts, there is danger of polluting the ground water. For these reasons, vertical outlets are not generally recommended.

14.10. Requirements of a Good Outlet. The importance of a good outlet is indicated by the fact that a high percentage of failures of drainage systems is due to faulty outlets. The requirements of a good outlet are to (1) provide a free outlet with minimum maintenance, (2) discharge the outflow without serious erosion or damage to the pipe, (3) keep out rodents and other small animals, (4) protect the end of the line against damage from the tramping of livestock as well as excessive freezing and thawing, and (5) prevent the entrance of flood water where the outlet is submerged for several hours.

DEPTH AND SPACING

A definite relationship exists between depth and spacing of drains. For soils of uniform permeability, the deeper the drains the wider the spacing; hence fewer drains are required and the cost is lower. The primary consideration in drainage design is to provide adequate root depth above the saturated zone. Because of inadequate data on crop requirements, the discussion which follows in regard to depth and spacing is based mostly on empirical data and field experience.

14.11. Depth. The depth of pipe drains should be such as to provide the desired water table depth midway between drain lines. Pipe depth is affected by soil permeability, outlet depth, spacing of laterals, depth to the impermeable layer in the subsoil, and limitations of trenching equipment. Pipe depth as here considered applies only to laterals and not to the depth of mains, since the depth of the main is governed primarily by outlet conditions and topography.

Pipe depth, defined as the distance from the surface to the bottom of the pipe, varies in different soils. Under no conditions should the amount of cover over the top of the pipe be less than 60 cm. This minimum is necessary to protect the pipe from heavy surface loads and to prevent shifting of the tile. In uniformly permeable mineral soils the depth of laterals usually varies from 80 to 250 cm. Unless limited by an impermeable layer, one should design for the maximum depth as this will permit a wide spacing. In deep organic soils after initial settlement has taken place, the minimum depth should not be less than 120 cm.

Where the subsoil is relatively impermeable, the pipe should be placed on or above the impermeable layer. If pipe must be placed below the impermeable layer, the trench should be backfilled with permeable soil.

In humid regions where the water table will rise to near the surface during heavy rainfall, the rate of drop is the important factor. However, in organic soils the water table may be maintained at a nearly uniform depth, which may be above the pipe. A few empirical criteria for drainage depths are presented in Table 14.2.

In arid regions under irrigation the drainage design criteria are determined more by the minimum depth of the water table for optimum crop growth than by the rate of drop. Depths of 200 to 300 cm are common. Water table depths approved by irrigation authorities and by financial institutions as a basis for long-time loans for improving irrigated land (Israelsen and Hansen, 1962) are as follows:

Classification	Range in Water Table Depth
Good	Static water table below 210 cm; up to 180 cm for about 30 days per year
Fair	Water table at 180 cm; up to about 120 cm for 30 days; no general rise
Poor	Some alkali on surface; water table 120 to 180 cm; up to 90 cm for 30 days
Bad	Water table less than 120 cm and rising

Selection of salt-tolerant crops, careful application of water, and proper soil management may permit successful cropping with shallower depths than indicated.

14.12. Spacing. In humid regions most drain lines are spaced 10 to 50 m apart and up to 90 m in very permeable soils. Where high-value crops are grown or under special conditions, spacings of 10 to 15 m are sometimes necessary. In irrigated areas of the West spacings ranging from 50 to 200 m are possible.

Table 14.2 Drainage Depth Requirements for Humid Areas

Crops	Depth and Rate of Drop of the Water Table	Location	References
Mineral Soils			
Field	Initial depth, 15 cm minimum; 30 cm/d through second 15 cm; 20 cm/d through third 15 cm	Minn.	Neal (1934)
Field	Drop from surface to 30 cm in 24 hr and 50 cm in 48 hr	Ill.	Kidder and Lytle (1949)
Field	Drop 20 cm/d	Va.	Walker (1952)
Grass	Constant depth 50 cm or less	Netherlands	Wesseling et al. (1957)
Arable	Constant depth 90 to >130 cm	Netherlands	Wesseling et al. (1957)
Organic Soils (controlled drainage)			
Grasses	Maximum depth, 50 cm	U.S.	U.S. SCS (1973)
Vegetables (shallow rooted)	Maximum depth, 60 cm	U.S.	U.S. SCS (1973)
Field (deep rooted)	Maximum depth, 80 cm	U.S.	U.S. SCS (1973)
Cereals, short grass, sugar beets	Optimum depth, 80–90 cm	England	Nicholson and Firth (1953)
Truck crops, grass, and sugar cane	Optimum depth, 30–60 cm	Fla.	Clayton, Neller, and Allison (1942)

14.13. Depth and Spacing Design Criteria. In general, the depth and spacing of drains vary largely with soil permeability, crop and soil management practices, kind of crop, and the extent of surface drainage. With good crop and soil management practices, the depth and spacing for normal conditions vary within the limits given in Table 14.3.

Many depth and spacing formulas have been proposed which are approximate mathematical solutions. With all of these the procedure is briefly: (1) determine a drainage rate or a water table height, (2) estimate or measure the hydraulic conductivity and other required soil characteristics, (3) select a suitable depth for the drains, and then (4) compute the spacing.

Steady State. For conditions where the rainfall rate or irrigation rate is constant, the drain discharge will also be constant if the water table does not change with time. The ellipse equation, which was one of the first mathematical solutions for drain spacing, will be derived by using the symbols and geometry shown in Fig. 14.8. Assuming that a constant rate of rainfall is removed equally well at all distances from the drain,

$$q_x = \left(\frac{S/2 - x}{S/2}\right)\frac{q}{2} \tag{14.1}$$

where q_x = rate of flow across a vertical plane at any $x(L^3/T)$,
   S = spacing between drains (L),
   q = total flow rate into a drain from both directions per unit length of drain (L^3/T).

From Darcy's law and the Dupuit assumption that the velocity is proportional to the water table slope,

$$q_x = -yv_x = Ky\left(\frac{dy}{dx}\right) \tag{14.2}$$

where v_x = velocity at x (L/T),
   K = hydraulic conductivity (L/T).

By equating Eqs. 14.1 and 14.2, the differential equation is

$$ydy = \left(\frac{q}{SK}\right)\left(\frac{S}{2} - x\right)dx \tag{14.3}$$

Integrating from $x = 0$ and $y = d$ to $x = x$ and $y = y$, where d is the height of the water table in the drain above the impervious layer, gives

$$y^2 - d^2 = \frac{q}{SK}(Sx - x^2) \tag{14.4}$$

Table 14.3   Average Depth and Spacing for Pipe Drains

Soil	Hydraulic class	Conductivity (mm/h)	Spacing (m)	Depth (cm)
Clay	Very slow	1	9–15	90–110
Clay loam	Slow	1–5	12–21	90–110
Avg. loam	Mod. slow	5–20	18–30	110–120
Fine sandy loam	Mod.	20–65	30–37	120–140
Sandy loam	Mod. rapid	65–130	30–60	120–150
Peat and muck	Rapid	130–250	30–90	120–150
Irrigated soils	Variable	25–25 000	45–180	150–240

Note: 25 mm/h = 1 iph; 1 m = 3.3 ft.

which is the equation of an ellipse. Substituting $x = S/2$ and $y = b$ for the midpoint, yields

$$S = \frac{4K(b^2 - d^2)}{q} = \left[\frac{4K(b^2 - d_e^2)}{i} \right]^{1/2} \tag{14.5}$$

The flow, q, may also be expressed as Si, where i is the rainfall or excess irrigation rate in m per day. The basic ellipse equation was first developed by Colding in 1872. In the Imperial Valley of California Eq. 14.5 is the recommended spacing equation.

The principal limitation of the ellipse equation is that the resistance due to convergence of flow lines near the drains is ignored. Thus, it is suitable only where the spacing is large compared to the depth to the impervious layer. The accuracy of the spacing can be improved by substituting the equivalent depth, d_e, given in Fig. 14.9 for d in the equation. The depth, b, is equal to $(d_e + m)$. Moody (1966) has developed equations for d_e for other drain diameters.

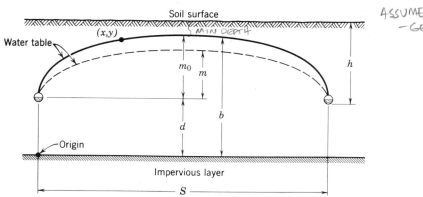

Fig. 14.8.   Symbols and geometry for drain spacing equations.

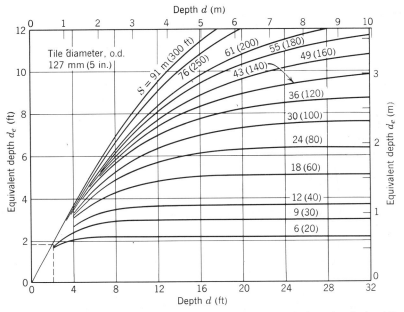

Fig. 14.9. Equivalent depth for the water conducting layer below the drain. (Compiled by van Schilfgaarde, 1963; original data from Hooghoudt.)

Example 14.1. For an irrigated area compute the drain spacing assuming a depth to the center of the drain as 1.8 m (6 ft) with the minimum depth to the water table of 1.5 m (5 ft). Subsurface explorations indicate a hydraulic conductivity 0.5 m/d (1.6 fpd) above an impervious layer at a depth of 6.7 m (22 ft). The excess irrigation rate is equivalent to a drainage coefficient of 1.2 mm/d (0.048 ipd).

Solution.   Assume a spacing of 55 m (180 ft) and for $d = 6.7 - 1.8 = 4.9$ m (16 ft), read from Fig. 14.9 $d_e = 2.9$ m (9.5 ft). With tile flowing half full as in Fig. 14.8, $m = 1.8 - 1.5 = 0.3$ m (1.0 ft). Therefore, $b = 2.9 + 0.3 = 3.2$ m (10.5 ft). Substituting in Eq. 14.5,

$$S = \left[\frac{4 \times 0.5(3.2^2 - 2.9^2)}{0.0012} \right]^{1/2} = 55.2 \text{ m (181 ft)}$$

Since 55.2 m (181 ft) is close to the assumed spacing of 55 m (180 ft), the solution is satisfactory. If the computed spacing is not close to the assumed value, assume a new spacing, which will change d_e, and recompute the spacing. Continue until the assumed and computed spacings are in close agreement.

Nonsteady State.   Mathematical analysis of the falling water table case is far more difficult than for the steady state. The Dupuit–Forchheimer theory, which assumes horizontal flow with the velocity proportional to the slope of the free water surface, has been applied to the nonsteady state by adjusting the continuity equation. This equation after appropriate substitutions for the soil porosity f and time t may be written

$$\frac{\partial^2 y}{\partial x^2} = \frac{f}{K(d + m_0/2)} \frac{\partial y}{\partial t} \tag{14.6}$$

It is often referred to as the heat flow equation since it is identical in form to the differential equation for solving heat flow problems. By assigning appropriate boundary conditions Eq. 14.6 can be solved in terms of a sine series, but only the first term is retained. The spacing equation as derived by Glover and reported by Dumm (1954) using symbols in Fig. 14.8 is

$$S = \pi \left[\frac{Kt(d + m_0/2)}{f \ln(4/\pi)(m_0/m)} \right]^{1/2} \tag{14.7}$$

Because of the assumptions made in its derivation, Eq. 14.7 also ignores the resistance due to the convergence of the flow lines near the drain and is based on a thickness of the water-conducting zone equal to $(d + m_0/2)$. Van Schilfgaarde (1963) modified the Glover equation to take these factors into consideration. This equation is

$$S = \left[\frac{9Ktd_e}{f\left[\ln m_0(2d_e + m) - \ln m(2d_e + m_0) \right]} \right]^{1/2} \tag{14.8}$$

where S = drain spacing (L),
K = soil hydraulic conductivity (L/T),
d_e = equivalent depth from Fig. 14.9 (L),
m = height of water table above the center of the drain at midplane after time, t (L),
m_0 = initial height of water table, (L),
t = time in days for water table to drop from m_0 to m,
f = drainable porosity of the water conducting soil expressed as a fraction (voids drained at 60 cm tension).

This equation is recommended for design purposes (see Example 14.2). It does not yield a practical solution where $d_e = 0$. For this case another form of the equation is available.

The soil hydraulic conductivity K should be the effective value for the water conducting zone, particularly for heterogeneous soils that may have horizontal strata of varying permeability and for soils with different permeability in the horizontal (K_h) and vertical (K_v) directions. In such soils $K = (K_h K_v)^{1/2}$. Auger hole, piezometer, core sample, and other methods have been proposed for measuring soil hydraulic conductivity, but these are time-consuming and costly. Skaggs (1975) developed a procedure that permits evaluation of the ratio K/f, where f is the drainable porosity. The water table drawdown may be taken from a single drain or from two adjacent drains. With this procedure neither K nor f need be determined individually, as only the ratio is required in the equation.

Example 14.2. Calculate the drain spacing for a falling water table assuming $K = 0.3$ m/d (1 fpd), $f = 0.02$, drain depth to center line is 0.9 m (3 ft), $m_0 = 0.9$ m (3.0 ft), $m = 0.6$ m (2.0 ft), $t = 1$ day for the water to drop from the soil surface to 0.3 m (1.0 ft) below, and $d = 0.6$ m (2.0 ft) depth of impervious layer below drains.

Solution. For purposes of evaluating d_e, assume $S = 20$ m (65 ft). From Fig. 14.9 read $d_e = 0.6$ m (1.9 ft). Substituting in Eq. 14.8,

$$S = \left\{ \frac{9 \times 0.3 \times 1 \times 0.6}{0.02 \left[\ln 0.9 (2 \times 0.6 + 0.6) - \ln 0.6 (2 \times 0.6 + 0.9) \right]} \right\}^{1/2} = 18 \text{ m (59 ft)}$$

Since 18 m is close to the estimated spacing of 20 m, d_e would not change and the spacing of 18 m is correct. Since the water table initially was at the soil surface, the drain depth h for a 100-mm (4-in.) drain is $(0.9 + 0.06) = 0.96$ m (3.2 ft), the depth to the bottom of the drain. The extra depth 0.06 m (0.2 ft) is one half the outside diameter of the 100-mm pipe.

Drain spacing is influenced by the amount of surface water to be removed by subsurface drains, although spacing equations do not take this factor into consideration. Computed spacings (Table 14.4) from experimental plots in northern Ohio with and without surface drainage demonstrate this effect. For a given rate of drawdown, spacings could be increased nearly 70 percent where surface drainage is provided. On plots having both surface and subsurface drainage

Table 14.4 Drain Spacings With and Without Surface Drainage

	Drain Spacing (m)	
Water Table Drawdown (mm/d)	With Surface Drainage	Without Surface Drainage
100	22	13
200	14	8
300	10	6

Note: Spacings computed from drain flow measurements for silty clay soil from drains with 12-m spacing, 0.9-m depth and initial water table near the surface.
Source: Hoffman and Schwab (1964).

about half of the flow from 10-year return period rainfall was removed in surface runoff and half from the tile.

14.14. Interceptor Drain Location. Typical interceptor drains were shown in Fig. 14.6d. In irrigated regions seepage from a canal as shown in Fig. 14.10 is often the cause of drainage problems. For this condition, the shape of the water table and the flow into the drain can be predicted from an equation and design curves reported by Keller and Robinson (1961). Schmid and Luthin (1964) have solved the problem of spacing of parallel ditches under steady state conditions. Specific problems of this type have been solved with digital computers and electric analogues. Depth and spacing of drains on sloping land is a complex problem and beyond the scope of this text.

SIZE OF PIPE DRAINS

14.15. Design Flow. The design flow for pipe drains is based on entirely different criteria for humid and for irrigated conditions. In either case the "drainage coefficient" is a convenient term for expressing the flow rate. It is defined as the depth of water to be removed from the drainage area in 24 hours. **Humid Conditions.** Normally, the drainage area is computed from the length and spacing of the drains. Where surface inlets are installed, the contributing watershed is the drainage area rather than the pipe-drained area. For example, if a surface inlet is located in a pothole and if the contributing watershed area is 5 ha, the outlet should be designed using the appropriate drainage coefficient for surface inlets with a drainage area of 5 ha. However, if no surface inlets are installed, the outlet drain should be designed using only the area actually drained by the pipe and the normal drainage coefficient as indicated in Table 14.5. The area drained by the pipe and the contributing watershed represent minimum and maximum areas, respectively. Where sidehill seepage occurs and surface inlets are not provided, the drainage area may be adjusted accordingly, but it would be somewhere between the minimum and maximum values. Another alternative would be to modify the drainage coefficient to provide for such unusual conditions.

In humid areas the drainage coefficient depends largely on rainfall. It is difficult to correlate rainfall with the drainage coefficient since the distribution of rainfall during the growing season and its intensity must be considered along with evaporation and other losses. For example, where the ground is dry, an intense storm of short duration produces rapid surface runoff and little infiltration. However, rains of low intensity over a long period of time may produce high rates of outflow from the drains.

The selection of a drainage coefficient is based primarily on experience and judgment. Recommended drainage coefficients are shown in Table 14.5. The drainage coefficient should be such as to remove excess water rapidly enough

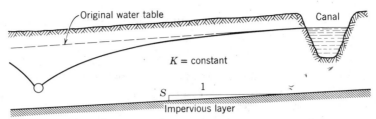

Fig. 14.10. An interceptor drain below a canal.

to prevent serious damage to the crop. In many instances removal of surface water necessitates doubling or increasing the drainage coefficient. Where the underlying stratum is sand or other porous material normal coefficients may be reduced. Where the design rate of drop of the water table is known, the drainage coefficient should be computed from the drainable porosity. For example, for 0.3 m drop per day of the water table and 3 percent porosity, the drainage coefficient is 9 mm/d.

Irrigated Conditions.   In irrigated areas the discharge from drain lines may be expected to vary from about 10 to 50 percent of the water applied. In these regions the drainage coefficient may vary with the size of the area contributing to the flow. Since not all of the area is irrigated at the same time, the design drainage area is not the same as the entire area but is estimated from the area being irrigated. Seepage water should also be considered in selecting the drainage coefficient, or it may be included by modifying the drainage area.

Table 14.5   Drainage Coefficients for Pipe Drains in Humid Regions

Crops and Degree of Surface Drainage	Drainage Coefficient			
	Mineral Soil[a] (clay and silt)		Organic Soil	
	(mm/d)	*(ipd)*	*(mm/d)*	*(ipd)*
Field Crops				
Normal[b]	10–13	⅜–½	13–19	½–¾
With blind inlets	10–19	⅜–¾	13–25	½–1
With surface inlets	19–25	¾–1	19–38	¾–1½
Truck Crops				
Normal[b]	13–19	½–¾	19–38	¾–1½
With blind inlets	13–25	½–1	19–51	¾–2
With surface inlets	25–38	1–1½	51–102	2–4

[a] These values may vary depending on special soil and crop conditions. Where available, local recommendations should be followed.
[b] Adequate surface drainage with outlets to other drains or ditches to be provided.
Source: U.S. SCS (1973).

The U.S. SCS (1973) reported that from surveys throughout the western states tile discharge varied from about 0.2 to 70 L/s per 100 m of tile with an average of about 9.3 L/s. Because of such wide variation, local recommendations must be developed. The following field procedure is a common method for determining the design flow. Make an estimate of deep percolation from irrigation, canal or ditch loss, and losses from other sources as a percentage of the amount of irrigation. From this estimate the design flow is computed. Pillsbury et al. (1965) developed the following discharge equation for the Coachella Valley of California:

$$q = 1.56 \, A^{0.75} \qquad\qquad (14.9)$$

where q = maximum flow in L/s,
        A = drained area in hectares.

14.16. Grades. Maximum grades are limiting only where pipes are designed for near maximum capacity or where pipes are embedded in unstable soil. Tile embedded in fine sand or other unstable material may become undermined and settle out of alignment unless special care is taken to provide joints that fit snugly against one another. Under extreme conditions it may be necessary to install bell and spigot tile, tongue and groove concrete tile, metal pipe, or plastic tubing. On mains steep grades up to 2 or 3 percent are not objectionable, provided the capacity at all points nearer the outlet is equal to or greater than the drain above.

A desirable minimum working grade is 0.2 percent. Where sufficient slope is not available, the grade may be reduced to that indicated in Table 14.6. In some soils these minimum grades should be higher to provide sufficient velocity to remove sediment from the drain. Minimum grades are sometimes based on a minimum velocity of about 0.5 m/s at full flow. The grade should not be less than the minimum, unless special precautions are taken during construction.

14.17. Hydraulic Design—Drain Size. The hydraulic capacity of drains can be determined from the Manning velocity equation. By equating the design flow to the hydraulic capacity at full flow, the required diameter is

$$d = 51.7 \, (D_c \times A \times n)^{0.375} \, s^{-0.1875} \qquad\qquad (14.10)$$

where d = inside drain diameter in mm,
        D_c = drainage coefficient in mm/d,
        A = drainage area in hectares,
        n = roughness coefficient in Manning's equation,
        s = drain slope in m/m.

Table 14.6 Minimum Grades for Tile Drains

Size Tile (Inside Diameter) (mm)	(in.)	Minimum Grade (%)	Velocity at Full Flow (m/s)	(fps)
75	(3)	0.2	0.29	(0.95)
100	(4)	0.1	0.25	(0.83)
125	(5)	0.17	0.24	(0.80)
150	(6 or >)	0.05	0.23	(0.77 or >)

After computing the required drain size, the next larger commercial size is selected. A graphical solution of this equation is given in Fig. 14.11.

The roughness coefficient in the Manning equation will increase with misalignment at tile joints and with irregularities of the drain surface, such as roughness of the walls and joints and corrugations in tubing. Design coefficients should be 0.013 for clay and concrete and 0.016 for corrugated plastic tubing. For drains in irrigated lands of the West a roughness coefficient of 0.015 for concrete and clay tile is recommended (see Appendix B).

For practical reasons a minimum tile size is usually specified. If the capacity were to match the design flow exactly, the tile should be gradually enlarged starting from the upper end of the line. Local custom, availability of tile, accuracy of installation, grade, and possible failure from sedimentation largely determine the minimum size. In most humid regions 75–100 mm is the minimum size recommended, but 125 and 150 mm are considered the minimum in organic soils and in irrigated regions. On flat slopes the minimum grade shown in Table 14.6 may determine the minimum size.

Example 14.3. Determine the diameter of corrugated plastic tubing and the flow rate where the slope is 0.4 percent, the drainage area is 12 ha (30 ac), and the drainage coefficient is 13 mm (½ in.).
Solution. Substituting in Eq. 14.10 using $n = 0.016$ (see Appendix B),

$$d = 51.7(13 \times 12 \times 0.016)^{0.375}(0.004)^{-0.1875} = 205 \text{ mm (8.1 in.)}$$

Select next largest size of 254 mm (10 in.). From Fig. 14.11 clay or concrete pipe size of 203 mm (8 in.) may be read directly from the nomograph ($n = 0.013$).

Converting 13 mm (½ in.) per day from 12 ha (30 ac),

$$q = \frac{12 \times 13 \times 10^4 \times 10^3}{1000 \times 60 \times 60 \times 24} = 18.1 \text{ L/s (0.64 cfs)}$$

$$(cfs)(28.3) = \ell/s$$

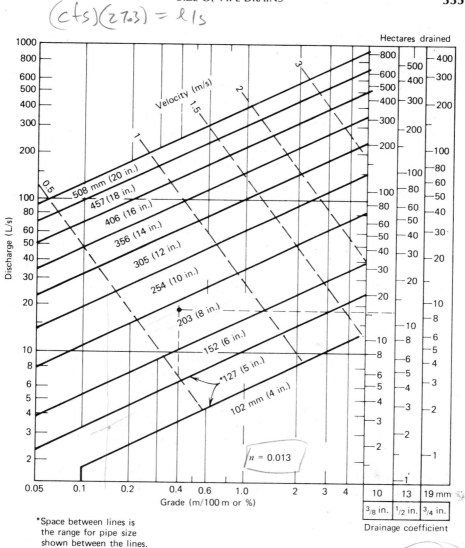

Fig. 14.11. Nomograph for size of concrete and clay pipe drains for $n = 0.013$.

Example 14.4. Determine the clay tile size to carry the design flow from 914 m (3000 ft) of tile spaced 30 m (100 ft) apart if the soil drainable porosity is 4 percent and the drainage requirement for optimum plant growth is a water table drop of 0.24 m/d (0.8 fpd). The grade for the tile is 0.15 percent.
Solution. Assume a uniform drop of the water table over the entire drained area. The design flow is

$$q = \frac{914 \times 30 \times 0.24 \times 0.04 \times 1000}{60 \times 60 \times 24} = 3.0 \text{ L/s } (0.11 \text{ cfs})$$

Using this flow rate enter Fig. 14.11 on the left side, move to the right to a slope of 0.15 percent, and read 127 mm (5 in.) as the tile size. Size may also be computed from Eq. 14.10 where D_c is $(0.24 \times 1000 \times 0.04) = 9.6$ mm. Computed diameter is 117 mm, but select 127 mm, nominal commercial size.

The size of the main is determined from the drainage area of all connected drains. It should not be computed from the maximum capacity of all laterals. The capacity at any point in the system should be adequate to carry the discharge of the drainage system above, allowance being made for surface inlets and future additions to the system.

14.18. Drain Openings. Adequate inflow area can be provided by crack or joint spacings between 0.3-m lengths of clay and concrete tile if the opening area is equal to about 1 percent of the outside (o.d.) surface of the drain (about 3500 mm²/m length for 100 mm drains). Experience has shown that such openings do not seriously restrict drainage. The usual practice is to provide joint cracks of 2 to 6 mm width for 0.3-m lengths of tile. As large a crack as possible is desired, provided the soil does not enter the drain. Doubling the crack width up to about 6 mm (maximum) will increase the inflow rate about 10 percent. For 100-mm diameter tubing about 88 1.6- by 25-mm slots or 125 6-mm diameter perforations per meter length are adequate. Doubling the area or number of uniformly distributed openings will increase inflow about 20 percent. With CPT, soil in the valley of the corrugation has a large effect on inflow. Inflow is slightly greater if the opening is located on the ridge rather than in the valley. Shape of openings is relatively unimportant if they are uniformly distributed. The restriction of openings is greatly affected by the presence of soil within the opening itself and the permeability of the soil within about 25 mm of it. Within this distance about 50 percent of the head is lost due to convergence and friction. If a drain is enclosed in a gravel envelope or surrounded by a permeable cover, the loss in head is usually less than 10 percent.

ACCESSORIES

Accessories for pipe drainage systems include special facilities, such as surface inlets, sedimentation basins, and blind inlets.

14.19. Surface Inlets. As shown in Fig. 14.12, a surface inlet, sometimes called an open inlet, is an intake structure for the removal of surface water from potholes, road ditches, other depressions, and farmsteads. Whenever practicable, surface water should be removed with surface drains (Chapter 12) rather than surface inlets.

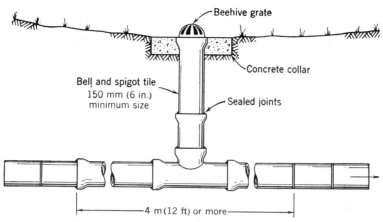

Fig. 14.12. Surface inlet for pipe drains.

Surface inlets should be properly located and constructed. They should be placed at the lowest point along fence rows or in land that is in permanent vegetation. Where the inlet is in a cultivated field, the area immediately around the intake should be kept in grass. The surface inlet should be constructed of bell and spigot tile (minimum size 150 mm) with sealed mortar joints on the vertical riser and extending at least 2 m as a short lateral on either side of the main line. Galvanized metal pipe or a manhole constructed of brick or concrete is also satisfactory. At the surface of the ground a concrete collar should extend around the intake to prevent the growth of vegetation and to hold it in place. On top of the riser a beehive cover or other suitable grate is necessary to prevent trash from entering the line.

14.20. Blind Inlet or French Drain. Where the quantity of surface water to be removed is small or the amount of sediment is too great to permit surface inlets to be installed, blind inlets may, at least temporarily, improve drainage. Though these inlets often do not function satisfactorily for more than a few years, they are economical to install and do not interfere with farming operations.

As shown in Fig. 14.13, a blind inlet is constructed by backfilling the tile trench with various gradations of material. The coarsest material is placed immediately over the line, and the size is gradually decreased toward the surface. These inlets seldom meet the filter design criteria given in Appendix H and for that reason are not permanently effective. Since the soil surface has a tendency to seal, the area should be kept in grass or permanent vegetation. Other materials, such as corncobs, sawdust, and straw, are considered less dependable than more durable materials.

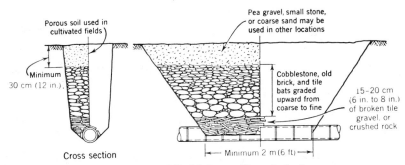

Fig. 14.13. Blind inlet or French drain.

14.21. Sedimentation Basin. Soils containing large quantities of fine sand frequently cause sedimentation since the particles enter the drains through the openings. A sedimentation basin is any type of structure that provides for sediment accumulation, thus reducing deposition in the drain.

A sediment basin may be desirable where the slope downstream is greatly reduced, where several laterals join the main at one point, or where surface water enters. The structure shown in Fig. 14.14 has a turned-down elbow on the outlet for retarding the outflow when the basin becomes filled with sediment. Where structures do not have this feature, the tendency is to neglect the clean-out of the basin. Sedimentation basins are seldom installed in small farm drainage systems.

Junction boxes may be installed where several lines join at different elevations. Except for the catch basin, they are similar to the structure shown in Fig. 14.14. To facilitate farming operations, the top of the manhole should be placed 30 cm or more below the surface.

14.22. Controlled Drainage Structures. Control structures in drain lines for maintaining the ground water at a specified level are similar to sedimentation basins except that crestboards are placed in the structure to keep the water at the desired level. The functioning of such structures is similar to that for control dams in open ditches described in Chapter 13. Controlled drainage is essential in the management of organic soils because of its effectiveness in controlling subsidence.

14.23. Relief Pipes and Breathers. Relief pipes and breathers are small-size vertical risers extending from the drain line to the surface. The riser should be made of steel pipe or sealed bell and spigot tile and should be located at fence lines where they are not likely to be damaged. Breathers are installed on long lines to prevent the development of a vacuum. In model studies Lembke et al. (1963) found that for slopes over 1 percent negative pressures could develop in a tile running full. Quantitative values could not be predicted in the field nor

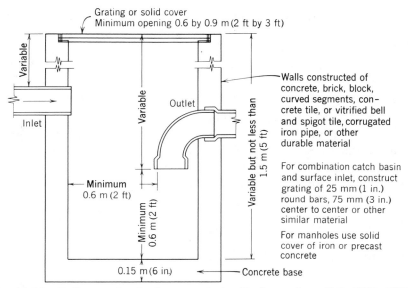

Fig. 14.14. Sedimentation basin or manhole. (Redrawn from U.S. SCS, 1973.)

were the effects of these negative pressures on flow determined. Many engineers do not feel that breathers are justified.

Relief pipes serve to relieve the excess water pressure in the drain during periods of high outflow, thus preventing blowouts. A relief pipe should be installed where a steep section of a main changes to a flat section, unless the capacity of the flat section exceeds the capacity of the steep section by 25 percent.

14.24. Envelope Filters. For drainage of irrigated land in the West gravel envelopes are placed completely around the drain. However, in humid areas this practice is not recommended. These envelopes are installed (1) to prevent the inflow of soil into the drain which may cause failure and (2) to increase the effective drain diameter which increases the inflow rate. Design criteria for these filters are given in Appendix H. Because of practical considerations and cost, the envelope is usually limited to one gradation of material, which is selected from local naturally occurring deposits. Some variation from filter standards may be justified. Envelopes should be fairly well-graded from the coarsest (38 mm dia.) to the finest. Sizes smaller than 0.7 mm tend to become selfclogging. Minimum thickness of envelope material is from 8 to 10 cm due to the physical difficulty of placing a uniformly small thickness.

Several types of sheet filters, called geotextiles, are available commercially. Strips can be placed above and below the drain during installation, and some filter materials are sealed around corrugated plastic tubing at the extruding plant. Filters are made from nylon, polypropylene, and other materials. The mesh size should allow some of the fine soil particles to pass through so as not to plug the filter, yet retain the larger particles that would tend to deposit in the drain. One recommendation is that the 50 percent particle size of the soil should be greater than or equal to the average diameter of openings in the filter.

14.25. Artesian Relief Wells. Where artesian aquifers are the cause of drainage problems, wells may be drilled into the aquifer to reduce the pressure and to lower the water table. Sometimes these wells are connected directly into a drain permitting the water to flow by gravity rather than by pumping. Water table contour maps or piezometric surface contours are often required prior to designing the drainage system.

DRAIN PIPE QUALITY

Only high-quality pipe should be installed in a drainage system. It is false economy to install second-grade or poor-quality materials. Drain tile are primarily of two kinds: clay and concrete. Clay tile are made of shale, fire clay, or surface clay; concrete tile are made of Portland cement and a suitable aggregate. Corrugated plastic tubing has been widely accepted since about 1970.

14.26. Characteristics of Good Drain Tile. Clay or concrete tile should have the following characteristics: (1) resistance to weathering and deterioration in the soil, (2) sufficient strength to support static and impact loads under conditions for which they are designed (see Chapter 15), (3) low water absorption, that is, a high density, (4) resistance to alternate freezing and thawing, (5) relative freedom from defects, such as cracks and ragged ends, and (6) uniformity in wall thickness and shape. Drain tile that meet current specifications of the American Society for Testing Materials have the essential qualities listed above (see Appendix D). Specifications have been prescribed for three classes of drain tile, namely, standard, extra-quality, and special-quality (concrete only) or heavy-duty (clay only). Standard-quality tile are satisfactory for drains of moderate size and depths found in most farm drainage work.

14.27. Concrete Tile. Concrete tile should be made with high-quality materials and be properly cured. Where concrete tile are to be placed in acid or alkali soils, the tile should be extra-quality and made with cements having specific chemical characteristics. Curing methods will also depend upon the degree of acidity or alkalinity of the soil. For nearly neutral soils where strength is the principal criterion, a steam-cure of 30 hr at 68°C or a water-cure of 28 days at 22°C is adequate.

Good-quality concrete tile are resistant to freezing and thawing but may be subject to deterioration in acid and alkaline soils. In these soils concrete tile should be used only if approved by local recommendations. The amount of corrosive action on concrete is proportional to the degree of acidity as measured by the pH scale. By visual inspection it is more difficult to determine the quality of concrete tile than that of clay tile.

14.28. Clay Tile. Clay tile should be well burned, with no checks or cracks, and should have a distinct ring when tapped with a metal object. Ordinary drain tile are not burned as hard as vitrified sewer tile. Clay tile made from shale are more durable and usually have less absorption than those made from surface clays. Evidence that freezing and thawing has some effect is indicated from a survey of drainage installations in the upper Mississippi Valley by Miller and Manson (1951). These tile had been in service from 30 to 40 years. Where the tile were less than 0.5 m in depth, there was some disintegration, but where the depth was greater than 0.5 m there was no evidence of frost action.

Clay tile are not generally affected by acid or alkaline soils. When subjected to frequent alternate freezing and thawing conditions, it is safer to use concrete tile, although most clay tile are resistant to frost damage. Where clay tile are laid with less than 0.7 m of cover, they should be extra-quality. In most mineral soils the kind of tile (clay or concrete) is not as important as the quality.

14.29. Corrugated Plastic Tubing. This tubing (CPT) is not damaged by soil chemicals, is light in weight, and is shipped in long lengths. Tubing should be uniform in color and density and free from visible defects. Parallel plate stiffness when deflected at 127 mm per minute should not be less than 0.17 N/mm/mm of length at 5 percent deflection and 0.13 N/mm/mm at 10 percent deflection for diameters up to 200 mm. In addition to the above standards CPT should meet other criteria specified by ASTM F405 and F667 (see Appendix D).

DESIGN AND LAYOUT PROCEDURE

The subsurface drainage system should be coordinated with existing and proposed surface ditches and other drains.

14.30. Preliminary and Location Surveys. The first step in design is to make a preliminary survey of the area, much as described for open ditches in Chapter 13. Soils should be investigated to determine whether subsurface drainage is practicable and economical. If so, sufficient information should be obtained to permit the selection of an adequate depth and spacing for the drains. The extent and depth of impermeable strata and ground water pressure and the source of seepage should be determined. On flat land, topographic maps may be prepared, especially where an extensive drainage system is planned. If the drainage system is not extensive, the preliminary survey and the location survey may be made at the same time.

Experience is desirable in the proper location of pipe drains, but a few general rules are (1) place the outlet at the best possible location; (2) provide as few outlets as possible; (3) lay out the system with short mains and long laterals; (4) use the available slope to best advantage, especially on flat land; (5) follow the general direction of natural waterways, particularly with mains and submains on land with considerable slope; (6) avoid routes that result in excessive cuts; (7) avoid crossing waterways except at an angle of 45 degrees or more; and (8) avoid soil conditions that increase installation and maintenance costs.

14.31. Alignment. Where a change in direction of pipe drains is necessary, a junction box, fitted tile, or a T- or Y-manufactured junction is required. The alignment of pipe at junctions is shown in Fig. 14.15. Where sufficient slope is available, the grade line of the lateral should intersect the main near the top of the main. However, if sufficient slope is not available, the grade line of the lateral may intersect the main at a lower elevation but should always be high enough (see y in Fig. 14.15) to permit the center line of the lateral to intersect the center line of the main. A difference in elevation (z in Fig. 14.15) between the extended grade line of the lateral and the main is adjusted by increasing the slope in the last few meters on the lateral.

14.32. Staking. The drain line is staked by placing hub and guard stakes at 15- or 30-m (50- or 100-ft) stations, starting at the outlet (Station 0+00). To prevent these stakes from being removed during construction, the stake line must be offset about 2 m from the center line of the trench. With the laser beam grade control system described in Chapter 15, staking is only necessary for location of the line and of points where grade changes are required.

Pipe drains like open ditches are arbitrarily designated in decreasing order of importance as mains, submains, and laterals. As shown in Fig. 14.16, the first lateral above the outlet to be connected to main A may be designated $A1$, the next $A2$, etc. The lines entering $A1$ may be indicated as $A1.1$, $A1.2$, etc.

Example 14.5 Determine the quantity of each size of pipe for the drainage system shown in Fig. 14.16. The spacing of all laterals is 30 m (100 ft.). The slope of Main A and A1 is 0.4 percent and all other laterals are 0.15 percent. The soil is fine sandy loam in a humid area. Field crops are to be grown. Minimum pipe size is 100 mm (4 in.), which is available only in corrugated plastic tubing (CPT). Larger sizes are to be clay tile.
Solution. From Table 14.5 select a drainage coefficient of 10 mm (⅜ in.) per day. Calculate drainage area (D.A.) from the length and spacing (30 m) of each line, but include the area drained by a lateral and the main only once. Determine pipe size from Fig. 14.11 or Eq. 14.10 using the appropriate n value.
Lengths required: 3002 m (9850 ft), 100 mm (4 in.) CPT; 183 m (600 ft), 150 mm (6 in.); and 85 m (280 ft), 200 mm (8 in.) clay tile plus 3 percent breakage allowance. Appropriate size and type junctions required as shown by plan. SI units are rounded.

	Slope (%)	D.A. [ha (ac)]	Nominal Pipe Diameter		Length	
			(mm)	(in.)	(m)	(ft)
Laterals						
A2 to A7	0.15	1.1 (2.8)	100	(4)	2195	(7200)
A1.1, A1.11,						
A1.12	0.15	1.3 (3.3)	100	(4)	442	(1450)
A1	0.4	2.4 (6.0)	100	(4)	366	(1200)
Main A from						
A1 to A7	0.4	7.0(17.3)	150[a]	(6)	183	(600)
0+00 to A1	0.4	9.7(24.0)	200	(8)	85	(280)

[a] Minimum size suggested for main, but size could be reduced to correspond to D.A. upstream.

Example 14.6. Determine the pipe size in Example 14.5 if laterals A1.1, A1.11, and A1.12 were all located within a 2.4 ha (6 ac) pothole and all of the surface water from this area was removed through a surface inlet at Sta. 1 + 98 m (6 + 50) on A1.1.

Solution. From Table 14.5 select a D_c of 19 mm (¾ in.) that applies to the entire pothole area. Since a flow rate for surface water is more than adequate for the three laterals within the pothole, the area drained by the three laterals need not be considered in computing outlet size. See next page.

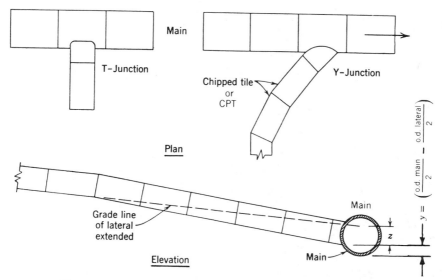

Fig. 14.15. Vertical and horizontal alignment at junctions.

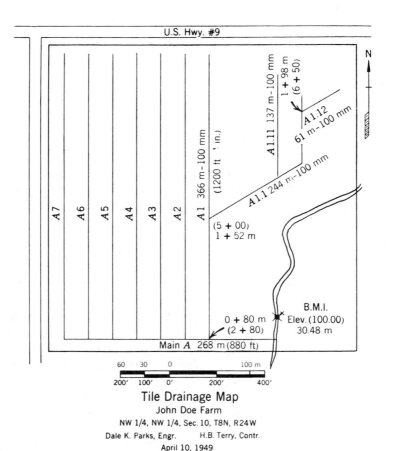

Fig. 14.16. A suitable subsurface drainage map.

	Slope (%)	D.A. [ha (ac)]	D_c (mm)	(in.)	Nominal Pipe Diameter (mm)	(in.)
A1.1 below inlet	0.15	2.4 (6.0)	19	(3/4)	150	(6)
A1 at Main A	0.4	6.0(14.8)[a]	10	(3/8)	150	(6)
Main A to Outlet	0.4	13.2(32.7)	10	(3/8)	200	(8)

[a] D.A. for A1 only, 1.1 ha (2.8 ac), plus 4.9 ha (12 ac) from A1.1 (2.4 ha at 19 mm is about equivalent to 4.9 ha at 10 mm). Conversion to a common D_c is necessary to read size from Fig. 14.11. Alternate procedure is to convert to flow rates in L/s (cfs).

Tile size on Main A is the same, but sizes should be increased on A1 and A1.1 to 150 mm (6 in.) to carry the increased flow from the pothole.

REFERENCES

Clayton, B. S. et al. (1942). "Water Control in the Peat and Muck Soils of the Florida Everglades." *Florida Agr. Expt. Sta. Bull.* 378.

Dumm, L. D. (1954). "Drain-Spacing Formula." *Agr. Eng.* **35**, 726–730.

Hoffman, G. J., and G. O. Schwab (1964). "Spacing Prediction Based on Drain Outflow." *Trans. Am. Soc. Agr. Eng.* **7**, 444–447.

International Institute for Land Reclamation and Improvement (1972-74). *Drainage Principles and Applications,* Vols. I, II, III, IV. The Institute, Wageningen, The Netherlands.

Israelsen, C. W., and V. E. Hansen (1962). *Irrigation Principles and Practices* (3rd ed.). Wiley, New York.

Kramer, P. J. (1949). *Plant and Soil Water Relationships.* McGraw-Hill, New York.

Kramer, P. J., and W. T. Jackson (1954). "Cause of Injury to Flooded Tobacco Plants." *Plant Physiol.* **29**, 241–245.

Keller, J., and A. R. Robinson (1961). "Laboratory Research on Interceptor Drains." *Am. Soc. Civil Eng. Trans.* **126**, Pt 3, 687.

Kidder, E. H., and W. F. Lytle (1949). "Drainage Investigations in the Plastic Till Soils of Northeastern Illinois." *Agr. Eng.* **39**, 384–386, 389.

Lembke, W. D., J. W. Delleur, and E. J. Monke (1963). "Model Study of Flow in Steep Tile Drains." *Am. Soc. Agr. Eng. Trans.* **6**(2), 142–144.

Luthin, J. N. (1966). *Drainage Engineering.* Wiley, New York.

Manson, P. W., and D. G. Miller (1948). "Essential Characteristics of Durable Concrete Draintile for Alkali Soils." *Agr. Eng.* **29**, 485–487, 489.

—— (1954). "Making Durable Concrete Drain Tile on Packer-Head Machines." *Minn. Agr. Expt. Sta. Bull.* 426.

Manson, P. W., and C. O. Rost (1951). "Farm Drainage—an Important Conservation Practice." *Agr. Eng.* **32**, 325–377.

Miller, D. G., and P. W. Manson (1948). "Essential Characteristics of Durable Concrete Drain Tile for Acid Soils." *Agr. Eng.* **29**, 437–441.

—— (1951). Long-Time Performance of Some Clay Draintile, *Agr. Eng.* **32**, 95–97, 100.

Moody, W. T. (1966). "Nonlinear Differential Equation of Drain Spacing." *Proc. Am. Soc. Civil Eng. IR2.* **92**, 1–9.

Neal, J. H. (1934). "Proper Spacing and Depth of Tile Drains Determined by the Physical Properties of the Soil." *Minn. Agr. Expt. Sta. Tech. Bull.* 101.

Nicholson, H. H., and D. H. Firth (1953). "The Effect of Ground Water Levels on the Performance and Yield of Some Common Crops." *J. Agr. Sci.* **43**, 95–105.

Pillsbury, A. F. et al. (1965). "Tile Drainage Performance in Coachella Valley, California." *Am. Soc. Civil Eng. Proc. IR2,* **91,** 1–10.

Schmid, P., and J. Luthin (1964). "The Drainage of Sloping Lands." *J. Geophys. Res.* **69,** 1525–1529.

Skaggs, R. W. (1975). "Determination of the Hydraulic Conductivity-Drainable Porosity Ratio from Water Table Measurements." *Trans. Am. Soc. Agr. Eng.* **19** (1), 73–80.

U.S. Bureau Reclamation (1978). "Drainage Manual." GPO, Washington, D.C.

U.S. Soil Conservation Service (1973). "Drainage of Agricultural Lands." Water Information Center, Port Washington, N.Y.

Van Hoorn, J. W. (1958). "Results of a Ground Water Level Experimental Field with Arable Crops on Clay Soil." Neth. J. Agr. Sci. 6 (1), 1–10.

Van Schilfgaarde, J. (1963). "Tile Drainage Design Procedure for Falling Water Tables." *Am. Soc. Civil Eng. Proc.* **89,** No. IR 2, June, and discussion Dec. 1963, March and Dec. 1964.

Van Schilfgaarde, J. (ed.) (1974). "Drainage for Agriculture," Monograph 17. *Am. Soc. Agron.,* Madison, Wisc.

Visser, W. C. (1959). "Crop Growth and Availability of Moisture." Inst. for Land and Water Mgt. Res., Tech. Bull. 6, Wageningen, The Netherlands.

——— (1954). "Tile Drainage in the Netherlands." *Netherlands J. Agr. Sci.* **2,** 69–87.

Walker, P. (1952). "Depth and Spacing for Drain Laterals as Computed from Core-Sample Permeability Measurements." *Agr. Eng.* **33,** 71–73.

Wessling, J., et al. (1957). "Land Drainage in Relation to Soils and Crops." Chapter V, in J. N. Luthin (ed.) *Drainage of Agricultural Lands,* Monograph 7. Am. Soc. Agron., Madison, Wisconsin.

PROBLEMS

14.1. Calculate the discharge of a 150 mm (6 in.) diameter corrugated plastic tube flowing full if the slope is 0.4 percent.

14.2. How many hectares (ac) will a 254-mm (10-in.) clay tile drain with a slope of 0.3 percent if the tile are in mineral soil in your area and field crops are to be grown?

14.3. Determine the quantity of each size of concrete tile required for a gridiron system, as shown in Fig. 14.6c, if the length of each of the five laterals is 400 m (1310 ft) and the main is 305 m (1000 ft). Slope in the laterals is 0.12 percent and in the main 0.3 percent. Spacing of the laterals is 30 m (100 ft). Design the system for field crops in your area.

14.4. What size clay tile is required to remove the surface water with a surface inlet if the runoff accumulates from 8.1 ha (20 ac) and the slope in the tile is 0.4 percent? Design for field crops in your area.

14.5. If a surface inlet draining 8.1 ha (20 ac) is added at the upper end of the main in Problem 14.3, what size tile are required from the surface inlet to the outlet?

14.6. What slope is required to provide a velocity of 0.46 m/s (1.5 fps) at full flow in a 100-mm (4-in.) clay tile? In a 300-mm (12-in.)?

14.7. If a tile drainage system draining 12.1 ha (30 ac) flows full for three days following a period of heavy rainfall, determine the effluent volume during this period if the system was designed for a drainage coefficient of 13 mm ($\frac{1}{2}$ in.) per day.

14.8. The grade elevation of a 400 mm (16 in.) main at the junction of a 100 mm (4 in.) lateral is 27.93 m (91.62 ft). Assuming a wall thickness of 10 percent of the nominal diameter, what should be the grade elevation for the lateral? Use a manufactured junction.

14.9. Determine the depth of flow as a percentage of the drain diameter to obtain maximum discharge. Note maximum discharge does not occur at full flow.

14.10. Determine the design flow from 3048 m (10 000 ft) of pipe at a spacing of 61 m (200 ft) following local recommendations for irrigated conditions. What size of concrete tile is required for the main at the outlet if the tile slope is 0.25 percent and $n = 0.013$?

14.11. Determine the depth and spacing of drains to maintain a constant water table not less than 1.5 m (5 ft) below the surface assuming that the height of the water table at the midplane is not more than 0.5 m (1.5 ft) above the center of 150 mm (6 in.) drains. The average hydraulic conductivity of the soil is 0.6 m/d (2 fpd), the depth d to the impervious layer is 3 m (10 ft) and the flow rate is 0.9 L/d per m length (0.01 cfd per ft length).

14.12. Compute the drain spacing for a clay loam soil having a drainable porosity of 0.06, a hydraulic conductivity of 0.6 m/d (2 fpd), and an impervious layer 2.4 m (8 ft) below the drains. Drainage requirements for the crop are such that the initial water table depth below the surface is 15 cm (6 in.) with a rate of drop of 21 cm (0.7 ft) for the first day. Because of outlet depth, the maximum depth of the drain is limited to 114 cm (3.75 ft) measured to the bottom of the 125-mm (5-in.) drains.

14.13. From a contour map supplied by your instructor design and layout the most efficient tile drainage system for the soil and cropping practices specified for the field. Determine the length of each size of tile and a list of all other materials required.

14.14. Determine the size at the outlet of 3658 m (12 000 ft) of clay tile if the spacing is 18.3 m (60 ft) and the grade is 0.1 percent. The soil porosity is 3 percent and the optimum rate of drop of the water table is 15 cm (0.5 ft) in the first 12 hr following heavy rainfall.

CHAPTER 15

Installation and Maintenance of Subsurface Drains

Although a subsurface drainage system is adequately designed and is staked out according to plan, it will not function satisfactorily unless properly installed and maintained. The trench should be dug to the specified grade, the bedding conditions and width of the trench should be such as to prevent overloading of the tile, good workmanship should be secured in laying the pipe and in making junctions, the cost of installation should be reasonable, and the drainage system should be mapped and properly recorded.

INSTALLATION PRACTICES

Installation should always begin at the outlet and progress upstream so that seepage water is free to move to the outlet. The installation includes digging to an established grade, laying the pipe, and backfilling.

15.1. Trenching Methods. Pipe are usually installed with a trenching machine or a plow: Trenching machines may be divided into three general classes: (1) wheel excavators, (2) endless-chain excavators, and (3) hoe excavators. The first two types are shown in Fig. 15.1 and the hoe in Fig. 13.7.

With wheel excavators, soil is carried to the top of the wheel and then dropped onto a conveyor belt. Endless-chain excavators may be subdivided into two groups: the vertical-boom and the slant-boom type. The vertical-boom type reduces the amount of handwork where obstructions, such as pipelines or cables, are encountered. On many hoe-type machines, described in Chapter 13, the digging bucket is simply a concave blade of the desired width. This class of excavator is suitable for making deep cuts, removing large stones, and making junctions. They also work satisfactorily in wet subsoil. Draglines are suitable for deep, wide trenches.

A trenching machine should dig to a uniform grade under a variety of soil conditions. A crumber (shoe) following the digging mechanism is helpful, as it obviates the necessity of hand finishing the ditch bottom. To ensure good bedding conditions and to align the pipe, the crumber should make a curved or

V-shaped bottom in the trench. If the shoe is curved, a keel for making a groove should be attached to the bottom. Machines not equipped with facilities for maintaining grades should not dig closer than about 15 cm to the grade line, the bottom of the ditch then being finished by hand.

The factors that influence the rate of installation for a trenching machine are: (1) soil moisture; (2) soil characteristics, such as hardness, stickiness, stones, and submerged stumps; (3) depth of trench; (4) condition of the trenching machine; (5) skill of the operator; (6) width of the trench; and (7) delays due to interruptions during operation. An extremely wet soil may stop machine operation; soil with a low moisture content may not affect the digging rate to any extent. Soils that stick to the buckets of the machine, sandy soils that fall apart easily, stones in the subsoil, and deep cuts reduce digging speeds. Increasing the depth from 90 to 150 cm decreased the digging speed by 56 percent under Iowa and Minnesota conditions. Although the width of the trench may not be important for machines with sufficient power, the rate of installation may be reduced especially for trenches wider than 0.4 m.

The average output for a trenching machine at depths from 90 to 150 cm may be as much as 700 to 300 m per day, respectively. A ditching machine operated by a contractor in northern states can be expected to install pipe only about 150 working days during the year. Studies showed that a total of 66 percent of the available working hours were lost. This time loss is accounted for as follows: weather 19 percent, repairs 14 percent, making junctions 10 percent, miscellaneous 9 percent, moving from job to job 8 percent, and servicing of the machine 6 percent.

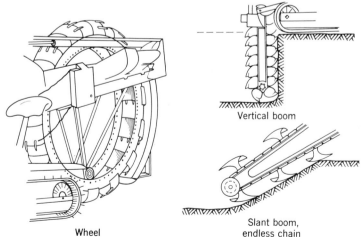

Vertical boom

Slant boom,
endless chain

Wheel

Fig. 15.1. Wheel and endless-chain trenching machines.

15.2. Methods of Establishing Grade. Since the late 1960s, laser beam equipment for grade control has become available commercially. One system consists of a tripod command post with a rotating laser beam that can give either a level or a sloping plane of reference. A detector on the trenching machine picks up the beam and automatically keeps the machine on the same grade as the plane produced by the laser beam. A system of electrically controlled hydraulic valves and cylinders keeps the trencher on grade. A grade modification device with a distance measuring wheel mounted on the trencher will change the drain grade from that made by the laser beam. The laser system has been adapted to many other earth-moving machines, such as the blade grader, scraper, bulldozer, etc. A special surveying rod with a detector to pick up the beam is available for topographic surveying and leveling.

For hand trenching or for trenching machines not equipped with an automatic grade-control system a sight or string line is a simple grade control method. The string line and the sight methods are essentially the same except that string is stretched between stations in the line method and sighting targets are set at each station in the sight method. As shown in Fig. 15.2, the depth of the trench is established by sighting over two targets. The line of sight or string line must be parallel to the grade line of the drain. The distance from the line of sight to the grade line is known as the *base distance,* which in Fig. 15.2 is 305 cm. The target is set near the hub stake, and the cut is subtracted from the *base distance* to obtain the difference in elevation between the top of the hub and the line of sight. For example, at the first station the target is 198 cm (305 − 107) above the hub stake. At least three consecutive targets should be set so that errors in design or calculation may be detected.

15.3. Laying the Drain. Pipe drains should be laid on true grade with the bottom of the trench shaped with a 90-degree groove angle to provide good alignment and bottom support. The crack spacing between individual tiles

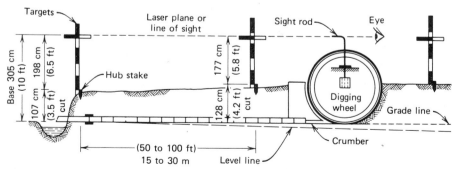

Fig. 15.2. Setting targets for trenching machines. (For laser systems the laser plane would be parallel to the line of sight as shown.)

should be about 3 mm, unless the soil is sandy (see Chapter 14). In unstable soils the crack spacing should be small to prevent inflow of fine sand. Where crack spacings are more than 3 mm in sandy soils and 6 mm in clay soils, the openings should be covered with pieces of broken tile, tubing, or other durable material. If the tile are not square on the ends, they may be rotated so that the widest opening will be on the bottom and the narrowest opening on top.

If for some reason the trench is excavated below grade, it should be refilled to the desired grade with well-graded gravel. If the trench does not have water in it, moist soil may be suitable if thoroughly compacted under the drain.

The upper ends of all drain lines should be closed tightly with flat slabs of concrete, stones, or prefabricated ends. Tile bats are not suitable unless sealed with concrete.

15.4. Junctions. A properly made junction is an essential part of the drain system. Either a T- or a Y-manufactured junction is satisfactory. With the Y-junction the flow should be directed downstream. Extensive tests conducted by Manson and Blaisdell (1956) showed that the energy loss for a 90-degree junction angle is insignificant for all practical purposes. For most conditions a 90-degree rather than a 45-degree junction would require an increase in slope on the main of much less than 0.01 percent to compensate for the increased head loss. Because of the ease and convenience in installation and the resulting improvement of workmanship, the 90-degree or the T-junction is preferred. Where a difference in elevation between the main and the lateral is allowed at the junction, the pipe are connected as described in Chapter 14.

Manufactured junctions are more satisfactory than those fabricated in the field. A homemade junction can be fitted together by chipping and breaking ordinary tile and by using concrete mortar completely around the connection. Plastic tubing can be connected in a similar manner using special fittings, usually of the snap-on type furnished by each manufacturer. Often they are not interchangeable.

15.5. Checking Grade. Immediately after installation and before backfilling, the drains should be checked for grade, alignment, and other design specifications. Allowable variation from true grade should be within reasonable limits. Some engineers allow a variation of 3 mm per 25 mm of drain diameter for sizes up to 250 mm. Under no circumstances should there be backfall in the drain. Somewhat greater variation may be allowed where slopes are greater than minimum (Chapter 14).

15.6. Blinding and Envelope Filters. As shown in Fig. 15.3, blinding is accomplished by placing loose, mellow topsoil or soil from another permeable layer in the trench immediately over the pipe. The purpose of blinding is to prevent the tubing or tile from being moved out of alignment, to prevent tile

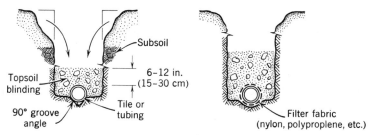

Fig. 15.3. Methods of protecting and stabilizing pipe drains in humid areas.

breakage or tubing deflection during the backfilling operation, and to provide permeable soil in contact with the pipe. If stones are present in the soil, the minimum depth of blinding should be about 15 cm. In heavy soils it may be desirable to backfill the trench with corncobs, cinders, well-graded gravel, and other porous materials. However, organic materials may not be effective for more than a few years.

In irrigated areas of the West, where gravel-envelope filters are recommended, trenching machines are often equipped with two gravel chutes as shown in Fig. 15.4. The first one places the gravel in the bottom of the trench, the pipe is laid on this layer, and the second chute places the gravel on the sides and top of the drain to the desired thickness. Gravel is normally deposited in the chute hoppers with motorized scoop loaders. Tar paper or filter mats may be placed on the drain in the manner shown. Experiments by Nelson (1960) indicate that fiber glass with random reinforcing fibers was superior to that with parallel fibers. Factory-wrapped and sealed fabrics placed completely around

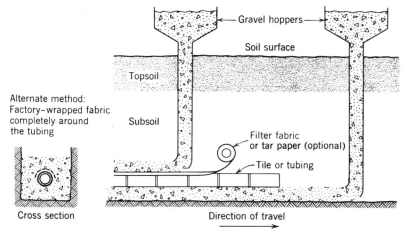

Fig. 15.4. Method of installing gravel envelope filters for irrigated land.

the tubing are a substitute for gravel-envelope filters. Such fabrics, called geotextiles, may be easier and more economical to install than gravel.

15.7. Backfilling. Backfilling of the trench can be accomplished with many machines, such as bulldozers, hoes, motor graders, manure loaders, and blades on small tractors. All soil should be replaced and heaped up so that after settlement the trench can be crossed with tillage equipment. If the soil is dry or contains stones, extreme care should be taken to prevent breakage of the tile or smashing of the plastic tubing.

15.8. Installation in Peat and Muck. Before drains are installed in newly developed peat and muck soils, the land should be drained with open ditches for several years to secure initial subsidence. The pipe should be laid on the underlying mineral soil, provided the mineral soil is not too deep and not impermeable. In peak or muck soils, 150-mm drains are often recommended as the minimum size because of differential subsidence. Perforated tile several feet in length or plastic tubing is recommended.

15.9. Installation in Quicksand. Quicksand is not a type of sand but rather a soil condition in which fine sand in a saturated, fluid condition is buoyed up by hydrostatic pressure from below. Some of the following recommendations may be helpful in planning and installing drains under quicksand conditions: (1) install pipe only during the driest season; (2) use plastic tubing with a suitable prefabricated covering; (3) install pipe that are at least 125 or 150 mm in diameter; (4) use grades of not less than 0.4 percent; (5) protect all cracks between individual tile with durable materials, covering the upper two-thirds of the tile circumference; (6) install each line as rapidly as possible and without interruptions to prevent settlement of the crumber on the machine; (7) blind the pipe with coarse material, such as gravel, sawdust, or sod; (8) prevent the caving-in of the ditch bank prior to and during backfilling operations; and (9) provide careful maintenance of the line, particularly if sinks or blowholes develop. A plastic stabilizing strip may be placed under the pipe to maintain alignment and grade. Under some situations, reducing the hydrostatic pressure by pump drainage may be practicable.

LOADS ON CONDUITS

Loads on underground conduits include those caused by the weight of the soil and by concentrated loads due to the passage of equipment or vehicles. At shallow depths concentrated loads largely determine the strength requirements of conduits; at greater depths the load due to the soil is the most significant. Loads for both conditions should be determined, particularly where the depth of the controlling load is not known. For pipe drains the weight of the soil usually determines the load. Concentrated loads may be calculated by methods described in Appendix E.

15.10. Types of Conduit Loading.   For purposes of analyzing loads, the two types of underground conduits are ditch conduits and projecting conduits, as shown in Fig. 15.5. Ditch conduit conditions apply to narrow trenches, and projecting conduit conditions occur in trenches wider than about 2 or 3 times the outside diameter of the pipe. However, the type of loading can be determined from Eqs. 15.1 and 15.2. Projecting conditions generally exist when the settlement of soil prisms A and C shown in Fig. 15.5b is greater than prism B. Since this condition applies to conduits placed under an embankment on undisturbed soil as in road fills or pond dams, the conduit projects above a relatively solid surface; hence its name.

15.11. Load Factors Based on Bedding Conditions.   The load factor is the ratio of the strength of a rigid conduit under given bedding conditions to its strength as determined by the three-edge bearing test. As shown in Fig. 15.6, four classes of bedding conditions are generally recognized.

For nonpermissible bedding conditions, shown in Fig. 15.6a, no attempt is made to shape the foundation or to compact the soil under and around the conduit. Such a bedding has a low load factor (1.1) and is not suitable for drainage work.

For ordinary bedding conditions (L.F. 1.5) the bottom of the trench must be shaped to the conduit for at least half its width. The conduit should be surrounded with fine, granular soil extending at least 15 cm above the top of the tile.

For first-class bedding the conduit must be placed in fine, granular material for 0.6 the conduit width and should be entirely surrounded with this material

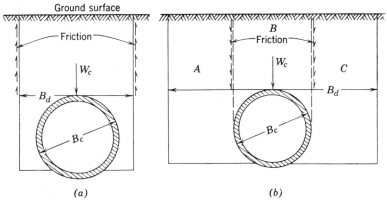

Fig. 15.5.   (a) Ditch-type and (b) projecting-type rigid conduit loading.

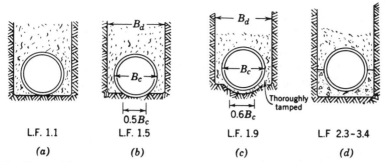

Fig. 15.6. Typical bedding specifications for rigid ditch conduits. (a) Not permissible. (b) Ordinary. (c) First class. (d) Concrete cradle.

extending 30 cm or more above the top. In addition, the blinding material should be placed by hand and thoroughly tamped in thin layers on the sides and above the conduit. Although the load factor for first-class bedding is 1.9, it is seldom necessary in drainage work.

The concrete cradle bearing is constructed by placing the lower part of the conduit in concrete. Such construction is not practical in conservation work except for earth dams or high embankments where excessive loads are encountered. This type of bedding provides the best conditions, with load factors varying from 2.2 to 3.4.

Based on a study of trench bottoms made by several ditching machines, load factors were found to vary from 0.9 to 2.5. With a keel on the bottom of a curved trencher shoe or a V-shaped bottom, a load factor of 1.5 or greater may be expected.

15.12. Soil Loads on Rigid Pipe. Loads on both ditch conduits and projecting conduits are based on studies by Marston (1930), Spangler (1947), and Schlick (1932) over a period of about 40 years. For loads on ditch conduits it is assumed that the density of the fill material is less than that of the original soil. As settlement takes place in the backfill, the sides of the ditch resist such movement (Fig. 15.5a). Because of the upward frictional forces acting on the fill material, the load on the conduit is less than the weight of the soil directly above it. In its simplest form the ditch conduit load formula is

$$W_c = C_d w B_d{}^2 \tag{15.1}$$

where W_c = total load on the conduit per unit length (FL^{-1}),
   C_d = load coefficient for ditch conduits,
   w = unit weight of fill material (FL^{-3}),
   B_d = width of ditch at top of conduit (L).

The backfill directly over a projecting conduit (prism B in Fig. 15.5b) will settle less than the soil to the sides of the conduit (prisms A and C). For projecting conditions the load on the conduit is greater than the weight of the soil directly above it, because shearing forces due to greater settlement of soil on both sides are downward rather than upward. The projecting conduit formula for wide ditches is

$$W_c = C_c w B_c^2 \tag{15.2}$$

where C_c = load coefficient for projecting conduits,
            B_c = outside diameter of the conduit (L).

The load coefficients C_d and C_c are functions of the frictional coefficient of the soil, the height of fill, and, respectively, the width of the ditch or the width of the conduit. Since these coefficients are rather complex, C_d and C_c curves, shown in Fig. 15.7, have been developed. The load on the conduit is the smaller value as computed from Eq. 15.1 or Eq. 15.2 (see Example 15.1). The width of trench for a given conduit that results in the same load when computed by both equations is known as the transition width (see Fig. 15.8).

Drain tile should be installed so that the load does not exceed the required average minimum crushing strength of the tile, as given in Appendix D. Whenever practicable, ditch conduit conditions should be provided rather than projecting conduit conditions. To prevent overloading in wide trenches, a narrow subditch can be excavated in the bottom of the main trench. The width of the subditch measured at the top of the tile determines the load, regardless of the shape of the trench above this point. The subditch need not extend above the top of the tile. A common method of construction in deep cuts is to remove the excess soil with a bulldozer and to dig the subditch with a trenching machine. For trench widths of 40 and 46 cm or less, standard-quality and extra-quality concrete or clay tile, respectively, may be installed at any depth.

Example 15.1.   Determine the static load on 254 mm (10 in.) tile (B_c = 305 mm or 1.0 ft) installed 213 cm (7.0 ft) deep in a ditch 46 cm (18 in.) wide if the backfill is ordinary clay weighing 1922 kg/m³ (120 pcf). What quality tile is required, assuming ordinary bedding conditions?
Solution. $H = d - B_c$, where H = depth to top of the tile and d = depth to bottom of the tile. $H = 213 - 30 = 183$ cm ($7.0 - 1.0 = 6.0$ ft). Read from Fig. 15.7 for $H/B_d = 183/46 = 4.0$, $C_d = 2.4$. Substitute in Eq. 15.1 to obtain load for ditch conditions:

$$W_c = 2.4 \times 1922 \times (0.46)^2$$
$$= 976 \text{ kg/m (648 lb/lin ft)}$$

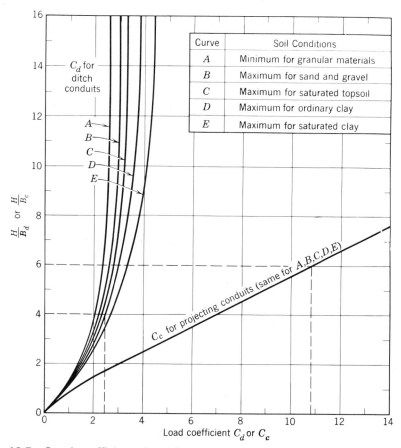

Fig. 15.7. Load coefficients for rigid drain tile. (Redrawn from Spangler, 1947.)

Read from Fig. 15.7 for H/B_c of 6.0, $C_c = 10.8$. Substitute in Eq. 15.2 to obtain load for projecting conditions:

$$W_c = 10.8 \times 1922 \times (0.305)^2$$
$$= 1931 \text{ kg/m (1296 lb/lin ft)}$$

The lower value 976 kg/m (648 lb/lin ft) is the actual load; hence ditch conditions apply. To allow for variation in tile strength and bedding conditions include a safety factor of 1.5, which results in a design load of $976 \times 1.5 = 1464$ kg/m (972 lb/lin ft).

From Fig. 15.6 the load factor for ordinary bedding is 1.5, and from Appendix D the supporting strength of standard-quality tile is 1190 kg/m (800 lb/lin ft), based on the three-edge bearing method. Strength for ordinary bedding condi-

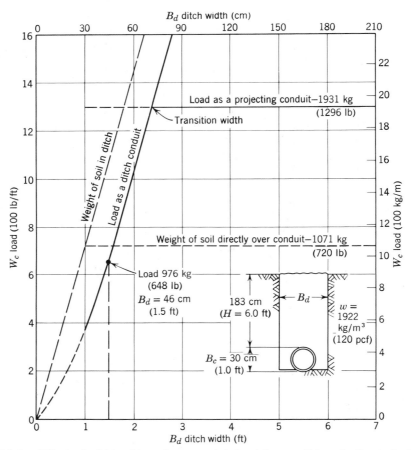

Fig. 15.8. Effect of width of trench on conduit load for conditions in Example 15.1.

tions is $1190 \times 1.5 = 1785$ kg/m (1200 lb/lin ft), which is sufficient to support a design load of 1464 kg/m (972 lb/lin ft).

15.13. Soil Loads on Flexible Pipe. For loads on flexible pipe Spangler and Handy (1973) assumed the soil beside the tubing had essentially the same stiffness as the tubing itself and multiplied Eq. 15.1 for ditch conditions by B_c/B_d to obtain

$$W_c = C_d w B_c B_d \tag{15.3}$$

This equation is valid only when thoroughly tamped soil on the sides of the tubing has essentially the same stiffness as the tubing itself. By knowing the load on the tubing the deflection can be computed from the modified Iowa formula (Spangler and Handy, 1973)

$$\Delta x = \frac{D_l \, K \, W_c}{(EI/r^3) + 0.061 \, E'}$$ (15.4)

where Δx = horizontal deflection ($\Delta x = 0.91 \, \Delta y$) (L),
D_l = deflection lag factor (usually 1.5),
K = bedding angle factor (0.096 for 90-degree angle),
W_c = soil load per unit length (FL^{-1}),
r = mean radius of tubing (L),
E = modulus of elasticity of tubing (FL^{-2}),
I = moment of inertia of tubing (L^4),
E' = modulus of soil reaction (FL^{-2}),

The two additive terms in the denominator represent the contributions of pipe stiffness and soil side support, respectively. The tubing stiffness term, EI/r^3, can be more easily computed from the parallel plate stiffness test, in which

$$EI/r^3 = 0.149C \, (F/\Delta y)$$ (15.5)

where $F/\Delta y$ = parallel plate stiffness for tubing as specified in ASTM F405,
C = correction factor for radius of curvature with deflection.

Field tests by Schwab and Drablos (1977) on tubing varying in diameter from 102- to 381-mm diameter showed that the load, when computed by the flexible pipe equation, was much too low probably because soil compaction and the modulus of soil reaction did not conform to the assumptions in Eq. 15.3. They found that tubing reaches a maximum deflection about 4 years after installation and that measured maximum deflections were about 17 percent. Deflections up to 20 percent in the field usually do not cause failure of corrugated HDPE tubing, but at 30 percent, collapse from buckling may occur.

Fenemor et al. (1979) found that the Iowa formula using $D_l = 3.4$, $E' = 345$ kN/m^2, and pipe stiffness at 20 percent deflection gave reasonable results that agreed with field measured deflections. Recommended maximum depths for tubing buried in loose, fine-grained soil are given in Appendix E. Howard (1977) developed E' values for a wide range of soils.

ESTIMATING COST

The three principal items of cost for a pipe drainage system are installation, engineering, and material costs, such as tile, tubing, outlet structures, and surface inlets. In the Midwest, the total cost for small laterals about 1 m in depth is approximately 45 percent for materials, 45 percent for installation, and 10 percent for engineering. In California, the total cost at a depth of 2 m is about three times the cost of 150-mm pipe; that is, labor and installation costs constitute about two-thirds of the total.

15.14.   Installation Costs.   The cost of installation depends primarily on the method of installation; depth of cut; size of pipe; presence of unusual soil conditions, such as stones and quicksand; and the number of junctions. Most ditching machine contractors install pipe at a fixed price per unit length, including trenching, laying the tile, blinding, and backfilling. An additional charge is usually made for each increment of overcut below a specified depth and for 200-mm pipe or larger. Frequently, contractors charge extra for removing stones and for making junctions.

15.15.   Material Costs.   The principal materials to be considered are: pipe, junctions, gravel, and material for constructing accessory facilities, such as outlets and surface inlets. The relative cost of drain pipe is given in Table 15.1. Sewer tile or metal pipe are suitable for outlet structures, surface inlets, and relief wells. Manufactured junctions, either T- or Y-shape, usually cost about five to ten times as much as 0.3 m of ordinary pipe.

15.16.   Engineering and Supervision Costs.   Engineering and supervision costs normally vary from 5 to 10 percent of the total cost. Engineering services should include the preliminary survey, location of the drains, staking the line, designing the system, and checking the grade and other specifications after installation.

15.17.   Other Costs.   Where the system is large and more than one property owner is involved, other costs may include the charge for the right-of-way,

Table 15.1   Relative Cost of Drain Pipe by Size

Nominal Dia.		Relative Cost	Clay or Concrete Tile	
(mm)	*(in.)*	*Corrugated Plastic Tubing*	*Std. Quality*	*Extra Quality*
75	3	82	90	95
100	4	100	100[a]	105
125	5	158	152	158
150	6	225	192	209
200	8	440	326	336
250	10	750	586	595
300	12	1320	788	820

[a] Arbitrarily taken as 100%, which was about $0.66 per m in Ohio in 1976.

clearing of trees, removal and replacement of fences and bridges, enlargement of outlet ditches, and so on. These costs vary widely for different installations.

MAPPING THE DRAINAGE SYSTEM

A suitable map should be made of the drainage system and filed with the deed to the property. Much time, effort, and money have been wasted because the location of old lines was not known. A record of the drainage system is also of considerable value to present and future owners for planning new lines, maintenance, and repair. The essential features of a drainage map are shown in Fig. 14.16. The drainage system may be mapped by aerial photography, which shows greater detail than sketch maps. Aerial photos should be taken shortly after installation, but the lines may show later because of drier soil or differences in vegetation over the drains.

MAINTENANCE

15.18. Causes for Tile Drain Failures. As a result of 30 years' experience in 36 states, Shafer (1940) classified all tile drain failures into five categories: (1) lack of inspection or maintenance, (2) improper design, (3) improper construction, (4) manufacturing processes and materials used, and (5) physical structure of the soil. As an indication of the relative importance of these classes of failures, a survey in Ohio showed that the percent of failure due to each of these causes was (1) 29, (2) 28, (3) 23, (4) 20, and (5) none, respectively. Although failures caused by poor physical structure of the soil are not very extensive, the installation of pipe in such soils might be properly classified as poor design. The principal causes of concrete and clay tile failures are the lack of resistance to freezing and thawing and inadequate strength. In design the major causes of failure are insufficient capacity, pipe placed at too shallow a depth, and lack of auxiliary structures, such as surface inlets. As shown in Fig. 15.9, failure to provide a cantilevered pipe or other structure at the outlet is a common cause of failure. Improper construction, which in the survey accounted for 23 percent of the failures, results from too wide crack spacings between the tile, improper bedding conditions, poor junctions, nonuniform grade, careless backfilling, and poor alignment. The lack of inspection and maintenance in the above survey was the major cause for failure, representing 29 percent of the total. Such failures are due mainly to the washout of surface inlets and outlet structures, piping over the lines, clogging from root growth, and failure to keep the outlet in good condition.

15.19. Preventive Maintenance. In comparison to open ditches pipe drains require relatively little maintenance. Open ditch outlets should be kept free of weed and tree growth, and the end of the outlet pipe should be covered

Fig. 15.9. Tile outlet failure due to lack of maintenance. (Courtesy U.S. Soil Conservation Service.)

with a flood gate, a screen, or a flap gate. The outlet ditch must not fill with sediment so as to obstruct the flow of the water from the pipe. Surface water should not be diverted into the ditch at or near the pipe outlet. Where shallow depths are required, drains should be protected from tramping of livestock and should not be crossed with heavy machinery, particularly during wet periods.

Pipe drains should be kept free of sediment and other obstructions. Roots of trees and certain other plant roots may grow into the drain and obstruct the flow, especially if the drain is fed by springs supplying water far into the dry season. Brush and trees, particularly willow, elm, cottonwood, soft maple, and eucalyptus, must be removed if within 30 m of the drain. Where it is impractical to remove this vegetation, nonperforated plastic tubing, sealed bell and spigot tile, or metal pipe should be installed. Sugarbeet and alfalfa roots have been known to enter drain lines, but these plants usually do not cause trouble because the roots die, decay, and are washed away.

During the first year after installation drain lines should be carefully watched to detect evidences of failure. Sinkholes over the line indicate a broken tile or too wide a crack or opening. Surface water should be diverted across the trench since this water may enter the drain or erode the backfill and wash out the pipe. Sediment basins should be cleaned at regular intervals. Surface inlets must be kept free of weed growth and sediment around the entrance.

15.20. Corrective Maintenance. Pipe drains that become filled with sediment or plant roots may be cleaned by digging holes along the line every 15 m or at shorter intervals. Between these holes the material in the drain may be removed with a suitable plug, swab, or sewer rod. Where sufficient water and grade are available, sediment can be washed out. Water can be forced at high pressure through a hose inserted from the outlet end. If the above methods are not practical, it may be more economical to re-install the entire line. Broken tile failures may be located by digging down to the drain at various points along the line. If water rises in any of the resulting holes, the failure is nearer the outlet.

In Europe and in California and Florida mineral deposits of predominantly iron oxide and manganese oxide have formed in pipe drains. MacKenzie (1962) reported that these minerals can be satisfactorily removed by injecting sulfur dioxide gas for a period of 24 hours. To obtain the desired chemical action water must also be in the lines during this time.

Especially in peat and muck soils, iron ochre may form a gelatinous mass that seals the drain openings. Ochre is generally a combination of bacterial and chemical sludge. Practical methods for eliminating the problem have yet to be developed.

REFERENCES

American Society of Agricultural Engineers. "Design and Construction of Subsurface Drains in Humid Areas," in *ASAE Yearbook* (See current issue for most recent recommendation).

Blaisdell, F. W., and P. W. Manson (1963). Loss of Energy at Sharp-Edged Pipe Junctions in Water Conveyance Systems, *U.S. Dept. Agr. Tech. Bull. 1283.*

Fenemor, A. D., B. R. Bevier, and G. O. Schwab (1979). "Predictions of Deflection for Corrugated Plastic Tubing." *Trans. ASAE,* 22 (6), 1338–1342.

Howard, A. K. (1977). "Modulus of Soil Reaction Values for Buried Flexible Pipe." *J. Geotech. Eng. Div., ASCE, GT1,* 103, 33–43.

MacKenzie, A. J. (1962). Chemical Treatment of Mineral Deposits in Drain Tile, *J. Soil Water Cons.* 17, 124–125.

Manson, P. W., and F. W. Blaisdell (1956). Energy Losses at Draintile Junctions, *Agr. Eng.* 37, 249–252, 257.

Marston, A. (1930). The Theory of External Loads on Closed Conduits in the Light of the Latest Experiments, *Iowa Eng. Expt. Sta. Bull. 96.*

Nelson, R. W. (1960). Fiberglass as a Filter for Closed Tile Drains, *Agr. Eng.* **41**, 690–693, 700.

Schlick, W. J. (1932). Loads on Pipe in Wide Ditches, *Iowa Eng. Expt. Sta. Bull. 108.*

Schwab, G. O. and C. J. W. Drablos (1977). "Deflection—Stiffness Characteristics of Corrugated Plastic Tubing." *Trans. ASAE,* **20** (6), 1058–1061, 1066.

Shafer, F. F. (1940). "Causes of Failure in Tile Drains." *Agr. Eng.* **21**, 17–18, 20.

Spangler, M. G. and R. L. Handy (1973). *Soil Engineering.* Intext Educational Publishers, New York.

Spangler, M. G. (1947). "Underground Conduits—An Appraisal of Modern Research." *Proc. Am. Soc. Civil Engrs.* **73**, 855–884.

PROBLEMS

15.1. Select the grade and compute cuts at each station for a 183-m (600-ft) drain line outletting into an open ditch so that cuts do not exceed 130 cm (4.2 ft). Design for an average depth of 120 cm (4 ft). Elevation of the water surface in the ditch is 25.15 m (82.50 ft) and hub elevations of successive 30-m (100-ft) stations are 26.46 m (86.81 ft), 26.61 (87.29), 26.80 (87.92), 26.87 (88.15), 26.94 (88.40), 27.01 (88.61), and 27.06 (88.79).

15.2. Determine the soil load on 300-mm (12-in.) drain tile installed at a depth of 305 cm (10 ft) in a trench 61 cm (24 in.) wide, assuming that the maximum weight of saturated topsoil in the backfill is 1761 kg/m³ (110 pcf). Calculate the design load, using a safety factor of 1.5.

15.3. What quality of drain tile are required to withstand the design load as determined in Problem 15.2 if ordinary bedding conditions are provided?

15.4. For the tile in Problem 15.3, what width of trench should be dug to permit the installation of standard-quality tile with ordinary bedding?

15.5. Based on the design load, what bedding conditions are necessary if only standard-quality tile were available for the installation described in Problem 15.2?

15.6. Estimate the cost of the drainage system in Fig. 14.16. The average cut in the main is 137 cm (4.5 ft) and for the laterals is 122 cm (4.0 ft). *Installation Cost:* 122 cm (4.0 ft) or less, $20 per 30 m (100 ft); overcut, $0.60 per 0.03 m (0.1 ft) per 30 m (100 ft). *Materials:* Tile price based on Table 15.1, using $200 per 1000 ft for 100-mm (4-in.) pipe and adding 5 percent for breakage; corrugated metal outlet pipe at $13 per m ($4 per ft); junctions at five times cost for 0.3 m length. *Engineering:* 5 percent of total materials and installation costs.

15.7. Determine the slope of a pipe drain in percent from the following field notes:

Sta	Hub Elev	Cut
0 + 00 m (0 + 00 ft)	30.00 m (98.40 ft)	122 cm (4.00 ft)
0 + 30 (1 + 00)	30.08 (98.70)	122 (4.00)
0 + 60 (2 + 00)	30.14 (98.90)	119 (3.90)
0 + 90 (3 + 00)	30.36 (99.60)	136 (4.45)

15.8. If the sight bar on a trenching machine is set 2.74 m (9.0 ft) above the bottom of the trench, compute the height above the hub stakes to set the target at each station along the drain line given in Problem 15.7.

CHAPTER 16

Pumps and Pumping

Although the agricultural engineer is not generally required to design pumping equipment, he should be able to determine the proper type of pump, the number and size of pumps required, and the size of power units. In addition, he may be called upon to design the plant installation, estimate the cost of operation, and supervise construction and operation of the plant.

Pumping plant installations are encountered principally in drainage and irrigation enterprises. Pumping plants for drainage provide outlets for open ditches and pipe drains. In irrigation, pumping from wells and storage reservoirs into irrigation canals or pipes or into other reservoirs is common practice.

TYPES OF PUMPS

Of the many types of pumps available, the centrifugal, propeller, and reciprocating pumps are by far the most common. From these three types, pumps may be selected for a wide range of discharge and head characteristics. Reciprocating pumps, sometimes called piston or displacement pumps, are capable of developing high heads, but their capacity is relatively small. They are not ordinarily suitable for drainage and irrigation, especially if sediment is present.

Centrifugal and propeller pumps both incorporate a rotating impeller. In centrifugal pumps the flow from the impeller is radial while in propeller pumps the flow from the impeller is axial. Many pumps discharge flow from their impellers on vectors which lie between radial and axial. These pumps are referred to as mixed flow pumps.

A commonly used index of pump type is the specific speed. Specific speed is calculated from the formula (Church, 1944)

$$n_s = \frac{51.65\, N\, q^{1/2}}{h^{3/4}} \tag{16.1}$$

where n_s = specific speed in rpm,
N = speed in rpm,
q = discharge in m³/s,
h = total head in m.

366

Specific speed is the speed in revolutions per minute at which an impeller would run if reduced in size to deliver 0.063 L/s (1 gpm) against a total head of 0.3 m (1 ft).

Figure 16.1 shows the efficiency of impeller pumps as a function of specific speed. Centrifugal pumps have low specific speeds, low discharge, and high head while propeller pumps are characterized by high specific speeds, high discharge, and low head. Low specific speeds apply to centrifugal pumps while high values are characteristic of propeller pumps. Mixed flow or turbine pumps have intermediate specific speeds.

CENTRIFUGAL PUMPS

Centrifugal pumps are economical in cost and simple in construction, yet they produce a smooth, steady discharge. They are small compared to their capacity, easy to operate, and suitable for handling sediment and other foreign material.

16.1. Principle of Operation. The centrifugal pump consists of two main parts: (1) the impeller or rotor that adds energy to the water in the form of increased velocity and pressure, and (2) the casing that guides the water to and from the impeller. As shown in Fig. 16.2, the water enters the pump at the center of the impeller and passes outward through the rotor to the discharge opening. By changing the speed of the pump, the discharge can be varied. Theoretically, the so-called pump laws for a centrifugal pump state that (1) the discharge is directly proportional to the speed of the impeller, (2) the head varies as the square of the speed, and (3) the power varies as the cube of the speed. For a pump of a given size, changing the diameter of the impeller, thereby the peripheral velocity has the same effect as changing the speed.

16.2. Classification. With regard to the construction of the casing around the impeller, centrifugal pumps are classified as volute or turbine, as shown in Fig. 16.3. The volute-type pump is so named because the casing is in the form of a spiral with a cross-sectional area increasing toward the discharge opening. Turbine-type pumps, sometimes called diffuser-type, have stationary guide vanes surrounding the impeller. As the water leaves the rotor, the vanes gradually enlarge and guide the water to the casing, resulting in a reduction in velocity with the kinetic energy converted to pressure. The vanes provide a more uniform distribution of the pressure. Turbine pump casings may be either circular as shown or spiral like volute casings.

Centrifugal pumps are built with horizontal or vertical drive shafts and with different numbers of impellers and suction inlets. The suction inlet may be either single or double acting, depending on whether the water enters from both sides or one side of the impeller. Single-suction, horizontal centrifugal pumps

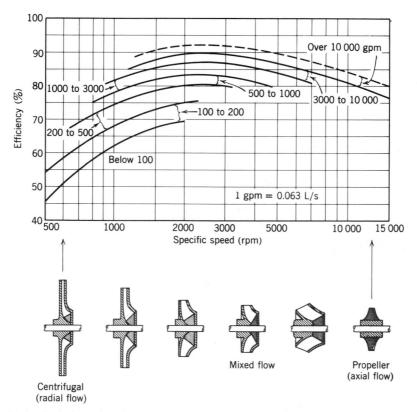

Fig. 16.1. Relationship of specific speed, impeller shape, efficiency, and pump type. (Courtesy Worthington Corporation.)

are frequently used where the suction lift does not exceed 4 to 6 m. Practically all turbine pumps are of the vertical type. Pumps with more than one impeller are known as multiple-stage pumps, sometimes referred to as deep-well turbine pumps. Both volute and turbine centrifugal pumps may have multiple impellers, but they are more common in the turbine type. Deep-well turbine pumps are here considered as centrifugal pumps with one of the following types of impellers to be described: centrifugal types, mixed flow, and propeller (see Fig. 16.6). Some authorities prefer to classify these pumps separately from ordinary centrifugals because all impellers do not function entirely on the centrifugal principle.

16.3. Centrifugal-Type Impellers. The design of the impellers greatly influences the efficiency and operating characteristics of the pump. Centrifugal-type impellers shown in Fig. 16.4 are classified as (a) open, (b) semienclosed, and (c)

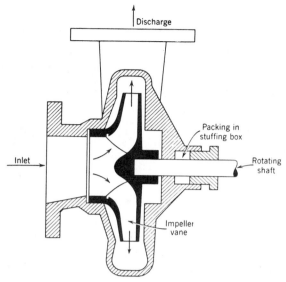

Fig. 16.2. Cross section of a horizontal centrifugal pump (single suction enclosed impeller).

enclosed. The open-type impeller has exposed blades that are open on all sides except where attached to the rotor. The semienclosed impeller has a shroud (plate) on one side; the enclosed impeller has shrouds on both sides, thus enclosing the blades completely. The open and semienclosed impellers are most suitable for pumping suspended material or trashy water. Enclosed impellers are generally not suitable where suspended sediment is carried in the water as this material greatly increases the wear on the impeller.

In deep-well turbine pumps mixed-flow impellers, which are a combination of axial-flow and centrifugal-type impellers, are often used. Such impellers have a higher capacity than the centrifugal type. Water movement in the pump is the result of both centrifugal force and direct thrust.

16.4. Performance Characteristics. In selecting a pump for a particular job the relationship between head and capacity at different speeds should be known as well as pump efficiency. Curves that provide this data are called *characteristic* curves, as shown in Fig. 16.5. The head-capacity curve shows the total head developed by the pump for different rates of discharge. At zero flow the head developed is known as the shut-off head. Head losses in pumps are caused by friction and turbulence in the moving water, shock losses due to sudden changes in momentum, leakage past the impeller, and mechanical friction. For a given speed the efficiency can be determined for any discharge from these characteristic curves. A pump should be selected that will have a high

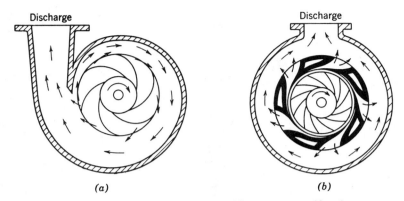

Fig. 16.3. (a) Volute-type and (b) turbine-type centrifugal pumps.

efficiency for a wide range of discharges. For example, from the upper curve in Fig. 16.5, 71 percent efficiency or more can be obtained for capacities varying from 0.027 to 0.046 m³/s (430 to 730 gpm) at heads of 99 to 75 m, respectively.

Characteristic curves can be obtained from the pump manufacturer. Such curves will vary in shape and magnitude, depending on the size of the pump, type of impeller, and over-all design. Performance characteristics are normally obtained by tests of a representative production line pump rather than by tests of each pump manufactured.

For operation of a centrifugal pump without cavitation, the suction lift plus all other losses must be less than the theoretical atmospheric pressure. The maximum practical suction lift can be computed by the equation,

$$H_s = H_t - H_f - e_s - NPSH - F_s \tag{16.2}$$

where H_s = maximum practical suction lift or elevation of the pump
                     center line minus the elevation of the water surface (L),
           H_t = atmospheric pressure at the water surface (L),

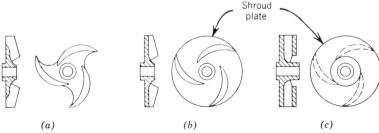

Fig. 16.4. (a) Open, (b) semienclosed, and (c) enclosed centrifugal-type impellers.

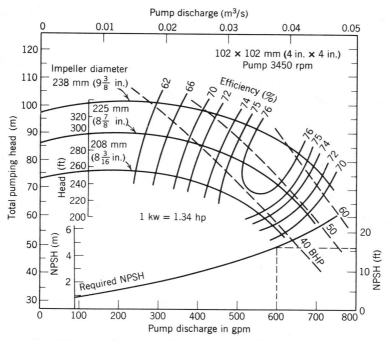

Fig. 16.5. Performance characteristics of a centrifugal pump.

H_f = friction losses in the strainer, pipe, fittings, and valves of the suction line (L),

e_s = saturated vapor pressure of the water (L),

$NPSH$ = net positive suction head of the pump including losses to the impeller and velocity head (L),

F_s = safety factor, which may be taken as 0.6 m (2 ft) with all other units in m (ft).

The approximate correction of H_t for altitude is a reduction of about 1.2 m/1000 m (1.2 ft/1000 ft) of altitude. Friction losses and suction lift should be kept as low as possible. For this reason the suction line is usually larger than the discharge pipe, and the pump is placed as close as possible to the water supply. The head loss equivalent due to the vapor pressure of the water must be considered to prevent cavitation, but it does not add to the total suction head when the pump is operating. The $NPSH$ is a characteristic of the pump and should be furnished with the characteristic curves as shown in Fig. 16.5. The following example will illustrate the calculation of suction lift.

Example 16.1. Determine the maximum practical suction lift for a pump having the characteristics shown in Fig. 16.5 if the discharge is 0.038 m³/s (600

gpm); the water temperature is 20°C (68°F), the total friction loss in the 102-mm (4-in.) suction line and fittings is 1.58 m (5.2 ft), and the pump is to be operated at an altitude of 305 m (1000 ft) above sea level.

Solution.   Substituting in Eq. 16.2 after reading $NPSH = 4.82$ m (15.8 ft) from Fig. 16.5, obtaining $e_s = 0.24$ m (0.8 ft) from Chapter 3, assuming $F_s = 0.61$ m (2.0 ft), and correcting for altitude $H_t = 10.36 - 0.36$ m (34.0 − 1.2 ft) = 10.00 m,

$$H_s = 10.00 - 1.58 - 0.24 - 4.82 - 0.61 = 2.75 \text{ m (9.0 ft)}$$

Pump operation at the maximum practical lift should be avoided in design since the collection of debris on the screen and increase of pipe friction with age may cause the pump to lose prime. In order to prime the pump the suction line and pump should be nearly full of water. This can be accomplished by filling the pump manually with water or by removing the air with a suction pump or an exhaust primer. A gate valve on the discharge side of the pump and a check valve at the lower end of the suction line are essential for priming, except at very low suction heads.

PROPELLER PUMPS

16.5.   Principle of Operation.   As distinguished from centrifugal pumps, the flow through the impeller of a propeller-type pump is parallel to the axis of the driveshaft rather than radial. These pumps are also referred to as axial-flow or screw-type pumps. The principle of operation is similar to that of a boat propeller except that the rotor is enclosed in a housing. As shown in Fig. 16.6, there is considerable radial flow in mixed-flow pumps. Many propeller pumps have diffuser vanes mounted in the casing, similar to the turbine-type centrifugal pump.

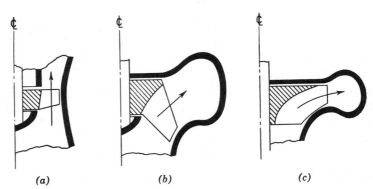

Fig. 16.6.   Flow through (a) propeller, (b) mixed flow, and (c) centrifugal pumps.

16.6.   Performance Characteristics.   Propeller pumps are designed principally for low heads and large capacities. The discharge of these pumps varies from 0.04 to 4.4 m³/s (600 to 70 000 gpm) with speeds ranging from 450 to 1760 rpm and heads usually not more than 9 m (30 ft). Although most propeller pumps are of the vertical type, the rotor may be mounted on a horizontal shaft.

Performance curves for a large propeller pump are shown in Fig. 16.7. In comparison to centrifugal pumps, the efficiency curve is much flatter, heads are considerably less, and the power curve is continually decreasing with greater discharge. With propeller pumps the power unit may be overloaded by increasing the pumping head. These pumps are not suitable where the discharge must be throttled to reduce the rate of flow.

PUMPING

Although in many respects pumping water for irrigation is similar to that for drainage, design requirements differ. For example, in pump drainage, heads are generally 6 m (20 ft) or less, whereas for irrigation heads up to 90 m (300 ft) are common.

16.7.   Power Requirements and Efficiency.   For a given discharge the power requirements for pumping are proportional to the total pumping head including velocity head, friction losses, static head—which is the difference between the free water surface at the inlet and at the discharge point—and the pressure head at the point of discharge. Where the end of the discharge pipe is higher than the water at the outlet, the head is measured to its center. Power requirements for pumping are computed by the formula

$$kW = 1.341 \text{ hp} = 17.64qh/E_p \tag{16.3}$$

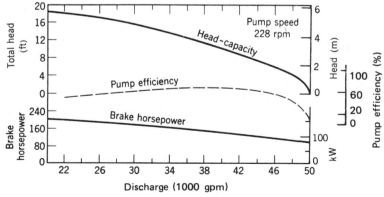

Fig. 16.7.   Performance curves for a propeller-type pump.

where kW (1.34 hp) = (input) power delivered to pump,

q = discharge rate, m³/s,

h = total head in m,

E_p = pump efficiency as a decimal fraction.

Considerable energy is utilized in overcoming friction losses in the pump and for velocity head losses. Because of such losses pumping plant efficiencies range from about 75 percent under favorable conditions to as low as 20 percent or less for unfavorable conditions. A well-designed pump should have an efficiency of 70 percent or more over a wide range of operating heads. Such efficiencies in the field are difficult to maintain because of wear in the pump and other factors. It is good practice to check the pump in the field to be sure that satisfactory efficiencies are obtained. If efficiencies are low, the source of difficulty should be located and corrected. The over-all efficiency of the installation, which includes the efficiency of the power plant, the pipe system, and the pump, should also be checked.

16.8. Power Plants and Drives. Power plants for pumping should deliver sufficient power at the specified speed with maximum operating efficiency. Internal combustion engines and electric motors are by far the most common types of power units. The selection of the type of unit depends on (1) the amount of power required, (2) initial cost, (3) availability and cost of fuel or electricity, (4) annual use, and (5) duration and frequency of pumping.

Internal combustion engines operate on a wide variety of fuels, such as gasoline, diesel oil, natural gas, and butane. Where the annual use is more than 800 to 1000 hours, the diesel engine may be justified. Otherwise, it may be more practical to use a gasoline engine. For continuous operation water-cooled gasoline engines may be expected to deliver 70 percent of their rated horsepower, diesel engines 80 percent, and air-cooled gasoline engines 60 percent. Where a vertical centrifugal or a deep-well turbine pump is to be driven with an internal combustion engine, right-angle gear drives are usually employed. In some instances internal combustion engines are adapted for mounting with their crankshafts vertical so they may be directly coupled to a vertical driveshaft. Belt drives may utilize either flat or V-belts suitable for driving vertical shafts or gear drives. In comparison with direct drives, which have a transmission efficiency of nearly 100 percent, the efficiency for gear drives ranges from 94 to 96 percent, V-belts from 90 to 95 percent, and flat belts from 80 to 95 percent.

Electric motors operate best at full-load capacity. They have many advantages over internal combustion engines, such at ease of starting, low initial cost, low upkeep, and suitability for mounting on horizontal or on vertical shafts. Direct drive motors are preferred because gears and belts are eliminated. Deep-well pumps are now available with watertight vertical motors. The motor

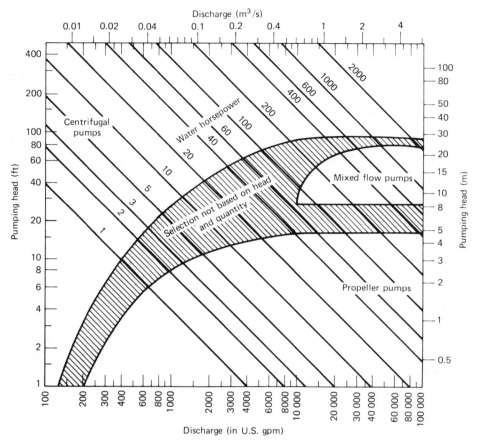

Fig. 16.8. Pump selection chart by discharge and head. (Redrawn from U.S. Soil Conservation Service, 1973.)

is submerged in the well near the impellers, thus eliminating the long shaft otherwise required.

16.9. Selection of Pump. Performance curves serve as a basis for selecting a pump to provide the required head and capacity for the range of expected operating conditions at or near maximum efficiency. The factors that should be considered include the head-capacity relationship of the well or sump from which the water is removed, space requirements of the pump, initial cost, type of power plant, and pump characteristics, as well as other possible uses for the pump. Storage capacity, rate of replenishment, and well diameter sometimes limit the pump size and type. For example, a large drainage ditch provides a nearly continuous source of water, whereas a well usually has a small storage capacity.

The initial cost that can be justified depends largely on the annual use and other economic considerations. The size of the power plant and type of drive should be adapted to the pump. In off-seasons pumps used intermittently may be made available for other purposes.

Figure 16.8 is a chart for selecting the type of pump. Their efficiencies may be estimated from Fig. 16.1. The centrifugal pump is suitable for low to high capacities at heads up to several hundred meters, the propeller pump for high capacities at low heads, and the mixed-flow pump for intermediate heads and capacities. Since impellers of these pumps may be placed near to or below the water level, the suction head developed may not be critical. Horizontal centrifugal pumps are best suited for pumping from surface water supplies, such as ponds and streams, provided the water surface does not fluctuate excessively. Where the water level varies considerably, a vertical centrifugal or a deep-well turbine pump is more satisfactory.

16.10. Pumping Costs. The cost of pumping consists of fixed and operating costs. Fixed costs include interest on investment, depreciation, taxes, and insurance. Included in the investment cost is the construction and development of the well, the pump, the power plant, pump house, and water storage facilities. Construction costs for wells depend largely on the size and depth of the well, construction methods, nature of the material through which the well is drilled or dug, type of casing, length and type of well screen, and the time required to develop and test the well.

Table 16.1 gives estimated service life of various components of pumping plants for estimating depreciation. Taxes and insurance are approximately 1.5 percent of the total investment. Operating costs include fuel or electricity, lubricating oil, repairs, and labor for operating the installation. Fuel costs for internal combustion engines are generally proportional to the consumption, whereas the cost of electrical energy generally decreases with the amount consumed. A demand charge or minimum is made each month regardless of the amount of energy consumed. This demand charge may or may not include a given amount of energy.

In a study of nine drainage pumping plants (electric power) in the upper Mississippi Valley, Sutton (1932) found that 33 percent of the total cost was for fixed costs, 53 percent for power, 7 percent for labor, and 7 percent for other expenses. The average over-all efficiency for these plants was 42 percent and the average lift 2.9 m.

The pumping cost per unit volume depends largely on the total quantity pumped per season. The effect of annual water use on the unit cost per unit height of lift is shown in Fig. 16.9. The unit cost for 12 units of annual water use

Table 16.1. Service Life or Pumping Plant Components

Item	Estimated Service Life
Well and casing	20 yr
Plant housing	20 yr
Pump turbine:	
Bowl (about 50% of cost of pump unit)	16 000 hr or 8 yr
Column, etc.	32 000 hr or 16 yr
Pump, centrifugal	32 000 hr or 16 yr
Power transmission:	
Gear head	30 000 hr or 15 yr
V-Belt	6000 hr or 3 yr
Flat-Belt, rubber and fabric	10 000 hr or 5 yr
Flat-Belt, leather	20 000 hr or 10 yr
Electric motor	50 000 hr or 25 yr
Diesel engine	28 000 hr or 14 yr
Gasoline or distillate engine:	
Air-cooled	8000 hr or 4 yr
Water-cooled	18 000 hr or 9 yr
Propane engine	28 000 hr or 14 yr

Source: U.S. Soil Conservation Service (1959).

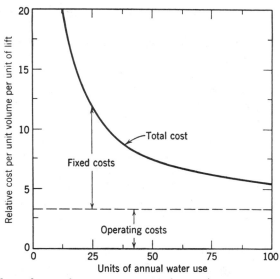

Fig. 16.9 Effect of annual water use on total pumping cost per unit of volume.

is about three times that for 100 units. Because of lower lifts and smaller investment, pumping for drainage is usually not as expensive per unit pumped as for irrigation.

16.11. Design Capacity. The first consideration in selection of a pump and design of a pumping plant must be the determination of the required capacity.

Irrigation pumping requirements depend upon peak evapotranspiration rates, irrigation efficiencies, and the planned frequency and duration of pumping plant operation. Design capacity of the pumping plant may also be dictated by characteristics of the water supply. A crop that has a peak consumptive use rate of 5 mm/day will require a pumping capacity of 0.58 L/s per hectare irrigated plus additional capacity to supply conveyance and other irrigation losses if 24-hour per day pumping is required.

Drainage coefficients for pumping plant design vary according to soil, topographic, rainfall, and cropping conditions. Recommended coefficients for the upper Mississippi Valley range from 7 to 25 mm/d. Requirements in Southern Texas and Louisiana are as high as 76 mm, including pumping discharge and reservoir storage. In Florida a drainage coefficient of 25 mm/d is suitable for field crops on organic soils. Truck crops may require and economically justify a drainage coefficient of as much as 76 mm/d.

16.12. Drainage Pumping Plants. Pumps offer a means of providing drainage where gravity outlets are not otherwise available. Typical applications of pump drainage include land behind levees, outlets for pipe drains and open ditches, and flat land where the gravity outlet is at considerable distance. Drainage pumping plants are normally required to handle large capacities at low lifts.

The design of small farm pumping plants may be simplified by using the design runoff rate recommended for open ditches plus 20 percent. Where electricity is available, electric motors with automatic controls are recommended for small installations. For drainage areas larger than 40 ha, storage capacity should be obtained by enlarging the open ditches or by excavating a storage basin. Constructed sumps as shown in Fig. 16.10 are generally too expensive where the diameter is greater than 5 m. Where storage is available in open ditches or other reservoirs, sumps about 2 m in diameter are satisfactory. Figure 16.11 illustrates a pumping plant for lifting drainage water over a levee. Most pumps operate only during the wet spring months, but some installations draining extremely low land may operate more or less continuously throughout the year. Usually, these plants operate intermittently and are used only 10 to 20 percent of the time.

For economical performance it is better to operate a small pump for several days than to run a large pump intermittently for short periods. Frequently, drainage installations lack storage capacity and the allowable head variation is

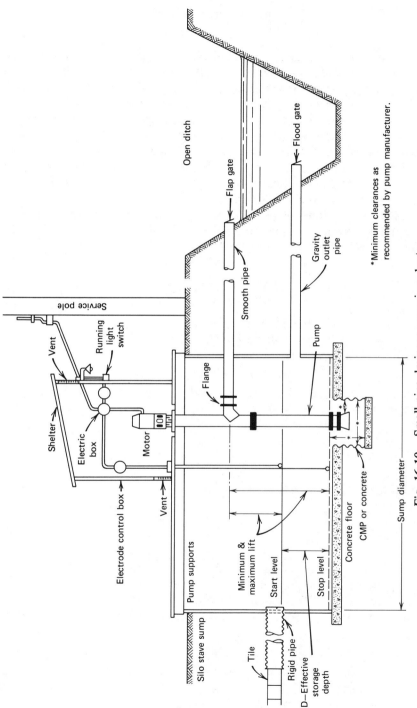

*Minimum clearances as recommended by pump manufacturer.

Fig. 16.10. Small pipe-drainage pumping plant.

379

small. Under these conditions automatically controlled electric motors are especially suitable. Larson and Manbeck (1961) have found that 10 to 15 cycles per hour are satisfactory for drainage pumping plants with automatic controls. Internal combustion engines are usually started manually, but an automatic shut-off can be provided. The number of stopping and starting cycles should not exceed more than two per day for nonautomatic operation.

Engine-operated plants require a larger storage capacity than those electrically operated. For large installations two pumps with different capacities may be necessary, one for handling surface runoff during wet seasons and the other for seepage or flow from pipe drains.

Sufficient storage must be provided in a sump to avoid excessive starting and stopping. The continuity equation for the total time per cycle is the sum of the running and the stopping time for the pump and may be written

$$\frac{3600}{n} = \frac{S}{q - q_i} + \frac{S}{q_i} \tag{16.4}$$

where n = number of cycles per hour,
         S = storage volume in m^3,
         q = average pumping rate in m^3/s,
         q_i = inflow rate in m^3/s.

By solving for Sn, assuming q constant, differentiating S with respect to q_i, and setting $dS/dq_i = 0$ to obtain maximum storage,

$$q_i = \frac{q}{2} \tag{16.5}$$

Substituting q_i above in Eq. 16.4 results in

$$S = \frac{900\,q}{n} \tag{16.6}$$

Thus, maximum storage occurs when the inflow rate is one half the discharge of the pump, and the storage requirements depend on the discharge rate and the frequency of cycling. When the inflow rate exceeds the pumping rate, the pump operates continuously. When the inflow rate is less than the pumping rate, cycling occurs.

Example 16.2.   A drainage pump has a design capacity of 0.0379 m^3/s (600 gpm) and is to be installed in a 2.44-m (8-ft) diameter sump with a maximum cycling frequency of 10 cycles per hr. What is the minimum difference in elevation between the start and stop levels in the sump?
Solution.   Substituting in Eq. 16.6,

$$S = \frac{900 \times 0.0379}{10} = 3.41 \text{ m}^3 \ (120 \text{ ft}^3)$$

$$\text{Depth of storage} = \frac{3.41 \times 4}{3.14 \times 2.44 \times 2.44} = 0.73 \text{ m} \ (2.4 \text{ ft})$$

Other combinations of depth and sump diameter are possible, provided the storage volume and other design requirements of the installation are met.

The pumping plant should be installed to provide minimum lift and located so that the pump house will not be flooded. A farm pumping plant for pipe drain or open ditch drainage of small areas is shown in Fig. 16.10. A circular sump is recommended for such an installation since less reinforcing is required and it is easier to construct. The electric motor is usually operated automatically by means of an electrode switch or start and stop collars on a float rod. Where open ditches provide storage, the sump can be reduced in size. The discharge pipe should outlet below the minimum water level in the outlet ditch, where practicable, to reduce the pumping head to a minimum. The gravity outlet pipe should be installed only when the water level in the ditch is lower than the pipe invert for at least 10 percent of the flow period. The saving in pumping cost should justify the cost of the gravity outlet pipe and flood gate.

16.13. Irrigation Pumping Plants. Pumping requirements for irrigation vary considerably, depending upon such factors as the source of water, method of irrigation, and the size of area irrigated. Water may be obtained from wells, rivers, canals, pits, ponds, and lakes. Pumping from surface supplies is similar to pump drainage except where pressure irrigation is practiced. Much higher heads are normally required where water is pumped from wells or pumped to pressure irrigation systems. Water pumped from wells may be discharged into open ditches or into pipe lines for distribution to irrigated areas. Wells located in the vicinity of the area to be irrigated have many advantages as compared to canals supplying water from distant sources. A typical deep-well turbine pump installation discharging into an underground pipe line is shown in Fig. 16.12.

Because of excessive pressure (water hammer), which may develop from starting or stopping the pump when pumping into pipe lines, surge tanks should be installed in the discharge line, as illustrated in Fig. 16.11. However, they are usually not required for portable sprinkler irrigation systems. Surge tanks that are open to atmospheric pressure are suitable for pumping into pipe lines made of thin metal, concrete, or vitrified clay pipe. The surge tank must be high enough to allow for any increase in elevation of the pipe line, for friction head, and for a reasonable freeboard. When the total head exceeds 6 m (20 ft), an enclosed metal tank having an air chamber is usually more practical than a surge tank. The maximum head at which irrigation pumping is practical varies with the locality, value of the crop, and other economic factors. In some areas pumping is carried out for heads up to 90 m (300 ft) or more.

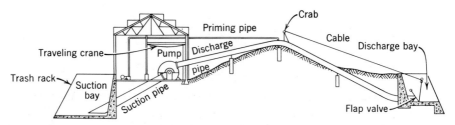

Fig. 16.11. Large drainage pumping plant with pipe over the top of levee. (Redrawn from Richey, 1961.)

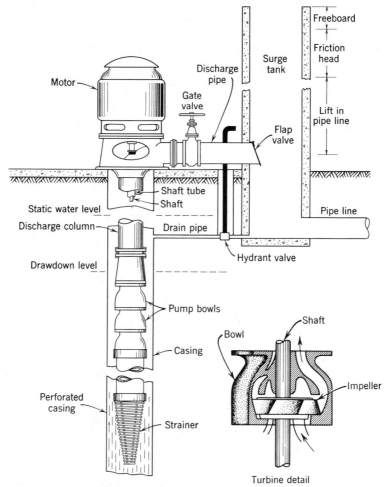

Fig. 16.12. Two-stage deep-well turbine pump installation with a surge tank for low heads. (Modified from Rohwer, 1943, and Wood, 1950.)

The pump should be fitted to the well. By plotting the head-capacity curve for the pump and the drawdown-discharge curve for the well (see Chapter 17) the intersection of the two curves gives the operating capacity of the plant. Such a set of curves for a typical field installation is shown in Fig. 16.13. The most desirable conditions for operating the pump occur when the capacity is 0.045 m³/s (720 gpm) and the head 13.4 m (44 feet). At this discharge the pump also operates at its maximum efficiency (65 percent) and the power unit at nearly its maximum power. If the discharge is changed, there is no danger of overloading the power unit since at other capacities the power requirement is always less. When maximum efficiency does not occur at the desired capacity, changing the pump speed may shift the efficiency curve to a higher value. If different speeds are not practicable, another size or type of pump must be selected. The above procedure for selecting a pump to fit the well emphasizes the importance of test-pumping the well to obtain the head-capacity curve before purchasing the pump.

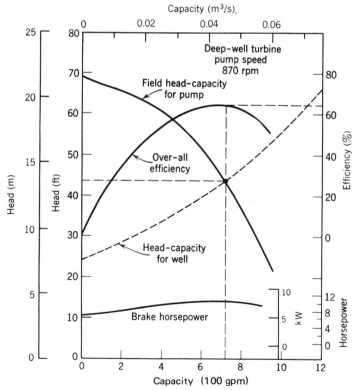

Fig. 16.13. Fitting the pump to the well using head-capacity curves. (Redrawn from Code, 1936.)

REFERENCES

Church, A. H. (1944). *Centrifugal Pumps and Blowers*. Wiley, New York.

Code, W. E. (1936). "Equipping a Small Irrigation Pumping Plant." Colo. Agr. Expt. Sta. Bull. 433.

Larson, C. L., and D. M. Manbeck (1961). "Factors in Drainage Pumping Efficiency." *Agr. Eng.* **42**, 296–297, 305.

Richey, C. B. (1961). *Agricultural Engineers Handbook*. McGraw-Hill, New York.

Rohwer, C. (1943). "Design and Operation of Small Irrigation Pumping Plants." U.S. Dept. Agr. Circ. 678.

Sutton, J. G. (1932). "Cost of Pumping for Drainage in the Upper Mississippi Valley." U.S. Dept. Agr. Tech. Bull. 327.

———— (1950). "Design and Operation of Drainage Pumping Plants." U.S. Dept. Agr. Tech. Bull. 1008.

U.S. Soil Conservation Service (1973). "Drainage of Agricultural Land." Water Information Center, Inc., Port Washington, N.Y.

U.S. Soil Conservation Service (1959). *"Irrigation Pumping Plants"* Irrigation, Section 15, Chapter 8. (Lithograph), Washington D.C.

Wood, I. D. (1950). "Pumping for Irrigation," U.S. Dept. Agr. SCS-TP-89.

PROBLEMS

16.1. A centrifugal pump discharging 0.0252 (400 gpm) against 20.7 m (68 ft) of head and operating with an efficiency of 60 percent requires 15.42 kW (11.5 hp) at 1000 rpm. What is the theoretical discharge if the speed is increased to 1500 rpm, assuming the efficiency remains constant? What is the theoretical head and horsepower at 1500 rpm?

16.2. From the performance curves for a centrifugal pump with a 238-mm (9⅜-in.) impeller shown in Fig. 16.5, what is the pump efficiency at a head of 91.5 m 130 psi)? What is the discharge? What is the efficiency and discharge if the head is reduced to 74.6 m (106 psi)?

16.3. What are the power requirements for pumping 0.063 m³/s (1000 gpm) against a head of 45.7 m(150 ft) assuming a pump efficiency of 65 percent? What size electric motor is required assuming an efficiency of 90 percent?

16.4. If the average discharge of a propeller pump is 0.063 m³/s (1000 gpm) what is the required depth between start and stop levels where the water is pumped from a sump 3.05 m (10 ft) in diameter? The maximum cycling rate is 12 cycles per hour. At what inflow rate would maximum cycling occur?

16.5. Determine the maximum practical suction lift for a centrifugal pump if the discharge is 0.063 m³/s (1000 gpm), the water temperature is 26.7°C (80°F), friction losses in pipe and fittings are 1.55 m (5.1 ft), the *NPSH* of the pump is 3.05 m (10 ft), and altitude for operation is 1524 m (5000 ft) above sea level.

CHAPTER 17

Water Supply and Quality

Precipitation is presently our only practical source of continuing fresh water supply for all agricultural, industrial, and domestic uses. Increasing attention being given to large scale desalinization of brackish or salty waters may eventually result in reasonable supplies of water for high value uses in some locations, but precipitation will remain the dominant source of water. Nearly 617 000 ha-m (5 000 000 ac-ft) of water fall on the 48 conterminous United States annually. Developed water supplies use only 4 percent of this amount, which is only 13 percent of the residual precipitation after allowing for evaporation and transpiration from natural plants and unirrigated crops. Thus, there is actually ample water for our needs. However, it is often not available at the time and place of need. The development of water resources involves storage and conveyance of water from the time and place of natural occurrence to the time and place of beneficial use.

This chapter emphasizes the development of water resources for agricultural use without intending in any way to minimize the importance of water for other uses. More detailed discussion is presented by Linsley and Franzini (1979).

SOURCES AND QUALITY OF WATER

17.1. Classification of Waters. Water in one or more of its three physical states—solid, liquid, or gaseous—is present in greater or lesser quantities in or on virtually all the earth, its atmosphere, and all things living or dead. Water, important from the standpoint of water-resource development, falls into the categories of atmospheric moisture, surface waters, and subsurface waters. Atmospheric moisture and resultant precipitation have been discussed in Chapter 2 and are the source of replenishment of surface and subsurface waters. Surface and subsurface waters are the direct source of our developable water resources.

17.2. Surface Water. Surface waters exist in natural basins and stream channels. Where minimum flows in streams or rivers are large in relation to water demands of adjacent lands, towns, and cities, development of surface waters is accomplished by direct withdrawal from the flow. On many streams and rivers, however, flow fluctuates widely from season to season and from year to year. Further, peak demands from many major rivers occur at seasons of minimum flow and in fact require that as much of the annual flow as possible be conserved and diverted for beneficial use.

This situation requires construction of reservoirs to hold flow during seasons or years of high runoff for later release to beneficial use. These reservoirs frequently incorporate hydroelectric, flood control, and recreational features in addition to their water supply function. Frequently, the income from electric power development is used to subsidize the water supply function.

Reservoirs range in size from several million hectare-meters for large multiple purpose reservoirs in the West to small ponds with less than a hectare-meter of storage. As examples of large reservoirs, the Grand Coulee and the Hoover Dam are 168 and 221 m high and have storage capacities of 0.74 and 3.46 million ha-m, respectively.

Surface water reservoirs are usually constructed to permit effective use of water downstream from the reservoir site. However, an interesting exception to this is the Glen Canyon Dam in northern Arizona, which was planned to permit use of Colorado River water in the upper basin of the Colorado River while the reservoir insures that yearly release of water from the upper basin will be adequate to meet lower basin appropriations.

Surface water supplies may be increased by water harvesting, which is any watershed manipulation carried out to increase surface runoff. In many arid areas, the majority of precipitation that infiltrates into the soil is lost for use either through direct evaporation or through transpiration by economically useless vegetation. For example, in the Colorado River Basin less than 6 percent of the precipitation appears as streamflow. Vegetation management has been shown to be effective as a means of increasing streamflow and it may be anticipated that it will receive increasing application.

More immediate in application is the practice of water harvesting by catchments as described by Lauritzen and Thayer (1966). Catchments are areas of concrete, sheet metal, asphalt, or otherwise treated or waterproofed soil specifically constructed to catch and collect precipitation. Successful water harvesting requires attention not only to collection of water but to the conveyance and storage of the water collected. Catchment treatment may vary from soil smoothing and removal of vegetation to application of plastic film or aluminum foil. Simple soil smoothing and vegetation removal has increased runoff by a factor of 3 in some tests.

Complete sealing will, of course, increase runoff to essentially 100 percent. Water harvesting is being applied to develop water supplies for wildlife, livestock and occasional domestic use.

17.3. Ground water. Subsurface water available for development is normally referred to as ground water. Ground water predominately results from precipitation that has reached the zone of saturation in the earth through infiltration and percolation. Ground water is developed for use through wells, springs, or dugout ponds.

In many areas where ground water is an important source of water supply it is being withdrawn much faster than it is being replenished from infiltration and percolation of precipitation. Figure 17.1 shows the severity of ground water lowering in a basin where the entire water supply is from ground water.

17.4. Water Quality. The quality of irrigation water depends on the amount of suspended sediment and chemical constituents in the water.

Sediment. The effect of sediment is influenced by the nature of the material and soil condition of the irrigated area. Where fine sediment is deposited on sandy soil, the textural composition and fertility may be improved. However, if the sediment has been derived from eroded areas, it may reduce fertility or decrease soil permeability. Sedimentation in canals or ditches may be serious, resulting in higher maintenance costs. Normally, ground water or water from reservoirs does not contain enough sediment to cause trouble in irrigation water. Sediment is the most important pollutant from agricultural land that gets

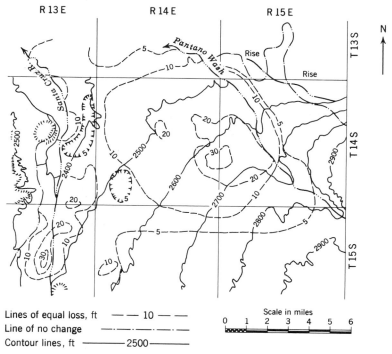

Fig. 17.1. Ground water changes in the Tucson basin of the Santa Cruz valley of Arizona, spring 1956-1961. (Adapted from Schwalen and Shaw, 1961.)

into streams and reservoirs causing eutrophication in lakes and increasing the cost of treatment for domestic and municipal supplies. Mean annual sediment concentration in U.S. streams is shown in Fig. 17.2. These concentrations are highest in the west-central states, but concentration is not necessarily related to total sediment loss because runoff is lower in the western than in the eastern states. Soil particles not only indicate an erosion loss, but they carry attached many chemical ions, such as phosphorus and potassium.

Chemical. Chemical properties of water affect its suitability for many uses. The discussion here is limited to chemical properties important in irrigation waters. A most important reference in relation to water quality in irrigation is USDA Handbook No. 60 (1954). The most important characteristics of irrigation water are (1) total concentration of soluble salts, (2) proportion of sodium to other cations, (3) concentration of potentially toxic elements, and (4) bicarbonate concentration as related to the concentration of calcium plus magnesium.

Total soluble salts are commonly indexed by the electrical conductivity of the water expressed in millimhos per centimeter. The relationship between conductivity and parts per million of soluble salts is illustrated in Fig. 17.3. The proportion of sodium to other cations or the sodium hazard of the water is indicated by the sodium-adsorption ratio, or SAR, calculated from

$$SAR = \frac{Na^+}{\sqrt{(Ca^{++} + Mg^{++})/2}}. \tag{17.1}$$

Where Na^+, Ca^{++}, and Mg^{++} represent the concentrations in milliequivalents per liter of the respective ions. The suitability of water for irrigation on the basis of conductivity and the sodium-adsorption ratio has been expressed diagrammatically in Fig. 17.4. Table 17.1 shows the suitability for irrigation of the various classes of water.

Boron, though essential to normal growth of plants, is toxic under some conditions in concentrations as low as ⅓ part per million. High concentrations of bicarbonate ions may result in precipitation of calcium and magnesium bicarbonates from the soil solution increasing the relative proportions of sodium and thus the sodium hazard. USDA Handbook No. 60 (1954), should be referred to if there are potential problems with toxic elements or bicarbonate ion concentration.

WATER MEASUREMENT

Effective use of water requires that flow rates and volumes be measured and expressed quantitatively. Water measurement is based upon application of the formula

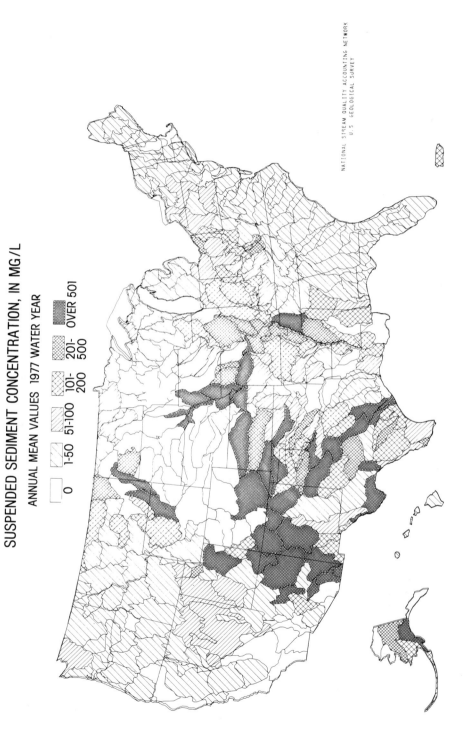

SUSPENDED SEDIMENT CONCENTRATION, IN MG/L

ANNUAL MEAN VALUES 1977 WATER YEAR

☐ 0
1-50
51-100
101-200
201-500
OVER 501

NATIONAL STREAM QUALITY ACCOUNTING NETWORK
U.S. GEOLOGICAL SURVEY

COUNCIL ON ENVIRONMENTAL QUALITY

Fig. 17.2. Average annual suspended sediment concentration, milligrams per liter. (From USDA, 1977.)

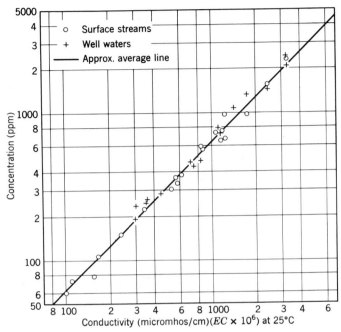

Fig. 17.3. Concentration of dissolved solids in irrigation waters in parts per million as related to conductivity. (USDA, 1954.)

$$q = av \tag{17.2}$$

where q = flow rate (L^3T^{-1}),
$\quad\quad a$ = cross-sectional area of flow (L^2),
$\quad\quad v$ = mean velocity of flow, (LT^{-1}).

Flow measurement thus involves determination of mean velocity and area of flow. Some techniques make each of these determinations separately and use them directly in Eq. 17.2. In others the calibration of the measurement device gives the flow directly. Measurement of flow volume requires integration of the flow rate q over the time period involved.

17.5. Units of Measurement. In agriculture, the common units of rate of flow in English units are the gallon per minute, cubic foot per second, and Miner's inch, and in SI units the liter per second and cubic meter per second. The Miner's inch is defined by state legislation and varies from state to state. Conversion factors among the various common units of rate and volumes are given in Appendix F and inside the back cover of the book.

17.6. Float. A crude estimate of the velocity of a stream may be made by determining the velocity of an object floating with the current. A straight uni-

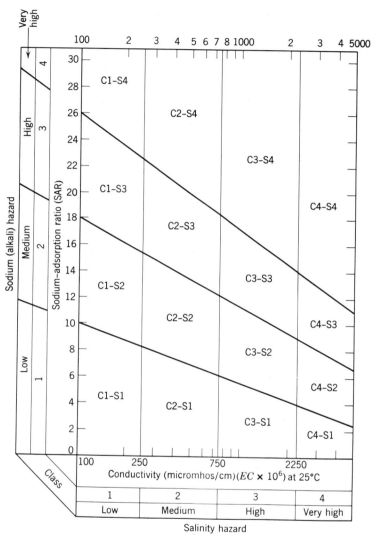

Fig. 17.4. Classification of irrigation waters with regard to sodium and salinity hazards. (USDA, 1954.)

form section of stream about 100 m long should be selected and marked by stakes or range poles on the bank. The time required for an object floating on the surface to traverse the marked course is measured and the velocity calculated. The average surface velocity is determined by averaging float velocities measured at a number of distances from the bank. Mean velocity of the stream is often taken as 0.8 to 0.9 of the average surface velocity.

Floats consisting of a weight attached to a floating buoy sometimes measure

Table 17.1   Suitability of Waters for Irrigation

Class	Salinity or Conductivity	Sodium-Adsorption Ratio[a]
Low 1	*Low-Salinity Water* (C1) can be used for irrigation with most crops on most soils with little likelihood that soil salinity will develop. Some leaching is required, but this occurs under normal irrigation practices except in soils of extremely low permeability.	*Low-Sodium Water* (S1) can be used for irrigation on almost all soils with little danger of the development of harmful levels of exchangeable sodium. However, sodium-sensitive crops such as stone-fruit trees and avocados may accumulate injurious concentrations of sodium.
Medium 2	*Medium-Salinity Water* (C2) can be used if a moderate amount of leaching occurs. Plants with moderate salt tolerance can be grown in most cases without special practices for salinity control.	*Medium-Sodium Water* (S2) will present an appreciable sodium hazard in fine-textured soils having high cation-exchange capacity, especially under low-leaching conditions, unless gypsum is present in the soil. This water may be used on coarse-textured or organic soils with good permeability.
High 3	*High-Salinity Water* (C3) cannot be used on soils with restricted drainage. Even with adequate drainage, special management for salinity control may be required and plants with good salt tolerance should be selected.	*High-Sodium* (S3) may produce harmful levels of exchangeable sodium in most soils and will require special soil management— good drainage, high leaching, and organic matter additions. Gypsiferous soils may not develop harmful levels of exchangeable sodium from such waters. Chemical amendments may be required for replacement of exchangeable sodium, except that amendments may not be feasible with waters of very high salinity.

Table 17.1 (Continued)

Class	*Salinity or Conductivity*	*Sodium-Adsorption Ratio*[a]
Very High 4	*Very High Salinity Water* (C4) is not suitable for irrigation under ordinary conditions, but may be used occasionally under very special circumstances. The soils must be permeable, drainage must be adequate, irrigation water must be applied in excess to provide considerable leaching, and very salt-tolerant crops should be selected.	*Very High Sodium Water* (S4) is generally unsatisfactory for irrigation purposes except at low and perhaps medium salinity, where the solution of calcium from the soil or use of gypsum or other amendments may make the use of these waters feasible.

[a] Sometimes the irrigation water may dissolve sufficient calcium from calcareous soils to decrease the sodium hazard appreciably, and this should be taken into account in the use of C1-S3 and C1-S4 waters. For calcareous soils with high pH values or for noncalcareous soils, the sodium status of waters in classes C1-S3, C1-S4, and C2-S4 may be improved by the addition of gypsum to the water. Similarly, it may be beneficial to add gypsum to the soil periodically when C2-S3 and C3-S2 waters are used.
Source: USDA, 1954.

directly mean velocity in a vertical. The weight is submerged to the depth of mean velocity, and the buoy marks its travel downstream. The float method has the advantage of giving an estimate of velocity with a minimum of equipment. The method is, however, obviously lacking in precision.

17.7. Impeller Meters. Instruments employing an impeller which rotates at speeds proportional to the velocity of flowing water are often used for velocity determination in open channels. In such applications they are called current meters. Figure 17.5 shows a typical current meter with accessories. The essential part of the meter is a wheel so arranged that it revolves when suspended in flowing water. An electrical circuit is incorporated so as to indicate the speed of revolution of the wheel. The revolving wheel actuates a set of breaker points in the electrical circuit and the revolutions are indicated by clicks in the earphone. The meter may be suspended by a cable for deep streams or attached to a rod in shallow streams. When supported by a cable a streamlined weight holds the meter against the current. A vane attached to the rear of the meter keeps the wheel headed into the stream. The revolutions are counted for a known period of time and the velocity of the current is determined from a calibration curve or equation for the particular meter.

When the mean velocity of a stream is determined with a current meter, the cross section of flow is divided into a number of subareas. Width of the sub-areas may be several meters depending on the size of the stream and the

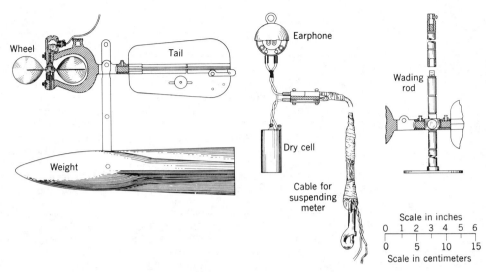

Fig. 17.5. Price current meter and attachments.

precision desired. This is illustrated in Fig. 17.6. The subareas may be indicated by marks on a tape or cable stretched across the stream or by marks on a bridge railing or other convenient structure. The average velocity at each station across the section is determined with the current meter. It has been found that the average of readings taken at 0.2 and 0.8 of the depth below the surface is an accurate estimate of the average velocity in the vertical. Where the stream is so shallow as to prevent the taking of a reading at 0.8 of the depth, the velocity at 0.6 of the depth below the surface may be taken as the average velocity. The area of the cross section may be determined by sounding with the current meter or other convenient device. Table 17.2 gives the calculation of the discharge for the section shown in Fig. 17.6. In using impeller meters for velocity measurements in closed conduits, the cross-sectional area of flow remains constant, and the meters are calibrated to read directly in cumulative volume or in flow rate. Figure 17.7 shows typical installations of impeller meters in closed conduits.

17.8. Slope Area. The Manning velocity equation for designing open channels described in Chapter 7 may be applied to stream flow measurement. A nomograph for calculating the velocity from the Manning formula is given in Appendix B. Application of the formula to estimation of flow in open channels requires measurement of the slope of the water surface and measurement of the properties of the cross section of flow. The reach of the channel selected should be uniform and if possible as much as 300 m long. The value of the roughness

Table 17.2 Calculation of Discharge from Current Meter Measurements

Gaging of Skunk River at Ames, Iowa.   Date: April 1, 1978
Meter No. SC5514394
Gage height at U.S.G.S.   Measurement began at 1:15 p.m.
Station—0.89 m   Measurement ended at 2:30 p.m.

Gaging by DeHart and Storm

Distance from Initial Point (m)	Width (m)	Depth (m)	Observation Depth (m)	Meter Revolutions	Time (sec)	Velocity (m/s) At Point	Velocity (m/s) Mean in Vertical	Area (m²)	Discharge (m³/s)
2	2	0.62	0.6	5	42	0.10	0.10	1.24	0.12
6	6	1.14	0.8	20	41	0.34	0.36	6.84	2.46
			0.2	25	45	0.38			
12	6	1.32	0.8	15	46	0.24	0.27	7.92	2.14
			0.2	20	45	0.30			
17	4.1	0.46	0.6	5	57	0.06	0.06	1.89	0.11
TOTALS	18.1							17.89	4.83

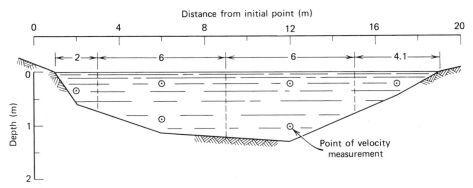

Fig. 17.6. Subdivision of a stream cross section for current meter measurements.

coefficient must be estimated and this is difficult to do accurately. Appendix B gives values of *n*, which are helpful in arriving at such estimates.

The slope-area method is sometimes used in estimating the discharge of past flood peaks. Cross-sectional area and flow gradient are measured from high-water marks along the channel. This method must be regarded as giving only a rough approximation of the peak flow.

17.9. Orifices. An orifice is an opening with a closed perimeter through which water flows. An orifice with prolonged sides, such as a pipe 2 or 3 diameters in length or an opening in a thick wall is called a tube. The orifice may be used as a measurement device because the velocity of discharge is a function of head on the orifice as discussed in Chapter 9. The ratio of orifice to pipe diameter should be between 0.5 to 0.83. The orifice coefficient is dependent on orifice configuration and must be determined for each design. Figure 17.8*a* illustrates an end-cap orifice for measurement of discharge from an irrigation pump. The pipe must be level and the manometer located about 0.6 m from the orifice. The coordinate method of measuring pipe flow, illustrated in Fig. 17.8*b*, requires the measurement of *X* and *Y* distances with the pipe level. The pipe should be long enough to produce a smooth flow from the pipe. Flow is determined from appropriate calibration tables.

17.10. Weirs and Flumes. For accurate measurement of flow in open channels structures of known hydraulic characteristics are required. As discussed in Chapter 9, these structures cause flow to pass through critical depth. They have a consistent relationship between head and discharge. The action of weirs and flumes in open channels is analogous to that of orifices and tubes for closed conduits.

Weirs. A weir consists of a barrier placed in a stream to constrict the flow and cause it to fall over a crest. The flow rate through such a structure is given in

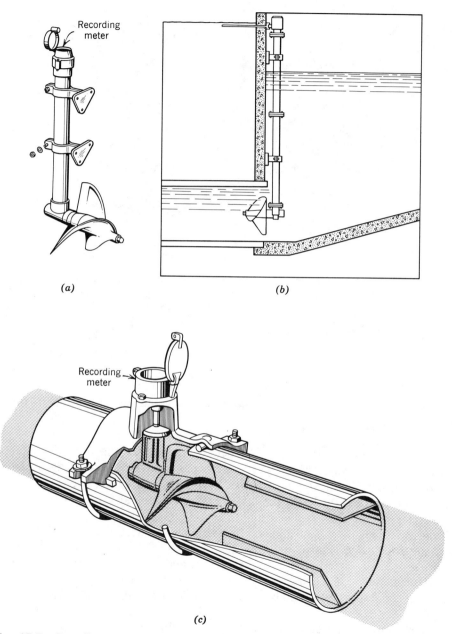

Fig. 17.7. Impeller meters: (a) Basic meter assembly; (b) location in an inverted siphon; and (c) low pressure pipe impeller meter. (Courtesy Sparling Div. Hersey-Sparling Meter Co., El Monte, California.)

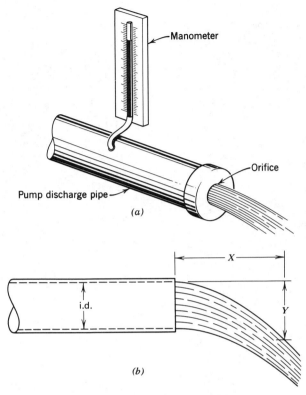

Fig. 17.8. (a) End-cap orifice and (b) coordinate method for measurement of flow from pump discharge.

Chapter 9. Weir openings may be rectangular, trapezoidal, or triangular in cross section, or they may take special shapes to give desired head-discharge relationships. Consult standard hydraulic handbooks and references such as King and Brater (1963) or Parshall (1950). A typical temporary weir for measuring stream flow is illustrated in Fig. 17.9.

Flumes. Specially shaped and stabilized channel sections such as a flume may also be used to measure flow. Flumes are generally less inclined to catch floating debris and sediment than are weirs, and for this reason they are particularly suited to measurement of runoff. One common type of measuring device is the flume developed by Parshall (1950). This flume is illustrated in Fig. 17.10. The Parshall flume has the advantage of requiring a very low head loss for operation. Discharge tables for all sizes of flumes are available in Parshall (1950). Other similar type flumes, which are simpler and easier to construct, have been developed, including the cutthroat flume described by Skogerboe (1973) and the critical-depth flume designed by Replogle (1971).

Fig. 17.9. Rectangular weir for measurement of flow in a small stream or irrigation ditch.

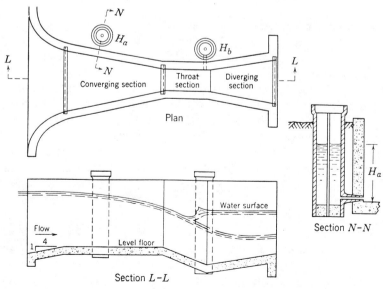

Fig. 17.10. Parshall measuring flume. (Redrawn from Parshall, 1950.)

17.11. Other Methods. Velocity determinations may be made in open channels or closed conduits with pitot tubes (King and Brater, 1963). Determination of mean velocity and calculation of discharge is done by methods similar to those applied to open channels when making current meter measurements. A pitot tube used in a closed conduit may be calibrated to read directly the flow rate from one velocity measurement. One such device, designed for insertion through the wall of a pipe, is known as a Cox meter.

For runoff measurement from small watersheds a special Type-H flume gives good accuracy at low flows as well as providing high capacity. The opening is V-shaped with the top of the V sloped toward the upstream side.

Pipe elbows offer an opportunity for measurement of flow as described by Langsford (1936). Water flowing through an elbow exerts different centrifugal forces at the inside and outside radii of the elbow. The resulting difference in pressure may be measured through a differential manometer. Discharge through the elbow is a function of the square root of the pressure differential with the coefficient determined by calibration for each size of elbow.

For open channels the vane or pendulum-type meter is growing in use. The vane that extends into the water from the surface is of variable width. Discharge is measured by calibrating the angular displacement of the vane. Another method is the injection of a fluorescent dye or substance that changes the water conductivity of the stream. It is detected with suitable equipment downstream so as to determine the velocity of flow.

GROUND WATER DEVELOPMENT

17.12. Ground water Hydrology. Ground water hydrology is the science of water below the ground water table. The classification of the earth's crust as a reservoir for water storage and movement of water is shown in Fig. 17.11. While other portions of this text are primarily concerned with the soil water zone, the earth's surface must be considered at much greater depths for the study of ground water. The profile of the earth is divided into two primary zones; the zone of rock fracture and the zone of rock flowage. Formations below the zone of rock fracture are of no concern in ground water hydrology. In the rock flowage zone the water is in a chemically combined state and is not available. In the zone of rock fracture interstitial water is contained in the pores of the soil or in the interstices of gravel and rock formations. This zone containing interstitial water is divided into the unsaturated and saturated zone. As here considered, ground water occurs only in the zone of saturation. Perched water tables shown in Fig. 17.11 are often encountered in the unsaturated zone. Good examples of perched and ground water tables can be observed in shallow and deep wells, respectively. Since the depth of perched water is shallow, ground water, which extends continuously downward, is of greater importance. While

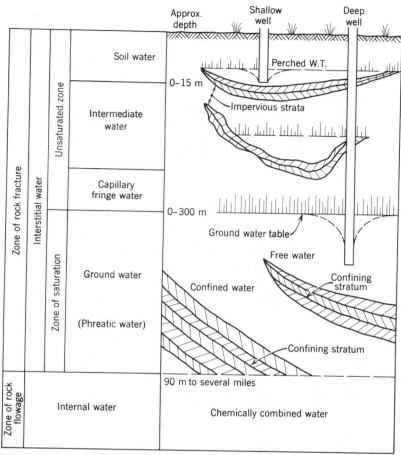

Fig. 17.11. Classification of the earth's crust and occurrence of subsurface water.

the ground water table is at the soil surface near lakes, swamps, and continuously flowing streams, it may be several hundred meters deep in drier regions. Ground water is often referred to as phreatic (a Greek term meaning "well") water. Capillary fringe water and intermediate water exist above the ground water table and are present above perched water tables as well. The occurrence, nature, and movement of soil water can be found in numerous texts on soil physics.

Formations from which ground water is derived in the zone of saturation have considerably different characteristics than the soil near the surface. The various types of deposits which furnish water supplies are shown in Fig. 17.12. A good water-bearing formation should have high hydraulic conductivity and

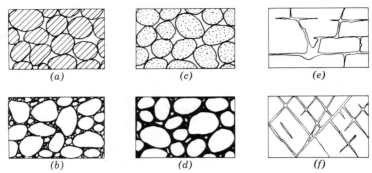

Fig. 17.12. Several types of rock interstices and the relationship of rock texture to porosity: (*a*) Well-sorted sedimentary deposit having high porosity. (*b*) Poorly sorted sedimentary deposit having low porosity. (*c*) Well-sorted sedimentary deposit with pebbles that are porous so that the deposit as a whole has a very high porosity. (*d*) Well-sorted sedimentary deposit whose porosity has been diminished by the deposition of mineral matter in the interstices. (*e*) Rock rendered porous by solution. (*f*) Rock rendered porous by fracturing. (Redrawn from Meinzer, 1923.)

high drainable porosity, called specific yield. As shown in Table 17.3, sand and gravel have these characteristics although fractured limestone and rock formations are also good aquifers.

Wells may be classed as gravity, artesian, or a combination of artesian and gravity, depending on the type of aquifer supplying the water. Gravity wells are those which penetrate the water table where water is not confined under pressure. Frequently, maps are prepared which show contour lines (depth) of the water table and indicate the type of formation.

Gravity water may be obtained from wells, springs, and dugout ponds. Wells, by far the most common, are either shallow or deep as shown in Fig. 17.11. Where the ground water table reaches the soil surface because the underlying strata are impervious, springs or seeps may develop. In areas where the ground

Table 17.3 Approximate Characteristics of Ground Water Aquifers

Soil Material	Total Porosity (%)	Specific Yield (%)	Relative Hydraulic Conductivity
Dense limestone or Shale	5	2	1
Sandstone	15	8	700
Gravel	25	22	5000
Sand	35	25	800
Clay	45	3	1

water table is near the surface, dugout ponds or open pits dug below the water level are practical for water storage.

Whenever the water level in a hole rises above the ground water table, artesian conditions are present. According to this definition, an artesian well is not necessarily a flowing well. The conditions under which artesian flow takes place are shown in Fig. 17.13. For artesian flow the following conditions must be present: (1) pervious stratum with an intake area, (2) impervious strata below and above the water-bearing formation, (3) inclination of the strata, (4) source of water for recharge, and (5) the absence of a free outlet for the water-bearing formation at a lower elevation. The level to which water rises in a pipe placed in the water-bearing formation is known as the piezometric head or, in three dimensions, the piezometric surface. Where the pervious stratum outcrops at the surface, artesian springs may develop. These are similar to water table springs except the flow is under pressure.

17.13. Hydraulics of Wells. A cross section of a well installed in homogeneous soil overlying an impervious formation is shown in Fig. 17.14. Under static

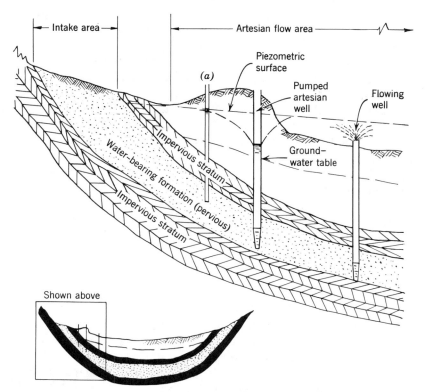

Fig. 17.13. Diagrammatic sketch of ideal conditions for artesian flow.

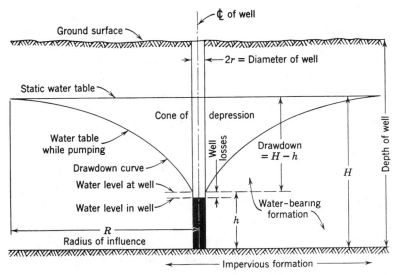

Fig. 17.14. Cross section of a typical gravity well in homogeneous soil.

conditions the water level will rise to the water table. When pumping begins, the water level in the well is lowered, thus removing free water from the surrounding soil. The distance from the well to where the static water table is not lowered by drawdown is known as the radius of influence. The water level at the edge of the well will be slightly higher than in the well because of friction losses through the perforated casing. The effect of the number and size of perforations on inflow has been investigated by Muskat (1942). For a given rate of pumping the water table surrounding a well in time reaches a stable condition. The shape of the drawdown curve depends on the soil permeability and the aeration porosity and is similar to an inverted cone. The base of the cone is at the water table or for artesian conditions the base is on the piezometric surface (Fig. 17.13).

The rate of flow into a gravity well, illustrated in Fig. 17.14, is

$$q = \frac{\pi K(H^2 - h^2)}{\log_e R/r} \tag{17.3}$$

where q = rate of flow (L^3T^{-1}),
K = hydraulic conductivity (LT^{-1}),
H = height of the static water level above the bottom of the water-bearing formation (L),
h = height of the water level at the well, measured from the bottom of the water-bearing formation (L),
R = radius of influence (L),
r = radius of the well (L).

The radius of influence given in Table 17.4 may be estimated from the texture and other characteristics of the aquifer. Since the discharge varies inversely as the logarithm of the radius of influence, an error in estimating this radius results in a much smaller error in the discharge. Equation 17.3 also indicates that the discharge is proportional to the logarithm of the well diameter. Assuming an R of 300 m and other conditions constant, a well 600 mm in diameter will yield about 10 percent more than a well 300 mm in diameter.

The flow into a well (artesian) completely penetrating an extensive confined aquifer, developed from Darcy's law, is

$$q = \frac{2\pi K d(H - h)}{\log_e R/r} \qquad (17.4)$$

where d = thickness of the confined layer (L),

H = height of the piezometric surface above the top of the confined layer (L),

h = height of the water in the well above the top of the confined layer (L).

If the aquifer is level, the flow is radial into the well and horizontal.

For a given well there is a definite relationship between drawdown and discharge. For thick water-bearing aquifers or artesian formations, the discharge-drawdown relationship is nearly a straight line. As shown in Fig. 17.15, the discharge-drawdown relationship can be obtained by pumping the well at various rates and plotting the drawdown against the discharge. Test pumping a well should be continued for a considerable length of time. Short pumping tests are often misleading. It has been found that even 24-hr tests are not long enough, and 30-day tests are more likely to indicate the true capacity of the well.

The theoretical water level for continuous pumping from a well having no recharge is shown in Fig. 17.16. These curves show the fallacy of using short-period tests for determining the discharge of a well. For example, 0.019 m³/s may be pumped for a period of 20 days without exceeding an assumed maxi-

Table 17.4 Radius of Influence of Wells

Soil Formation and Texture	*Radius of Influence* *[m (ft)]*
Fine sand formations with some clay and silt	30–90 (100–300)
Fine to medium sand formations fairly clean and free from clay and silt	90–180 (300–600)
Coarse sand and fine gravel formations free from clay and silt	180–300 (600–1000)
Coarse sand and gravel, no clay or silt	300–600 (1000–2000)

Source: Bennison (1947).

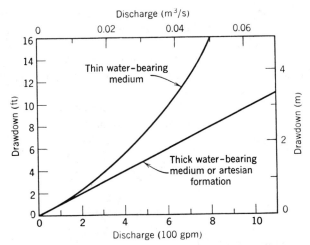

Fig. 17.15. Relationship between drawdown and discharge of a well.

mum economical lift of 46 m. However, to maintain the same drawdown, the discharge must be reduced if pumping is to be continued for more than 20 days.

17.14. Construction of Wells. The three general types of wells are: dug, driven, and drilled. The dug well consists of a pit dug to the ground water level. Often it is lined with masonry, concrete, or steel to support the excavation. Because of difficulty in digging below the water level, dug wells do not penetrate the ground water to a depth sufficient to produce a high yield.

Wells up to 76 mm in diameter and 18 m deep may be constructed by driving a well point into unconsolidated material. A well point is a section of perforated

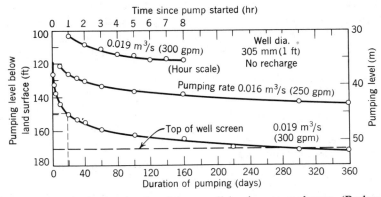

Fig. 17.16. Theoretical pumping level in a well having no recharge. (Redrawn from Ferris, 1959.)

pipe pointed for driving and connected to sections of plain pipe as it is driven to the desired depth. Sometimes penetration of well points is aided by discharging a high-velocity jet of water at the tip of the point as it is driven.

Deeper and larger diameter wells are drilled with cable-tool or rotary equipment. With cable tools a heavy bit is repeatedly dropped onto material at the bottom of the well. Crushed material is removed periodically with a bailer. Cable-tool wells have been drilled to depths of 1500 m. Deep wells are also drilled with rotary tools consisting of a bit rotated by a string of pipe. A mud slurry pumped through the drill pipe brings cuttings to the surface as it flows up the outside of the drill pipe. In unconsolidated materials wells may be cased as drilling progresses. Casings may also be installed after drilling is completed.

Well casings are perforated where they pass through water-yielding strata. In some situations perforated casing may be formed in place by ripping or shooting holes through a solid casing. Better results are achieved by placing a corrosion-resistant screen at the water-bearing strata. Screens are made of brass, bronze, or special alloys to resist corrosion. Screen openings are selected to permit 50 to 70 percent of the particles in the aquifer to pass the screen. The open screen area should keep entrance velocities below 0.15 m/s to minimize head loss. In aquifers of uniformly fine, unconsolidated material a gravel pack may be placed around the screen. Figure 17.17 shows a cross section of a gravel-packed well.

17.15. Development of Wells. After the screen is placed or the casing is perforated, a well should be developed by pumping at a high discharge or surging with a plunger. This practice develops higher velocities through the screen and in the aquifer adjacent to the screen than will be developed in normal pumping from the well. This action brings fine materials into the well where they are removed by pumping or bailing. As a result the aquifer is opened for freer flow of water and a stabilized filter varying from coarse to fine material is developed.

WATER CONSERVATION

17.16. Evaporation Surpression. Reduction of evaporation from free water surfaces is an important water conservation measure. Two broad approaches are employed, reduction of the free water surface area and protection of free water surfaces.

Reduction of free water surface is accomplished by minimizing the surface-area to volume ratio of reservoirs. Storage of water in natural ground water reservoirs rather than in surface reservoirs also reduces evaporation losses.

Protection of free-water surfaces has been uneconomical except in special situations. In recent years, however, much attention has been given to application of monomolecular films for evaporation suppression. These methods have

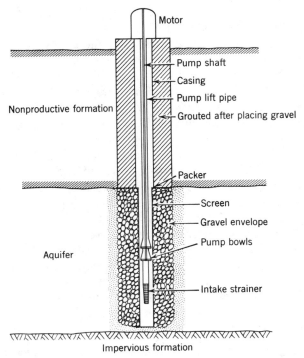

Fig. 17.17. Cross section through a gravel packed well. (Modified from Linsley and Franzini, 1979.)

been discussed by Harbeck (1958) and Crow (1961). Fatty alcohols such as hexadecanol will form on a water surface, a monomolecular film that is resistant to the action of dust and wind. The film inhibits the escape of water molecules. The film material must be continuously supplied to maintain the film against break up by wind action and deterioration by biological processes. Evaporation reductions in excess of 25 percent have often been achieved. The method is receiving limited practical application on stock reservoirs in the Southwest.

17.17. Artificial Recharge. Ground water reservoirs are supplied by water percolating to them from the surface. Under natural conditions only a small fraction of rainfall reaches the ground water. Since ground water reservoirs provide evaporation free storage and since surface runoff waters are often wasted, a logical water conservation measure is to attempt to increase the recharge of ground water from surface runoff.

There are four general methods of artificial recharge: basin, furrow or ditch, flooding, and pit or shaft. The basin method of spreading water consists of a series of small basins formed by dikes or banks. The dikes often follow contour lines, and they are so arranged that the water flows from one basin to the next. In the furrow or ditch method the water flows along a series of parallel ditches placed closely together. The flooding method consists of ponding a thin layer of water over the land surface. Pits or shafts as a method of recharge are used primarily in municipal areas and industrial centers. Regardless of the method, it is desirable to spread water which is relatively free of sediment. It is not unusual to use a combination of several methods.

In artificial recharge basins layers of accumulated sediment are periodically removed and replaced with sand. Soil conditioners such as organic residues, grasses, or chemical treatments are effective in increasing infiltration rates. Some waters require desilting or biological control treatment before they can be recharged without clogging the infiltration area or the aquifers.

17.18. Control of Seepage. Water conveyance losses from canals and ditches can be greatly reduced through reduction or elimination of seepage (see Chapter 13). Concrete linings are frequently used in irrigation canals and ditches. Asphalt and plastic linings are also used in canals and ponds. Chemical additives are often successful in canals, ponds, and ditches for seepage reduction.

17.19. Phreatophyte Control. The term, phreatophyte, includes plants that habitually obtain their water supply from the zone of saturation or from the overlying capillary fringe. Examples are tamarisk, cottonwood, willow, and mesquite. Eighty species of phreatophytes have been identified. Thompson (1958) reports that phreatophytes cover approximately 7 million hectares in the western United States and use an estimated 3 million ha-m of water annually. Phreatophyte growth is largely concentrated along the lower valleys of major rivers. Consumptive use of water by phreatophytes varies with species, climate, and depth to the ground water table. Under high water table conditions, water used by phreatophytes will approach open pan evaporation. Maximum use of water by tamarisk in the Rio Grande Valley of New Mexico has been measured at 3.3 m per year with a 0.6 m depth to the water table.

Control of phreatophytes thus offers a great potential for water conservation. Control can be effected either by chemical or mechanical means, however, the costs of control have limited its application. Thompson (1958) concludes that channelization is the most effective means of salvaging water that would otherwise be lost to phreatophytes. Through channelization and drainage, ground water can be lowered in phreatophyte infested areas and conveyed to downstream reservoirs. The accompanying lower water table greatly reduces the consumptive use by phreatophytes.

WATER RIGHTS

Much of the confusion regarding water rights stems from the failure to make a distinction among the several types of naturally occurring waters. From the legal standpoint water may be classified as (1) diffused surface water, (2) water in well-defined surface channels, (3) water in well-defined underground aquifers, and (4) underground percolating water. Diffused surface water and underground percolating water, because of their diverse nature, are normally regulated by common or civil law rather than by legislative action. In some western states, however, diffused surface water is treated the same as water in well-defined channels. In most states diffused surface water is considered the property of the landowner and he may use it in any way without regard to its effect on the water supply of other owners. Especially in the eastern states the law of diffused surface water has been concerned with the damage caused by such water and the fixing of responsibility for such damage. Underground percolating water is defined as that subsurface water which flows in small pores or filters through the soil in such a way that its course or direction cannot be easily determined.

Two basic divergent doctrines regarding the right to use water exist, namely, riparian and appropriation. They are recognized either separately or as a combination of both doctrines in different states. A comparison of the salient features of these two doctrines is made in Table 17.5.

17.20 Riparian Doctrine. This doctrine may have originated in early English water law that was borrowed in part from Roman civil law. The riparian doctrine in its American form recognizes the right of a riparian owner to make reasonable use of the stream's flow, provided the water is used on riparian

Table 17.5 Comparison of Water Rights Laws[a]

Characteristic	Riparian	Doctrine of Appropriation
Acquisition of water right	By ownership of riparian land	By permit from state (state ownership)
Quantity of water	Reasonable use	Restricted to that allowed by permit
Types of use allowed	Domestic, livestock, etc., but not precisely defined	Some beneficial use required
Loss of water by nonuse	No	Yes, but continued use not always required
Location where water may be used	On riparian land, but some exceptions	Anywhere, unless specified in permit

[a] Generally applicable only for surface water in well-defined channels and for water in well-defined underground aquifers. Some state laws on ground water deviate from the above.

land. Riparian land is that which is contiguous to a stream or other body of surface water. Land ownership accompanies the right of access to and use of the water, and this right is not lost by nonuse. Reasonable use of water generally implies that the landowner may use all that he needs for drinking, for household purposes, and for watering livestock. For percolating ground water most states recognize absolute ownership, and use of water from wells is not restricted. This principle is known as "common law riparian." In some states the riparian doctrine is modified by applying the principle of "reasonable use." The California Supreme Court under this principle definitely established that no proprietor can absorb all the water of the stream so as to allow none to flow down to a neighbor. In California, the riparian doctrine was further modified by establishing "correlative rights." Under this doctrine the landowner's use of ground water not only must be reasonable in consideration of the similar rights of others, but it must be correlated with the uses of others in times of shortage.

In states that do not have statutory laws governing water rights, the riparian doctrine is based on previous court decisions. Many of the eastern states have modified the riparian doctrine by regulating use through the issuance of permits for specified amounts of water. Others restrict use in certain areas of the state.

17.21. Doctrine of Appropriation. This doctrine is normally applied to prior rights. It is based on the priority of development and use; that is, the first to develop and put water to beneficial use has the prior right to continue his or her use. The right of prior appropriation is acquired mainly by filing a claim in accordance with the laws of the state. The water must be put to some beneficial use, but the appropriator has the right to water required to satisfy his needs at the given time and place. This principle assumes that it is better to let individuals, prior in time, take all the water rather than to distribute inadequate amounts to several owners. Water rights are not limited to riparian land and may be lost by nonuse and abandonment.

17.22. Water Rights Law by States. Water rights doctrine by states for water in well-defined surface channels is shown in Fig. 17.18a. The riparian doctrine is recognized in all but the eight western interior states and Alaska. Doctrine by states for percolating ground water is shown in Fig. 17.18b. The riparian doctrine in either its common law or reasonable use form for well-defined underground aquifers applies to all of the eastern states, Arizona, California, Hawaii, Nebraska, and Texas. Otherwise, either the appropriation or correlative rights doctrine applies.

REFERENCES

Bennison, E. W. (1947). *Ground Water, Its Development, Uses and Conservation.* E. E. Johnson, St. Paul, Minn.

Bouwer, H. (1978). *Groundwater Hydrology.* McGraw-Hill, New York.

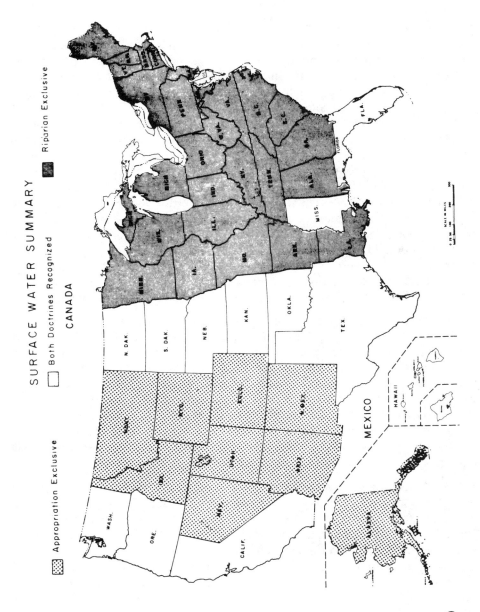

(a)

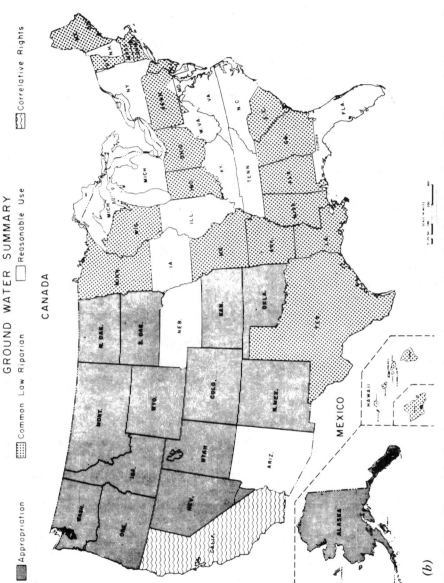

Fig. 17.18. Water rights law by states (a) for surface water in well-defined channels and (b) for percolating ground water. (From Garrity and Nitzschke, 1967.)

Crow, F. R. (1961). "Reducing Reservoirs Evaporation." *Agr. Eng.* **42**, 240–243.

Ferris, J. G. (1959). In C. E. Wisler and E. F. Brater, *Hydrology*. Wiley, New York.

Garrity, T. A., Jr., and E. T. Nitzschke, Jr. (1967). *Water Law Atlas. Socorro, N. Mexico*. State Bureau of Mines and Mineral Resources and N. Mexico Inst. Mining and Tech., Cir. 95.

Harbeck, G. E., Jr. (1958). "Can Evaporation Losses be Reduced?" Am. Soc. Civil Eng. Proc., J. Irrig. Drainage Div. Paper no. 1499.

Hutchins, W. A. (1942). "Selected Problems in the Law of Water Rights in the West." U.S. Dept. Agr. Misc. Publ. 418.

—— (1939). "Water Rights for Irrigation in Humid Areas." *Agr. Eng.* **29**, 431–432, 436.

Israelsen, O. W., and V. E. Hansen (1962). *Irrigation Principles and Practices*. Wiley, New York.

Johnson, E. E. Inc. (1966). *Ground Water and Wells*. E. E. Johnson Inc., St. Paul, Minn.

King, H. W., and C. F. Brater (1963). *Handbook of Hydraulics* (5th ed.), McGraw-Hill, New York.

Langsford, W. M. (1936). "The Use of an Elbow in a Pipe Line for Determining the Rate of Flow in the Pipe." *Univ. of Ill. Eng. Exp. Sta. Bull. 289.*

Linsley, R. K., and J. B. Franzini (1979). *Water-Resources Engineering* (3rd ed.). McGraw-Hill, New York.

Lauritzen, C. W., and A. A. Thayer (1966). "Rain Traps for Interception and Storage of Water for Livestock." *U.S. Dept. Agr., Agr. Inf. Bull.*

Meinzer, O. E. (1923). "The Occurrence of Ground Water in the United States." *U.S. Geological Survey, Water Supply Paper 489.*

Muskat, M. (1942). "The Effect of Casing Perforations on Well Productivity. *Am. Inst. Mining Met. Engrs. Tech. Publ. 1528.*

Parshall, R. L. (1950). "Measuring Water in Irrigation Channels with Parshall Flumes and Small Weirs." U.S. Dept. Agr. Cir. 843.

Replogle, J. A. (1971). "Critical-Depth Flumes for Determining Flow in Canals and Natural Channels." *Trans. Am. Soc. Agr. Eng.* **14**, 428–433.

Schwalen, H. C., and R. J. Shaw (1961). "Water in the Santa Cruz Valley, Arizona." Ariz. Agr. Expt. Sta. Report no. 205.

Skogerboe, G. V. (1973). "Selection and Installation of Cutthroat Flumes for Measuring Irrigation and Drainage Water." Colo. St. Univ. Eng. Expt. Sta. Tech. Bull. 120.

Thomas, R. O. (1959). "Legal Aspects of Ground Water Utilization." *Am. Soc. Civil Eng. Proc.* **85**, IR-4.

Thompson, C. B. (1958). "Importance of Phreatophytes in Water Supply." *Am. Soc. Civil Eng. Proc.* **84** IR-1, Jan.

U.S. Department of Agriculture (1954). "Handbook No. 60: Diagnosis and Improvement of Saline and Alkali Soils." U,S. GPO, Washington, D.C.
———— (1977). "Handbook of Agriculture Charts: Agr. Hbk. 524." U.S. GPO, Washington D.C.

PROBLEMS

17.1. Determine the stream discharge for the velocities and gage widths of the stream shown in Fig. 17.6 if the depths at the point of gaging from left to right were changed to 0.5 m (1.64 ft), 2.0 m (6.56 ft), 1.5 m (4.92 ft), and 0.5 m (1.64 ft), respectively. Record and tabulate data as shown in Table 17.2 for the velocities as given. Compute the average stream velocity.

17.2. Determine the discharge of a stream having a cross-sectional area of 18.59 m² (200 ft²) by the float method. Trial runs for surface floats to travel 91.4 m (300 ft) were 122, 128, 123, 124, and 128 seconds each.

17.3. Determine the discharge of a stream having a cross-sectional area of 9.29 m² (100 ft²) and a wetted perimeter of 9.14 m (30 ft) using the slope-area method. The channel has some weeds and stones with straight banks and is flowing at full stage. The difference in elevation of the water surface at points 122 m (400 ft) apart is 8.53 cm (0.28 ft).

17.4. Determine the capacity of a Parshall flume having a throat width W of 0.38 m (1.25 ft) for $H_a = 0.40$ m (1.30 ft) and $H_b = 0.27$ (0.90 ft). See Parshall, (1950).

17.5. Determine the discharge rate of a 10-cm (0.33-ft) square sharp-edged orifice if the head to the center of the submerged orifice is 0.2 m (0.66 ft) and the discharge coefficient is 0.6.

17.6. Compute the flow rate into a gravity well 610 mm (24 in.) in diameter if the depth of the water-bearing stratum is 24.38 m (80 ft), the drawdown is 9.14 m (30 ft), soil hydraulic conductivity is 76 mm/h (3 iph), and the radius of influence is 183 m (600 ft).

17.7. Compute the flow rate into a well completely penetrating a 6.10-m (20-ft) depth confined aquifer in which the piezometric surface is 15.24 m (50 ft) above the top of the aquifer. Diameter, drawdown, K, and R are the same as in Problem 17.6.

17.8. An irrigation reservoir has a storage capacity of 9.87 ha-m (80 ac-ft). If the irrigation requirement for the crop is 610 mm (24 in.) and the seepage and evaporation losses are 60 percent of the stored water, how many hectares (acres) can be irrigated?

17.9. Derive the equation for the flow rate into a gravity well completely penetrating the aquifer.

17.10. Derive the equation for the flow rate into a well completely penetrating a confined horizontal aquifer with a uniform constant depth.

$SM_f = SM_i + Prec + Irrig - Evap - Trans - Drainage - Runoff$

CHAPTER 18

Irrigation Principles

Man's dependence upon irrigation can be traced to earliest biblical references. Irrigation in very early times was practiced by the Egyptians, the Asians, and the Indians of North America. For the most part, water supplies were available to these people only during periods of heavy runoff. Modern concepts of irrigation have been made possible only by the application of modern power sources to deep well pumps and by the storage of large quantities of water in reservoirs. Thus, by using either underground or surface reservoirs it is now possible to bridge over the years and even out water excesses and deficiencies.

18.1. Estimating Evapotranspiration. With increasing demands for water and with limited supplies available, more effective use of water is becoming essential. Both the planning and the operation of irrigation systems can be most intelligently accomplished if predictions can be made of water use. Several approaches to estimation of evapotranspiration have been discussed in Chapter 3. The Penman method is considered by many to be the most accurate indicator, but its application is limited by lack of needed data for many locations. The Blaney–Criddle method is widely used, but in the western portions of the United States there is increasing acceptance of the Jensen–Haise method. Variations of these methods, such as the one described by Doorenbos and Pruitt (1977), are used for special circumstances.

18.2. Irrigation Requirements. The irrigation requirement is the quantity of water, exclusive of precipitation, to be supplied by artificial means. Irrigation requirements are dependent not only on evapotranspiration but also on water application efficiency, water supplied by percolation, by capillary movement from the ground water table, and by effective rainfall. Where rainfall is a significant part of the plant water supply, a probability procedure for estimating the contribution of rainfall developed by the U.S. Soil Conservation Service (1967) is recommended.

18.3. Crop Needs. Water requirements and time of maximum demand vary with different crops. Although growing crops are continuously using water, the rate of transpiration depends on the kind of crop, degree of maturity, and atmospheric conditions, such as humidity, wind, and temperature. Where sufficient water is available, the moisture content should be maintained within the limits for optimum growth. The rate of growth at different soil moistures varies with different soils and crops. Some crops are able to withstand drought or high moisture content much better than others. During the later stages of maturity, the water needs are generally less than during the maximum growing period. When crops are ripening, irrigation is usually discontinued.

From soil moisture measurements or from the appearance of the soil or the crop, the irrigator is able to determine when and how much water should be applied. A recommended procedure involves taking soil samples from the root zone. With experience, it is possible to estimate the need for irrigation from the "feel" of the soil. Tensiometers or neutron probe sites located in representative areas of the field provide a more reliable basis for determining water needs. In practice, however, much irrigation water is applied on the basis of a routine schedule based on the experience of the farm manager.

Table 18.1 illustrates evapotranspiration estimates and water requirements for crops grown near Deming, New Mexico, based on the Blaney–Criddle method. It should be noted that a correction is made for expected effective rainfall in determining the field irrigation requirements.

18.4. Seasonal Use of Water.   To make maximum use of available water, the irrigator should have a knowledge of the water requirements of crops at all times during the growing season. It may be possible to select the crop to fit the water supply. Figure 18.1 shows evapotranspiration for three different crops grown in the Salt River Valley of Arizona. Wheat, being a fairly short season crop, has the lowest seasonal use of 655 mm (25.8 in.) with the high water requirement occurring during March and April. Alfalfa, a long season crop, has a seasonal use of 1888 mm (74.3 in.) and in this area grows during the entire year with the exception of December and January. Cotton, a tropical crop, has its highest seasonal use during the hottest portion of the summer and a seasonal use of 1046 mm (41.2 in.). Similar data are available in other regions.

18.5. Moisture Deficiency Recurrence.   The duration and length of dry periods during the growing season in humid and semihumid areas largely determine the economic feasibility of irrigation. In the Northern Hemisphere moisture deficiency during the months of June, July, and August is more serious than in earlier or later months.

18.6. The Soil Moisture Reservoir.   In planning and managing irrigation it is helpful to think of the soil's capacity to store available moisture as the soil moisture reservoir. The reservoir is filled periodically by irrigations. It is slowly depleted by evapotranspiration. Water application in excess of the reservoir capacity is wasted unless it is used for leaching (Section 18.9). Irrigation must be scheduled to prevent the soil moisture reservoir from becoming so low as to inhibit plant growth.

Table 18.1 Seasonal Evapotranspiration and Irrigation Requirements For Crops Near Deming, New Mexico[a]

Crop	Length of Growing Season (Days)	Evapo-transpiration Depth (mm)	Effective Rainfall Depth (mm)	ET Less Rainfall (mm)	Water Application Efficiency (%)	Irrigation Requirement Depth (mm)
Alfalfa	197	915	152	763	70	1090
Beans(dry)	92	335	102	233	65	358
Corn	137	587	135	452	65	695
Cotton	197	668	152	516	65	794
Grain						
(spring)	112	396	33	363	65	558
Sorghum	137	549	135	414	65	637

[a] Average frost-free period is April 15 to October 29. Irrigation prior to the frost-free period may be necessary for some crops.
Source: Jensen (1973).

Irrigation can raise the soil moisture to the field capacity. In either sprinkler or surface irrigation the infiltration capacity and the permeability of the soil will determine how fast water can be applied. In sprinkler irrigation the water should be applied at a rate lower than the infiltration capacity. In surface irrigation, the soil surface must be flooded to allow water to enter the soil.

In some cases where early spring runoff is available beyond that which may be stored in surface reservoirs, fields are sometimes flooded with surface runoff channeled through the irrigation ditches in order to fill the soil moisture reservoir and to conserve other water supplies for use later in the season.

18.7. Irrigation Scheduling. In many cases irrigation districts, government agencies, or consulting engineering firms are providing irrigation water management services to farm operators. With information on the water holding capacity of the soil and evapotranspiration rates, the time and amount of irrigation can be predicted. Computer programs may be utilized to facilitate the process. This service provides the farmer with a basis for effective management of his irrigation activities.

18.8. Salinity. The presence of soluble salts in the root zone can be a serious problem especially in arid regions. In subhumid regions, where irrigation is provided on a supplemental basis, salinity is usually of little concern because rainfall is sufficient to leach out any accumulated salts. However, all water from surface streams and underground sources contains dissolved salts. The salt applied to the soil with irrigation water remains in the soil unless it is flushed out in drainage water or is removed in the harvested crop. Usually the quantity of salt removed by crops is so small that it will not make a significant contribution to salt removal or enter into determinations of leaching requirements.

Salt-affected soils may be classified as saline, sodic, or saline-sodic soils.

Saline soils contain sufficient soluble salt to interfere with the growth of most plants. Sodium salts are in relatively low concentration in comparison with calcium and magnesium salts. Saline soils often are recognized by the presence of white crusts on the soil, by spotty stands, and by stunted and irregular plant growth. Saline soils generally are flocculated, and the permeability is comparable with that of similar nonsaline soils.

Sodic soils are relatively low in soluble salts, but contain sufficient exchangeable (adsorbed) sodium to interfere with the growth of most plants. Exchangeable sodium is adsorbed on the surfaces of the fine soil particles. It is not leached readily until displaced by other cations, such as calcium or magnesium.

Saline-sodic soils contain sufficient quantities of both total soluble salt and adsorbed sodium to reduce the yields of most plants. As long as excess soluble salts are present, the physical properties of these soils are similar to those of

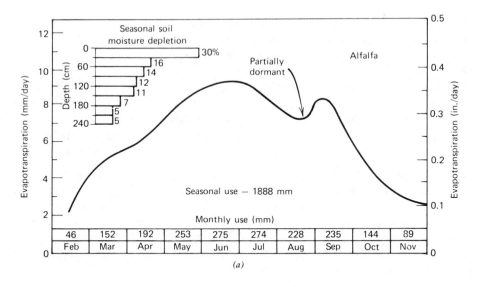

(a)

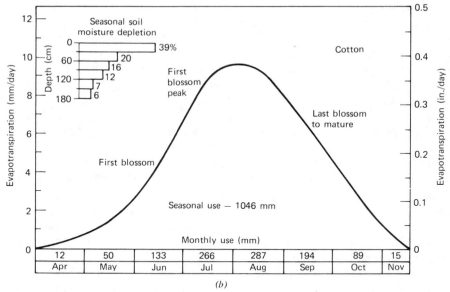

(b)

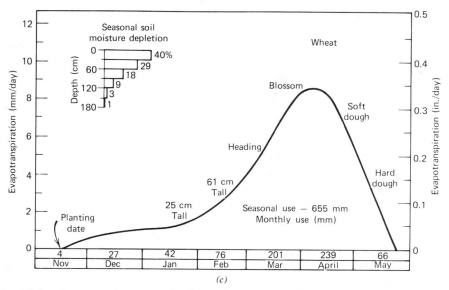

Fig. 18.1. Average evapotranspiration and seasonal moisture depletion with depth for (a) alfalfa, (b) cotton, and (c) wheat at Mesa and Tempe, Arizona. (Redrawn and adapted from Erie, French, and Harris, 1968.)

saline soils. If the excess soluble salts are removed, these soils may assume the properties of sodic soils.

The principal effect of salinity is to reduce the availability of water to the plant. In cases of extremely high salinity, there may be curling and yellowing of the leaves, firing in the margins of the leaves, or actual death of the plant. Long before such effects are observed, the general nutrition and growth physiology of the plant will have been altered.

As the proportion of exchangeable sodium increases, soils tend to become dispersed, less permeable to water, and of poorer tilth. High-sodium soils usually are plastic and sticky when wet, and are prone to form clods and crusts on drying. These conditions result in reduced plant growth, poor germination, and, because of inadequate water penetration, poor root aeration and soil crusting.

Both sodic and saline-sodic soils may be improved by the replacement of the excessive adsorbed sodium by calcium and magnesium. This usually is done by applying soluble amendments that supply these cations. Acid-forming amendments, such as sulfur or sulfuric acid, may be used on calcareous soils since they react with limestone (calcium carbonate) to form gypsum, a more soluble calcium salt.

Leaching is the only way by which the salts added to the soil by the irrigation water can be removed satisfactorily. Sufficient water must be applied to dissolve the excess salts and carry them away by subsurface drainage.

With the necessity of using additional water beyond the needs of the plant to provide sufficient leaching, it is imperative under irrigation that there be adequate drainage of water passing through the root zone. Natural drainage through the underlying soil may be adequate. In cases where subsurface drainage is inadequate, open or pipe drains must be provided.

Water will rise 0.6 to 1.5 m or more in the soil above the water table by capillarity. The height to which water will rise above a free-water surface depends on soil texture, structure, and other factors. Water reaching the surface evaporates, leaving a salt deposit typical of saline soils.

Some crop plants can tolerate relatively large amounts of salt. Others are more easily injured. The relative salt tolerance of a number of crop plants is shown in Table 18.2. The tolerance of crops listed may vary somewhat, depending on the particular variety grown, the cultural practices used, and climatic factors. The term "ECe" in the table denotes the electrical conductivity of the saturated extract of the soil reported as determined in millimhos per centimeter at 25°C. "ECw" is the electrical conductivity of the irrigation water in millimhos per centimeter at 25°C. Each millimho per centimeter represents about 670 ppm of dissolved salts. In general, a yield reduction of 10 percent due to salinity is considered acceptable since there are many other factors that might well be limiting in determining the maximum yield of a given crop. Similar data for a number of additional crops are available from Ayers and Westcot (1976).

18.9. Leaching. The traditional concept of leaching involves the ponding of water to achieve more or less uniform salt removal from the entire root zone. However, Ayers and Westcot (1976) show that salt accumulation can take place for short periods of time in the lower root zone without adverse effects. As can be noted in Fig. 18.1, most of the water transpired by the plant is taken from the upper portion of the root zone. This area will be leached to a considerable degree by normal applications of irrigation water and by rainfall that may come at any time of the year. The same amount of water when applied with more frequent irrigations is more effective in removing salts from this critical upper portion of the root zone than from the lower root zone. Thus, high-frequency sprinkler, trickle, or surface irrigation should be effective for salinity control. The methods of determining the leaching requirement are summarized by Ayers and Westcot (1976). Other concepts that may be helpful in controlling salinity are the use of soil or water amendments, deep tillage, and irrigation before planting.

18.10. Irrigation Efficiencies. Efficiency is an output divided by an input and is usually expressed as a percentage. An efficiency figure is only meaningful when the output and input are clearly defined. Three basic irrigation efficiency concepts are the following.

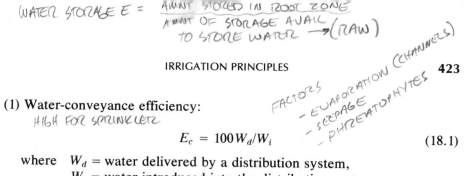

WATER STORAGE E = AMNT STORED IN ROOT ZONE / AMNT OF STORAGE AVAIL TO STORE WATER → (RAW)

FACTORS
- EVAPORATION (CHANNELS)
- SEEPAGE
- PHREATOPHYTES

(1) Water-conveyance efficiency:

HIGH FOR SPRINKLER

$$E_c = 100 W_d/W_i \tag{18.1}$$

where W_d = water delivered by a distribution system,
W_i = water introduced into the distribution system.

The water-conveyance efficiency definition can obviously be applied along any reach of a distribution system. For example, a water-conveyance efficiency could be calculated from a pump discharge to a given field or from a major diversion work to a farm turn-out.

(2) Water-application efficiency:

70 - 95% SPRINKLER
15 - 75% SURFACE

$$E_a = 100 W_s/W_d \tag{18.2}$$

LOSSES
- EVAP, WIND
DRIFT, LEAKS, SUBSURFACE
NONUNIFORMITY CAUSES
WHICH DRAINAGE

where W_s = water stored in the soil root zone by irrigation,
W_d = water delivered to the area being irrigated.

This efficiency may be calculated for an individual furrow or border, for an entire field, or for an entire farm or project. When applied to areas larger than a field, it overlaps the definition of conveyance efficiency.

(3) Water-use efficiency:

- IN WI ISC SAME AS WATER APPLICATION
SINCE NO LEACHING REQ

$$E_u = 100 W_u/W_d \tag{18.3}$$

where W_u = water beneficially used,
W_d = water delivered to the area being irrigated.

The concept of beneficial use differs from that of water stored in the root zone in that leaching water would be considered beneficially used though it moved through the soil moisture reservoir. Sometimes water-use efficiency is based on dry plant weight produced by a unit volume of water.

Another useful measurement of the effectiveness of irrigation is the uniformity coefficient,

$$C_u = 1 - y/d \tag{18.4}$$

where y = average of the absolute values of the deviations in depth of water stored from the average depth of water stored,
d = average depth of water stored.

This coefficient indicates the degree to which water has penetrated to a uniform depth throughout a field. Note that when the deviation from the average depth is zero, the uniformity coefficient is 1.0.

Table 18.2 Crop Salt Tolerance Levels for Crops[a]

| Crop | Yield potential | | | | | | | | Max. ECe |
| | 100% | | 90% | | 75% | | 50% | | |
	ECe	ECw	ECe	ECw	ECe	ECw	ECe	ECw	
Field crops									
Barley[b]	8.0	5.3	10.0	6.7	13.0	8.7	18.0	12.0	28
Corn	1.7	1.1	2.5	1.7	3.8	2.5	5.9	3.9	10
Cotton	7.7	5.1	9.6	6.4	13.0	8.4	17.0	12.0	27
Sorghum	4.0	2.7	5.1	3.4	7.2	4.8	11.0	7.2	18
Soybeans	5.0	3.3	5.5	3.7	6.2	4.2	7.5	5.0	10
Wheat[b]	6.0	4.0	7.4	4.9	9.5	6.4	13.0	8.7	20
Vegetable crops									
Beans	1.0	0.7	1.5	1.0	2.3	1.5	3.6	2.4	7
Cantaloupe	2.2	1.5	3.6	2.4	5.7	3.8	9.1	6.1	16
Carrot	1.0	0.7	1.7	1.1	2.8	1.9	4.6	3.1	8
Lettuce	1.3	0.9	2.1	1.4	3.2	2.1	5.2	3.4	9
Potato, sweet potato	1.6	1.1	2.5	1.7	3.8	2.5	5.9	3.9	10

Forage crops

Alfalfa	2.0	1.3	3.4	2.2	5.4	3.6	8.8	5.9	16
Bermuda grass	6.9	4.6	8.5	5.7	10.8	7.2	14.7	9.8	23
Orchard grass	1.5	1.0	3.1	2.1	5.5	3.7	9.6	6.4	18
Sudan grass	2.8	1.9	5.1	3.4	8.6	5.7	14.4	9.6	26
Wheat grass	7.5	5.0	9.0	6.0	11.0	7.4	15.0	9.8	22

Fruit crops

Apple, pear	1.7	1.0	2.3	1.6	3.3	2.2	4.8	3.2	8
Date palm	4.0	2.7	6.8	4.5	10.9	7.3	17.9	12.0	32
Fig, olive, pomegranate	2.7	1.8	3.8	2.6	5.5	3.7	8.4	5.6	14
Orange, grapefruit, lemon	1.7	1.1	2.3	1.6	3.3	2.2	4.8	3.2	8
Plum, peach	1.6	1.0	2.1	1.4	2.9	1.9	4.2	2.8	7

[a] All values are in mmhos/cm at 25°C.
[b] During germination and seedling stage ECe should not exceed 4 or 5 mmhos/cm. Data may not apply to new semidwarf varieties of wheat.
Source: Ayers and Westcot (1976).

Example 18.1.   If 42.48 m³/s (1500 cfs) are pumped into a farm distribution system and 38.23 m³/s (1350 cfs) are delivered to a turn-out 3.2 km (2 mi) from the well, what is the conveyance efficiency of the portion of the farm distribution system used in conveying this water?

Solution.   Substituting in Eq. 18.1,

$$E_c = 100 \frac{W_d}{W_i} = 100 \frac{38.23}{42.48} = 90 \text{ percent}$$

Example 18.2.   Delivery of 10.20 m³/s (360 cfs) to a 32.4-ha (80-acre) field is continued for 4 hours. Tail-water flow is estimated at 0.28 m³/s (10 cfs). Soil probing after the irrigation indicates that 0.31 m (1 ft) of water has been stored in the root zone. Compute the application efficiency.

Solution.   Apply Eq. 18.2,

$$W_d = \frac{(10.20) \ (3600 \text{ sec/hr}) \ (4 \text{ hr})}{10\ 000} = 14.69 \text{ ha-m (119 ac-ft)}$$

$$W_s = (0.31) \ (32.4) = 10.04 \text{ ha-m (80 ac-ft)}$$

$$E_a = 100 \frac{W_s}{W_d} = 100 \frac{10.04}{14.69} = 68 \text{ percent}$$

Example 18.3.   A uniformity check is taken by probing at 30-m (100-ft) stations down one border. The depths of penetration recorded were as follows:

Station	Penetration (cm)	Deviation from Mean (cm)
0 + 00 m	195	10
0 + 30	198	13
0 + 60	198	13
0 + 90	192	7
1 + 20	189	4
1 + 50	189	4
1 + 80	183	−2
2 + 10	177	−8
2 + 40	174	−11
2 + 70	168	−17
3 + 00	177	−8
3 + 30	183	−2
3 + 60	186	1
Sum	2409	Sum of absolute values 100
Mean	185	7.7

Compute the uniformity coefficient.

Solution. Apply Eq. 18.4,

$$C_u = 1 - \frac{y}{d} = 1 - \frac{7.7}{185} = 0.96$$

IRRIGATION METHODS

The methods of applying irrigation water may be classified as subsurface, surface, sprinkler, and trickle irrigation.

18.11. Subsurface Irrigation. In unique situations, water may be applied below the surface of the soil. There are two types of subsurface irrigation (also called subirrigation). The most common system develops or maintains a water table allowing the water to move up through the root zone by capillary action. This is essentially the same practice as controlled drainage discussed in Chapters 13 and 14. Controlled drainage becomes subirrigation if water must be supplied to maintain the desired water table level. Water may be introduced into the soil profile through open ditches, mole drains, or pipe drains. The open-ditch method is most widely used. Water table maintenance is suitable where the soil in the plant root zone is quite permeable and there is either a continuous impermeable layer or a natural water table below the root zone. Since subirrigation allows no opportunity for leaching and establishes an upward movement of water, salt accumulation is a hazard; thus the salt content of the water should be low.

A second method of subirrigation introduces water into the soil through perforated pipes. Water introduced into these pipes moves throughout the root zone by capillary action. The system has been successfully applied to irrigation of turf.

18.12. Surface Irrigation. By far the most common method of applying irrigation water, especially in arid regions, is by flooding the surface. Surface methods include wild flooding where the flow of water is essentially uncontrolled and surface application where flow is controlled by furrows, corrugations, border dikes, contour dikes, or basins. Except in the case of wild flooding, the land should be carefully prepared before irrigation water is applied. In order to conserve water, the rate of water application should be carefully controlled and the land properly graded.

18.13. Sprinkler Irrigation. In recent years increasing use has been made of irrigation pipe systems for distributing water to sprinkler heads. Lightweight portable pipes with slip joint connections are common. However, in view of the high labor cost in moving these systems, such applications are becoming more limited to high value crops. Mechanical-move systems are now widely accepted. These may be either intermittent or continuous mechanical move. Solid set and permanent systems are suitable for intensively cultivated areas growing a high income crop, such as flowers, fruits, or vegetables. Sprinkler irrigation systems provide reasonably uniform application of water. On coarse-textured soils, water application efficiency may be twice as high as with surface irrigation.

18.14. Trickle Irrigation. Increasing use is being made of trickle (drip) systems that apply water at very low rates, often to individual plants. Such rates are accomplished through the use of specially designed emitters or porous tubes. A typical emitter might apply water at from 2 to 10 liters per hour. Other techniques for applying water at low rates may also be called trickle irrigation. These systems provide an opportunity for efficient use of water because of minimum evaporation losses and because irrigation is limited to the root zone. Due to high cost, their use if generally limited to high value crops. Since the distribution pipes are usually at or near the surface, operation of field equipment is difficult. Both sprinkler and trickle systems are well-adapted to application of agricultural chemicals, such as fertilizers and pesticides with the irrigation water.

18.15. Comparison of Irrigation Methods. Table 18.3 compares different types of irrigation systems in relation to various site and situation factors.

Efficient surface irrigation requires grading of the land surface to control the flow of water. The extent of grading required depends upon the topography. In some soil and topographic situations, the presence of unproductive subsoils may make grading for surface irrigation unfeasible. The utilization of level basins where large streams of water are available generally provide high irrigation efficiencies.

Sprinkler irrigation is particularly adaptable to hilly land where grading for surface irrigation is not feasible. It is appropriate for most circumstances where the infiltration rate exceeds the rate of water application. With sprinkler irrigation the rate of water application can be easily controlled. Sprinkler irrigation systems usually have a relatively high cost of installation. With mechanical-move systems labor can be substantially reduced. In some cases disease problems have resulted from moistened foliage. Evaporation losses with sprinkler irrigation are not excessively high even in arid regions. A well-designed sprinkler irrigation system can provide a high efficiency of water application.

A well-designed trickle irrigation system probably provides the highest efficiency of water application. It is especially well-suited to tree fruit and high value crops, but not to annual row crops. Water must be clean and uncontaminated, usually achieved by a filtration system. Trickle irrigation lends itself well to automation and has a low labor requirement. Since trickle systems usually operate at low pressure, energy requirements are generally lower than with sprinkler systems. Some low pressure sprinklers operate at pressures comparable to trickle systems.

18.16. Automation. With the current high cost of labor, much attention is given to minimizing labor requirements. Mechanical-move and solid-set sprinkler systems lend themselves well to automatic controls. The application of automatic controls to surface irrigation is also feasible in some cases. Electrically triggered, mechanically operated control gates and valves can provide a considerable reduction in labor for most irrigation systems. In some cases water pressure may be reduced with continuous mechanical-move systems by aligning crop rows parallel with the wheel movement and by using drop pipes attached to the moving pipe to place the water into each furrow. By decreasing pressure, total energy requirements are reduced with a resultant saving in operating costs.

18.17. Temperature Control. Irrigation systems are often used as a means of frost protection. Irrigation water, especially if supplied from wells, is often considerably warmer than the soil and the air near the surface under frost conditions. The heat of fusion released by water freezing on plant parts keeps the temperature from falling below 0°C as long as freezing continues.

Conversely, irrigation may be used for cooling, particularly when germination occurs under high temperatures. Sprinkler irrigation has also been used to delay premature blossoming of fruit trees when warm weather occurs before the frost danger has passed. The cooling effect of evaporation effectively lowers the temperature of the plant parts. See Chapter 20 for more details.

REFERENCES

Ayers, R. S., and D. W. Westcot (1976). "Water Quality for Agriculture," Irrigation and Drainage Paper No. 29. Food and Agriculture Organization of the United Nations, Rome.

Doorenbos, J., and W. O. Pruitt (1977). "Crop Water Requirements" (revised), Irrigation and Drainage Paper No. 24. Food and Agriculture Organization of the United Nations, Rome.

Erie, L. J., O. F. French, and K. Harris (1968). "Consumptive Use of Water by Crops in Arizona." Arizona Agricultural Experiment Station Technical Bulletin 169.

Fangmeier, D. D. (1977). "Alternative Irrigation Systems." Agricultural Engineering and Soil Science Series (lithographed). University of Arizona, Tucson, Ariz.

Hansen, V. E., O. W. Israelsen, and G. E. Stringham (1980). *Irrigation Principles and Practices* (4th ed.). Wiley, New York.

Table 18.3 Comparison of Irrigation Systems in Relation to Site and Situation Factors

| Site and Situation Factors | Improved Surface Systems | | Sprinkler Systems |
	Redesigned Surface Systems	Level Basins	Intermittent Mechanical Move
Infiltration rate	Moderate to low	Moderate	All
Topography	Moderate slopes	Small slopes	Level to rolling
Crops	All	All	Generally shorter crops
Water supply	Large streams	Very large streams	Small streams nearly continuous
Water quality	All but very high salts	All	Salty water may harm plants
Efficiency	Average 60–70%	Average 80%	Average 70–80%
Labor requirement	High, training required	Low, some training	Moderate, some training
Capital requirement	Low to moderate	Moderate	Moderate
Energy requirement	Low	Low	Moderate to high
Management skill	Moderate	Moderate	Moderate
Machinery operations	Medium to long fields	Short fields	Medium field length, small interference
Duration of use	Short to long	Long	Short to medium
Weather	All	All	Poor in windy conditions
Chemical application	Fair	Good	Good

Source: Fangmeier (1977).

Jensen, M. E. (1973). *Consumptive Use of Water and Irrigation Water Requirements*. American Society of Civil Engineers, New York.

U.S. Soil Conservation Service (1967). "Irrigation Water Requirements." Tech. Release No. 21, Washington, D.C.

PROBLEMS

18.1. Determine the evapotranspiration and irrigation requirement for small grain where the monthly evapotranspiration coefficient $k = 0.65$ in the Blaney–Criddle formula. The average monthly temperatures for the three-month growing season are 21.6, 23.4, and 24.6°C, respectively. The percentages of daytime hours for the same period are 8.7, 9.3, and 9.5, and the average rainfall is 33, 49, and 54 mm (1.30, 1.91, and 2.13 in.), respectively. Assume the water application efficiency is 60 percent.

18.2. A 76-mm (3-in.) application of water measured at the pump increased

Table 18.3 Comparison of Irrigation Systems in Relation to Site and Situation Factors

Sprinkler Systems		Trickle Systems
Continuous Mechanical Move	*Solid Set and Permanent*	*Emitters and Porous Tubes*
Medium to high	All	All
Level to rolling	Level to rolling	All
All but trees and vineyards	All	High value required
Small streams nearly continuous	Small streams	Small streams, continuous and clean
Salty water may harm plants	Salty water may harm plants	All—can potentially use high salt waters
Average 80%	Average 70–80%	Average 80–90%
Low, some training	Low to seasonal high, little training	Low to high, some training
Moderate	High	High
Moderate to high	Moderate	Low to moderate
Moderate to high	Moderate	High
Some interference circular fields	Some interference	May have considerable interference
Short to medium	Long term	Long term, but durability unknown
Better in windy conditions than other sprinklers	Windy conditions reduce performance; good for cooling	All
Good	Good	Very good

the average moisture content of the top 0.6 m (2 ft) of soil from 18 to 24 percent (dry weight basis). If the average dry density of the soil is 1201 kg/m³ (75 pcf), what is the water application efficiency?

18.3. A flow of 5.66 m³/s (200 cfs) is diverted from a river into a canal. Of this amount 4.25 m³/s (150 cfs) are delivered to farm land. The surface runoff from the irrigated area averages 0.71 m³/s (25 cfs) and the contribution to ground water is 0.42 m³/s (15 cfs). What is the water-conveyance efficiency? What is the water-application efficiency?

18.4. Determine the water-application efficiency and the uniformity coefficient if a stream of 0.085 m³/s (3 cfs) is delivered to the field for 2 hours, runoff averaged 0.042 m³/s (1.5 cfs) for 1 hour, and the depth of penetration of the water varied linearly from 1.68 m (5.5 ft) at the upper end to 1.06 m (3.5 ft) at the lower end of the field. The root zone depth is 1.68 m (5.5 ft).

DESIGN
- SIZE OF FLOW
- RATE OF ADVANCE
- LENGTH OF BORDER
- DEPTH TO SUPPLY
- INTAKE RATE
- LAND SLOPE
- EROSION HAZARD
- DEPTH OF FLOW

CHAPTER 19

Surface Irrigation

In spite of the many advantages of applying water through sprinkler and trickle systems, surface irrigation is the predominant method in the United States. The 1974 U.S. census reported that almost three-fourths of the nation's irrigation was accomplished by surface methods. In the western states, where surface irrigation is especially predominate, the major portion of the water comes from surface runoff usually stored in reservoirs. Since this water must be conveyed for considerable distances over rough terrain, conveyance canals and control structures are key parts of most irrigation systems in arid regions. The hydraulic principles involved in the design of control structures are presented in Chapter 9 and the basis for the design of canals in Chapter 13. Wells also provide an important source of water for surface irrigation (see Chapter 17).

DISTRIBUTION OF WATER ON THE FARM

The farm water supply is normally delivered either from surface storage by conveyance ditches or from irrigation wells. Sometimes surface storage and underground supplies are combined in order to provide an adequate water supply at the farm.

19.1. Surface Ditches. A system of open ditches often distributes the water from the source on the farm to the field as shown in Fig. 19.1. These ditch systems should also be carefully designed as discussed in Chapter 13 so as to provide adequate head (elevation) and capacity to supply water at all areas to be irrigated. The amount of land that can be irrigated is often limited by the head of water available as well as by the design and location of the ditch system. Where irrigation is used as an occasional supplement to rainfall, these ditches may be temporary. In order to minimize water losses there is an increasing tendency to line ditches with impermeable materials. This practice is particularly applicable in the more arid regions where irrigation water supplies are limited and crop needs are largely dependent on irrigation water.

- open channel
resures - desirele
- head (elev.)
- capacity

19.2. Underground Pipe. Since surface distribution systems provide continuing problems of maintenance, constitute an obstruction to farming operations, and provide a water surface subject to evaporation losses, underground

432

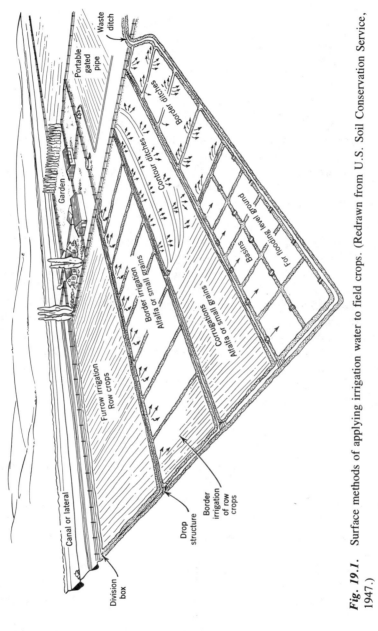

Fig. 19.1. Surface methods of applying irrigation water to field crops. (Redrawn from U.S. Soil Conservation Service, 1947.)

UNIT STREAM

— FLOW DIVERTED TO IRRIGATE
STRIP 1'WIDE × 100' LONG

$$\frac{MAX\ VEL}{MAX\ DEPTH} = \frac{ALLOW\ FLOW}{UNIT\ STREAM} = \frac{MAX\ Q}{OF\ 100'\ SEGMENTS}$$

BOUNDARY CONDITIONS

— MAX VELOCITY TO PREVENT EROSION
— MAX DEPTH — CROP DAMAGE

$$Q = VD$$

$$MANNING\ V = \frac{1.49}{n}R^{2/3}S^{1/2}$$

ft/sec

pipe distribution systems shown in Fig. 19.2a are becoming increasingly popular. In these systems water flows from the distribution pipe upward through riser pipes and irrigation valves to appropriate basins, borders, or furrows. In Fig. 19.2b a multiple-outlet riser controls the distribution of water to several furrows. Alfalfa valves regulate the flow to a header ditch from which water can be distributed to the field.

19.3. Portable Pipe. Surface irrigation with portable pipe or large diameter plastic tubing may be advantageous particularly in circumstances where water is applied infrequently. Light-weight gated pipe, Fig. 19.2c, provides a convenient and portable method of applying water in furrow irrigation. Portable flumes are sometimes used, particularly with high value crops, where they may be removed from the fields during certain field operations.

19.4. Devices to Control Water Flow. Control structures, such as illustrated in Fig. 19.3 are essential in open ditch systems to (1) divide the flow into two or more ditches, (2) lower the water elevation without erosion, and (3) raise the water level in the ditch so that it will have adequate head for removal. Various devices are used to divert water from the irrigation ditch and to control its flow to the appropriate basin, furrow, or border. Valves may be installed in the side or bottom of the ditch during construction. Other devices shown in Fig. 19.4 are spiles, gate take-outs for border irrigation, and siphon tubes. These siphons, usually aluminum or plastic, carry the water from the ditch to the surface of the field and have the advantage of metering the quantity of water applied. Figure 19.5 gives the rate of flow that can be expected from siphons of various diameters with different head differences between the inlet and outlet.

APPLICATION OF WATER

The various surface methods of applying water to field crops are illustrated in Fig. 19.1.

19.5. Flooding. Ordinary flooding is the application of irrigation water from field ditches that may be nearly on the contour or up and down the slope. After the water leaves the ditches, no attempt is made to control the flow by means of levees or by other methods of restricting water movement. For this reason ordinary flooding is frequently referred to as "wild flooding." Although the initial cost for land preparation is low, labor requirements are usually high and the efficiency of water application is generally low. Ordinary flooding is most suitable for close-growing crops, particularly where slopes are steep. Contour ditches are usually spaced from 15 to 45 m apart, depending on the slope, texture and depth of the soil, size of the stream, and crop to be grown. This method may be used on rolling land where borders, basins, and furrows are not feasible and adequate water supply is not a problem.

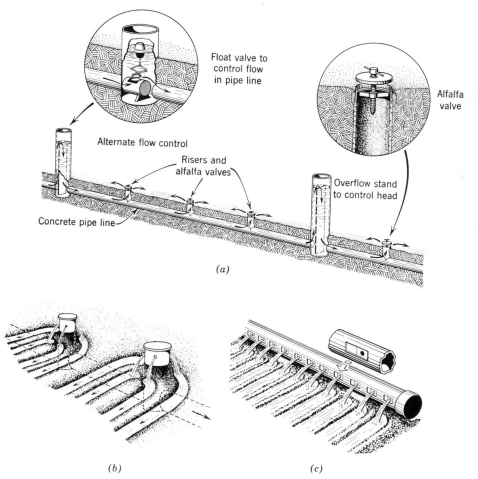

Fig. 19.2. Methods of distribution of water from (a) low-pressure underground pipe, (b) multiple-outlet risers, and (c) portable gated pipe. (Adapted from U.S. SCS and USBR, 1959.)

19.6. Graded Borders. The graded border method of flooding consists of dividing the field into a series of strips separated by low ridges. Normally, the direction of the strip is in the direction of greatest slope, but in some cases, the borders are placed nearly on the contour. The strips usually vary from 10 to 20 m in width and are 100 to 400 m in length. Ridges between borders should be sufficiently high to prevent overtopping during irrigation. To prevent water from concentrating on either side of the border, the land should be level perpendicular to the flow. Where row crops are grown in the border strip, furrows confine the flow and eliminate this difficulty.

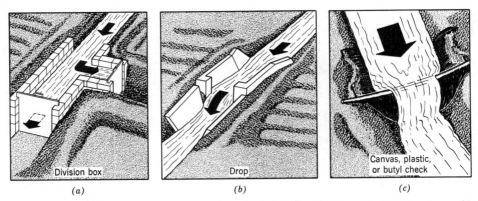

Fig. 19.3. Devices to control water flow in irrigation ditches. (a) Division box. (b) Drop. (c) Canvas, plastic, or buytl check. (Adapted from U.S. SCS and USBR, 1959.)

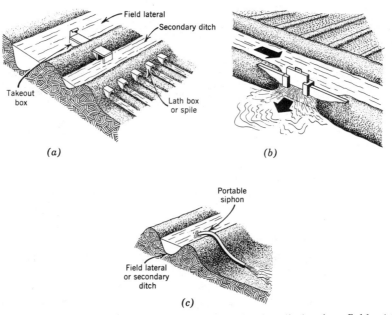

Fig. 19.4. Devices for distribution of water from irrigation ditches into fields. (a) Spile or lathe box. (b) Border takeout. (c) Siphons. (Adapted from USBR, 1951; and U.S. SCS and USBR, 1959.)

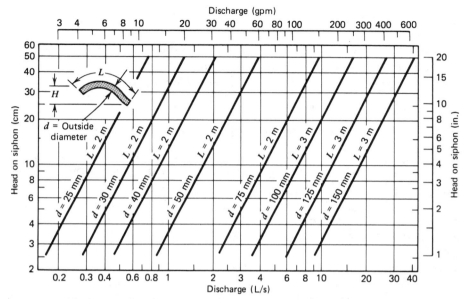

Fig. 19.5. Discharge of aluminum siphon tubes as a function of head. (Adapted from U.S. SCS.)

19.7. Level Borders. The layout of level borders is similar to that described for graded borders, except that the surface is leveled within the area to be irrigated. These areas may be long and narrow or they may be nearly square (often called basins). Modern laser leveling techniques make possible the preparation of smooth level surfaces required for this method of irrigation. Where relatively large rates of flow are available, the field can be quickly covered resulting in high field irrigation efficiencies. Level borders also lend themselves well to preirrigation, utilizing stream flow diverted during periods of high runoff. In orchard irrigation small leveled basins may include as few as one to four trees.

19.8. Furrow. While in the flooding methods water covers the entire surface, irrigating by furrows submerges only from one fifth to one half the surface, resulting in less evaporation, less puddling of the soil, and permitting cultivation sooner after irrigation. Furrows vary from large to small and are up and down the slope or on the contour. Small, shallow furrows, called corrugations, are particularly suitable for relatively irregular topography and close-growing crops, such as meadow and small grains. Furrows 80 to 200 mm deep are especially suited to row crops since the furrow can be constructed with normal tillage. Contour furrow irrigation may be practiced on slopes up to 12 percent, depending on the crop, erodibility of the soil, and the size of the irrigation stream.

The length of furrows depends on the infiltration capacity, the slope, and size of the stream. Excessively long furrows result in deep percolation losses and erosion in the upper ends of furrows. Suggested lengths of furrows for various slopes and soil textures are given in Table 19.1.

DESIGN AND EVALUATION

A recognition and understanding of the variables involved in the hydraulics of surface irrigation is essential to effective design. Hansen (1960) has listed, with reference to Fig. 19.6, the pertinent variables as (1) size of stream, (2) rate of advance, (3) length of run and time involved, (4) depth of flow, (5) intake rate, (6) slope of land surface, (7) surface roughness, (8) erosion hazard, (9) shape of flow channel, (10) depth of water to be applied, and (11) fluid characteristics.

Since the hydraulics of surface irrigation is complex and since some of the variables involved have not been evaluated and their relationships have not been determined, empirical procedures are often employed in design of surface irrigation systems. In designing a system it is recommended that local practices be explored and locally available data be considered. For example, extension services, experiment stations, and the Soil Conservation Service have prepared "Irrigation Guides" suggesting design procedures for many states. References, such as Hansen, Israelsen, and Stringham (1980) should be consulted. Selection of design values, such as border width, depends upon judgment and experience in managing water under specific soil, slope, and crop conditions.

19.9. Surface Irrigation Design. Probably the most commonly used reference in the United States for the design of graded and level border irrigation

Table 19.1 Lengths of Run for Furrows and Corrugations

| Slope (%) | *Length of Furrows or Corrugations (m)* | | | |
	Loamy Sand and Coarse Sandy Loams	*Sandy Loams*	*Silt Loams*	*Clay Loams*
0–2	75–120	90–200	200–400	270–400
2–5	60–90	60–90	90–200	120–270
5–8	45–60	45–75	60–90	75–120
8–15	30–45	30–60	30–60	60–90

Source: U.S. Bureau Reclamation (1951).

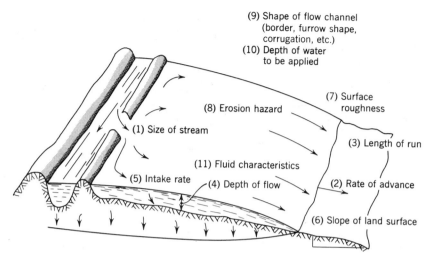

Fig. 19.6. Schematic view of flow in surface irrigation indicating the variables involved. (From Hansen, 1960.)

systems is Chapter 4 of the National Engineering Handbook on Irrigation published by the U.S. Soil Conservation Service (1974). This publication presents the derivation of equations relating the infiltration, roughness, irrigation efficiency, length of run, time of application and stream size.

The length of run and slope are generally defined by the field situation. The field efficiency will depend on management practices of the irrigator and also on the preparation of the field for irrigation. On gently sloping, well-leveled, uniformly graded fields an efficiency of 60–75 percent is usually feasible. The publication gives suggested design efficiencies for various situations.

Since Manning's equation is used in design, the roughness coefficient must be considered. In general, an n value of 0.04 is used for a smooth, bare soil surface. The value 0.10 is usually accepted for small grain and similar crops if the rows run lengthwise with the border strip. A value of 0.15 is suggested for alfalfa and broadcast small grain while dense sod and small grain crops drilled across the border strip have an n of about 0.25.

The determination of the infiltration rate of the field to be irrigated is an important consideration. In this procedure the soils are classified by intake families based on their ability to accept water. Usually these intake families are determined either by experience or by completing a series of cylinder infiltrometer runs. Figure 19.7 shows the relationship of time for infiltration against the accumulated intake. Results of infiltrometer runs can be plotted on this chart and the most appropriate intake family selected.

Since the equations must be solved by trial and error procedures, a series of

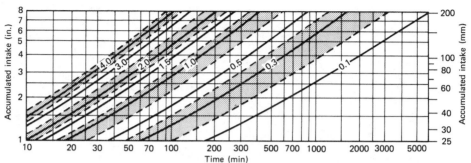

Fig. 19.7. Intake families for border irrigation design. (Redrawn from U.S. SCS, 1974.)

charts is presented to cover a wide variety of field situations. One of these for the design of a graded border system is given as an example in Fig. 19.8. This chart presents three families of curves; time curves, depth curves, and stream-size curves with each family involving several slopes. By entering the upper abscissa, dropping to the appropriate time curve, and reading from the right-hand ordinate, the time of application in minutes is obtained. At the intersection of the vertical line with the stream-size curve, read on the left ordinate the required stream size per unit width of border. By moving to the right to the appropriate depth curve and dropping to the lower abscissa, read the depth of flow.

Example 19.1. Determine the required time of application, stream size per unit border width, and the depth of flow for a smooth, unplanted graded 192-m (630-ft) border with a slope of 0.1 percent and a field application efficiency of 70 percent. The intake family is 1.0 and the depth of application is 50 mm (2 in.).
Solution. The slope of 0.1 percent requires the use of the 0.001 curve in each family of curves. The abscissa is

$$\frac{\text{Length}}{\text{Efficiency}} = \frac{192}{70} = 2.7 \text{ m (9 ft)}$$

Enter the upper abscissa at 2.7, drop to the 0.001-time curve and read the required time as 56 minutes. Enter the upper abscissa at 2.7, drop to the 0.001 stream-size curve and read the required stream size as 4.18 L/s per meter (0.045 cfs/ft) of border width. Enter the upper abscissa at 2.7 and drop to the 0.001 stream-size curve, move to the right to the 0.001-depth curve, and read the depth of flow as 41 mm (0.135 ft) on the lower abscissa.

The SCS publication presents additional curves for the solution of level-border design problems. Also, design charts and computer programs for design of furrow irrigation systems are available.

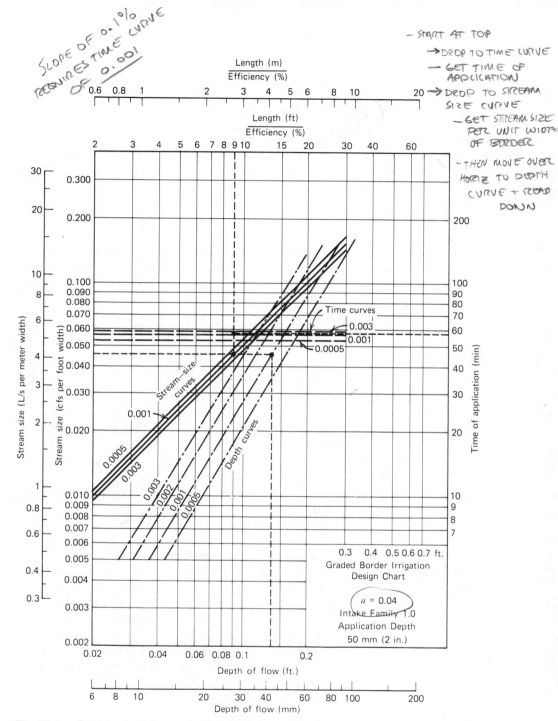

SLOPE OF 0.1%
REQUIRES TIME CURVE
OF 0.001

- START AT TOP
  → DROP TO TIME CURVE
- GET TIME OF
  APPLICATION
- → DROP TO STREAM
  SIZE CURVE
- GET STREAM SIZE
  PER UNIT WIDTH
  OF BORDER
- THEN MOVE OVER
  HORIZ TO DEPTH
  CURVE + READ
  DOWN

Length (m)
―――――――――
Efficiency (%)

Length (ft)
―――――――――
Efficiency (%)

Time curves
0.003
0.001
0.0005

Stream–size curves

0.001

0.0005
0.003

Depth curves

0.003
0.002
0.001
0.0005

0.3   0.4 0.5 0.6 0.7 ft.
Graded Border Irrigation
Design Chart

$n = 0.04$
Intake Family 1.0
Application Depth
50 mm (2 in.)

Stream size (L/s per meter width)

Stream size (cfs per foot width)

Time of application (min)

Depth of flow (ft.)

Depth of flow (mm)

Fig. 19.8. Sample chart for graded border irrigation design. (Adapted from U.S. SCS, 1974.)

441

19.10. Evaluation of Existing Systems. Although the design of new irrigation systems constitutes an important engineering activity, it should be noted that even greater opportunities are often available in the evaluation and improvement of existing systems. Many of these older systems were designed before much was known about intake rates and water-holding capacities of soils, and when water supplies were not as limited as they are now becoming. Fields or portions of fields that do not receive enough water have limited production potential. Excessive irrigation not only wastes water, but leaches water-soluble nutrients and may cause drainage problems.

Evaluation of existing systems can be approached in a number of ways. A simple method of determining underirrigation is by use of the soil auger or tube sampler. Observation of the *opportunity time* for infiltration in various parts of the field may be helpful. More sophisticated methods involve application of such equipment as portable flumes, meters, and water intake rings in carefully conducted diagnostic procedures, such as described by Merriam (1968).

REFERENCES

Hansen, V. E. (1960). "Mathematical Relationships Expressing the Hydraulics of Surface Irrigation." Proc. ARS-SCS Workshop Hydraulics Surface Irrig., ARS 41-43.

Hansen, V. E., O. W. Israelsen, and G. E. Stringham (1980). *Irrigation Principles and Practices,* (4th ed.). Wiley, New York.

Merriam, J. L. (1968). *Irrigation System Evaluation and Improvement.* Blake Printing & Publ. Co., San Luis Obispo, CA.

Turner, J. H., and C. L. Anderson (1971). *Planning for an Irrigation System.* Am. Assoc. for Voc. Instr. Materials, Athens, GA.

U.S. Bureau Reclamation (1951). "Irrigation Advisor's Guide." U.S. Dept. Interior, Washington, D.C.

U.S. Soil Conservation Service (1974). "Border Irrigation," in *National Engineering Handbook,* Sect. 15, Chap. 4. Washington, D.C.

———— (1947). "First Aid for the Irrigator." U.S. Dept. Agr. Misc. Publ. 624.

———— (1957). "Instructions and Criteria for Preparation of Irrigation Guides." Engineering Handbook for Western States and Territories, Sect. 15, Part I. Eng. and Watershed Planning Unit, Portland, OR.

U.S. Soil Conservation Service and U.S. Bureau Reclamation (1959). "Irrigation on Western Farms." Agr. Inform. Bull. 199, Washington, D.C.

PROBLEMS

19.1. Determine the time of application and stream size for a 10-m (33-ft) width border, and the depth of flow for a 140-m (460-ft) border length with a slope of 0.2 percent. The field application efficiency is 70 percent, the depth of application is 50 mm (2 in.), the intake family is 1.0, and the roughness coefficient is 0.04.

19.2. Determine the time of application, stream size, and depth of flow if the slope in Problem 19.1 is decreased to 0.05 percent.

19.3. Determine the discharge of a 25-mm (1-in.) and a 75-mm (3-in.) diameter siphon tube for a head of 20 cm (7.9 in.) from the graph. Compare the relative flow using the theoretical flow rates assuming the same friction loss in both tubes and that the Manning equation applies.

19.4. If the flow into the border described in Problem 19.1 is discharged by 50-mm (2-in.) diameter, 2-m (6.3-ft) length siphon tubes under a head of 20 cm (7.9 in.), how many siphon tubes are required to irrigate the 10-m (33-ft) width border?

Hand-move — not too practical in corn (too tall)

$TAW = \left(\frac{MOIST}{HOLD\ CAP}\right)\left(\frac{ROOT}{DEPTH}\right)$

$RAW \equiv \frac{1}{2} TAW$

$ED \leq RAW\ (SAY =)$

$PD = \frac{ED}{e}$

$IP = \frac{ED}{Peak\ rate}$

$IF \leq IP$

$AREA/DAY = \frac{TOTAL}{IF}$

SPRINKLER

$\begin{array}{c} sprinkler \\ lateral \quad lateral \\ spac \quad main \\ spac \end{array}$

$\underset{GPM}{CAPACITY} = \frac{S_l * S_M \times iph}{96.3}$

CHAPTER 20

Sprinkler Irrigation

Sprinkler irrigation is a versatile means of applying water to any crop, soil, and topographic condition. It is particularly popular in humid regions because surface ditches and prior land preparation are not necessary and because pipes are easily transported and provide no obstruction to farm operations when irrigation is not needed. Sprinkling is suitable for sandy soils or any other soil and topographic condition where surface irrigation may be inefficient or expensive, or where erosion may be particularly hazardous. Low rates and amounts of water may be applied, such as are required for seed germination, frost protection, delay of fruit budding and cooling of crops in hot weather. Fertilizers and soil amendments may be dissolved in the water and applied through the irrigation system. The major disadvantages of sprinkling systems are high investment cost and high labor requirements.

20.1. Sprinkler Systems. In general, systems are described according to the method of moving the lateral lines, on which are attached various types of sprinklers. These systems are identified and compared in Table 20.1 and Fig. 20.1. During normal operation, laterals may be hand-moved or mechanically moved. The sprinkler system may cover only a small part of a field at a time or be a solid-set system in which sprinklers are placed over the entire field. With the solid-set system, all or part of the sprinklers may be operated at the same time. Most sprinklers rotate a full circle, but some may be set to operate for any portion of a circle (part circle). Perforated pipes are suitable for distributing water to small acreages of high-value crops where a rectangular pattern is desired.

As shown in Table 20.1, hand-move laterals have the lowest investment cost but the highest labor requirement. With giant sprinklers, the spacing can be increased, thereby reducing labor, but higher pressures are required, which increase the pumping cost. These sprinklers are generally pulled or transported from one location to another or moved continuously. Hand-move laterals with standard sprinklers are most suitable for low-growing crops, and are impractical in tall corn because of adverse conditions for moving the pipe.

With the end-pull system shown in Fig. 20.1a, the lateral is moved to the next position by pulling the line with a tractor to the opposite side of the main. In

Table 20.1 Comparison of Sprinkler Irrigation Equipment

Type of System	Relative Investment Cost[a]	Relative Labor Cost	Practical Hours of Operation per Day
Hand-move laterals (standard sprinklers)	0.4	5.0	16
Hand-move laterals (giant sprinklers)	0.5	4.0	12–16
End-pull laterals (tractor tow)	0.5	1.4	16
Boom-type sprinklers (trailer-mounted)	0.6	3.7	12–16
Side-roll laterals (powered-wheel move)	0.7	1.7	18–20
Self-propelled (center-mounted)	1.0	1.0	24
Solid set	3.0–5.0	1.0	24

[a] Based on a 65-ha field, 63 L/s from pump, and 80 percent application efficiency.
Source: Berge and Groskopp, 1964.

moving the lateral, it is pulled at a slight diagonal so as to move the entire length down the field, one half the lateral spacing with each direction pulled. This system is suitable for low-growing crops and in places where adequate moving space is available. It is also practical for tree crops. Labor is greatly reduced (Table 20.1) compared to hand-move systems.

The rotating boom-type system (Fig. 20.1*b*) operates with one trailer unit per lateral at spacings up to 110 m. The trailer and boom unit is moved to the next position along the lateral with a tractor or winch. The lateral line is added or picked up as the trailer is moved progressively through the field away from or toward the main line. The boom-type unit will cover about the same area as a giant sprinkler, but the pressure required is not as high.

The side-roll lateral system (Fig. 20.1*c*) utilizes the irrigation pipe as the axle of large diameter wheels that are spaced about 9 m apart. The lateral is moved to the next position by a small gasoline engine mounted at the midpoint of the line or by hand with a lever and ratchet. The side-roll lateral is limited to crops that will not interfere with the movement of the pipe. Unless a flexible pipe is attached to the main, the lateral must be disconnected for each move. Labor requirements are about the same as for the end-pull system, but much less than for the hand-move systems. Self-aligning sprinklers, which stay in a vertical position regardless of the position of the wheels, eliminate the necessity of having the sprinkler on top as required with the fixed-position sprinklers.

The self-propelled center pivot system (Fig. 20.1*d*) consists of a radial pipe line supported at a height of about 2 m at intervals of about 30 m. The radial line

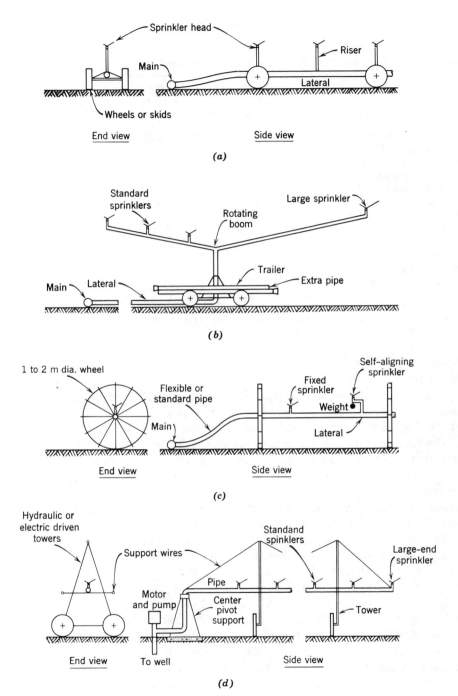

Fig. 20.1. Mechanical-move sprinkler systems. (a) End-pull lateral. (b) Rotating boom-type unit. (c) Side-roll lateral (hand or mechanical move). (d) Self-propelled (center-mounted) lateral.

rotates slowly around a central pivot by either water pressure or electric motors. The towers are supported by wheels or skids and are kept in alignment with supporting wires. The nozzles increase in size from the pivot to the end of the line, at which is placed a large sprinkler in order to obtain the maximum diameter of coverage. The nozzles are selected to provide a uniform depth of application, varying from 13 to 100 mm per revolution. The depth of application is determined by the speed of rotation. The system is best suited to sandy soils, but it will operate in heavier soils if the depth of application is greatly reduced. A common-size system designed for 65 ha (160 acres) is 392 m in length. Such a system will irrigate about 55 ha of the 65 ha; the remaining 10 ha in the corners are not covered. Some manufacturers provide a fold-back extension arm that will irrigate nearly all of the corners of a square field and odd-shape fields. Normally, the system, once set up, remains permanently in place; but some can be towed to another field. The major advantage of the center-pivot system is the saving of labor; the major disadvantage is high investment cost.

Solid-set systems may be operated by changing the flow from one lateral to the next or by sequencing sprinklers (one operating on each line) along the lateral lines. For frost protection, all sprinklers are operated at the same time. Operation can be made automatic with timing devices and solenoid valve controls. As with the self-propelled system, the only labor required is for setup and maintenance. Because of extremely high investment costs (Table 20.1), solid-set systems are practical only for high-value crops.

20.2. Components of Sprinkler Systems.

The major components of a hand-move portable sprinkler system are shown in Fig. 20.2. Most components are basically the same as required for the end-pull, rotating boom, and side-roll systems shown in Fig. 20.1. With the rotating boom or giant sprinkler systems, the lateral spacing and the distance between sprinkler setups along the lateral are much greater than that for standard sprinklers. Latches, seals, and other details of these components vary considerably, depending on the manufacturer.

A pump is required to overcome the elevation difference between the water source and the sprinkler nozzle, to counteract friction losses, and to provide adequate pressure at the nozzle for good water distribution. As discussed in Chapter 16, the type of pump will vary with the discharge, pressure, and the vertical distance to the source of water. The pump should have adequate capacity to meet future needs and to allow for wear. For most farm irrigation systems capacity generally varies from 6 to 50 L/s.

The second component of the system, the main line, may be either movable or permanent. Movable mains generally have a lower first cost and can be more easily adapted to a variety of conditions; permanent mains offer a saving in labor and reduced obstruction to field operations. Water is taken from the main either through a valve placed at each point of junction with a lateral or in some

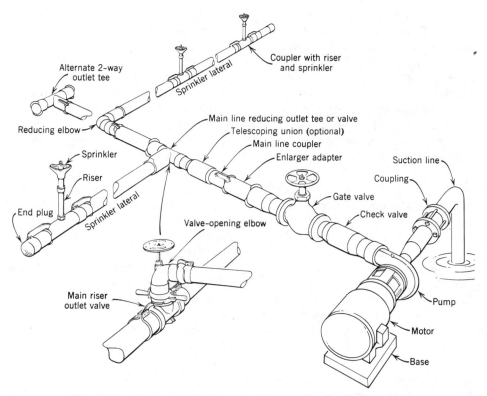

Fig. 20.2. Components of a portable sprinkler irrigation system.

cases through either an L- or a T-section that has been supplied in place of one of the couplings on the main.

The laterals are usually 6- or 9-m (20- or 30-ft) lengths of aluminum or other lightweight metal pipe connected with couplers. In some cases the couplers are permanently attached to the pipe (see Fig. 20.3a).

For rotating sprinklers, the sprinkler heads most often used have two nozzles, one to apply water at a considerable distance from the sprinkler and the other to cover the area near the sprinkler center (Fig. 20.3b). Of the devices to rotate the sprinkler, the most usual, also shown in Fig. 20.3b, taps the sprinkler head with a small hammer activated by the force of the water striking against a small vane connected to it. Sprinklers designed to cover a considerable area have a slow rate of rotation, about one revolution per minute.

A number of sprinkler heads are available for special purposes. Some provide a low-angle jet for use in orchards. Some work at especially low heads, of say 35 kPa, and others operate only in a part circle. Giant sprinkler units discharge 20 to 30 L/s at pressures of 550 to 700 kPa and throw the water many

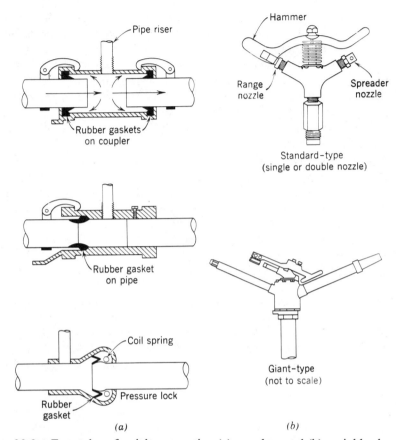

Fig. 20.3. Examples of quick-connecting (a) couplers and (b) sprinkler heads.

meters. In general, these work at higher pressures than the smaller units and result in greater pumping costs. On the other hand, because these large units cover a much greater area, a smaller number of moves is required.

20.3. Evaporation Losses. In sprinkler application of irrigation water, evaporation occurs from the spray and from the wet foliage surfaces. Frost (1963) developed relationships between spray losses and atmospheric parameters. He has compared the evapotranspiration rates of various crops under nonsprinkling conditions to the evapotranspiration rates of the same foliage as it is being wetted by sprinkler water. It was found that the evaporation from wet foliage was essentially the same as for dry foliage. In those situations where wet-leaf loss was less than dry-leaf loss, energy available for evaporation of moisture was not reaching the wet leaves at the rate at which it reached the dry leaves. The energy difference was absorbed by the spray.

20.4. Design for Equipment Requirements. Not only should the sprinkler system be properly designed hydraulically and economical in cost, but selection of the system should consider the availability of labor for moving the sprinklers and the pipe. The design of a sprinkler irrigation system involves the maximum rate of application, the irrigation period, and the depth of application. Application at rates in excess of the soil infiltration capacity results in runoff with accompanying poor distribution of water, loss of water, and soil erosion. Maximum water application rates for various soil conditions are given in Table 20.2. These values may serve as a guide where reliable local recommendations are not available.

Applications at rates well below the maximum have been found beneficial. Rates of one half the infiltration rate of the soil combined with nozzle pressures that provide a fine spray have resulted in improved maintenance of soil structure and minimization of soil compaction.

The depth of application and the irrigation period are closely related. Irrigation period is the time required to cover an area with one application of water. The depth of application will depend on the available moisture-holding capacity of the soil.

Under humid conditions rains may bring the entire field up to a given moisture level. As the plants use this moisture, the moisture level for the entire field decreases. Irrigation must be started soon enough to enable the field to be covered before plants in the last portion to be irrigated suffer from moisture deficiency.

Table 20.2 Suggested Maximum Water Application Rates for Sprinklers for Average Soil, Slope, and Cultural Conditions

Soil Texture and Profile Conditions	*Maximum Water Application Rate for Slope and Cultural Conditions (mm/h)*			
	0% Slope		*10% Slope*	
	w/cover	*bare*	*w/cover*	*bare*
Light sandy loams uniform in texture to 2 m	43	25	25	15
Light sandy loams over more compact subsoils	30	18	18	10
Silt loams uniform in texture to 2 m	25	13	15	8
Silt loams over more compact subsoil	15	8	10	3
Heavy-textured clays or clay loams	5	3	3	2

Source: Soil Conservation Service (1949).

SO THAT WHEN STARTING SPOT GETS DOWN
TO 10% YOU ARE ABLE TO START AGAIN

One recommended system is to commence irrigation when the moisture level of the field reaches 55 percent of the available moisture capacity. The net depth of application under this plan is equal to 45 percent of the available moisture capacity. The irrigation period is set so that the entire irrigated area will be covered before the finishing end of the field reaches a moisture level below 10 percent of the available moisture. Typical moisture-holding capacities are:

sandy loam	130 mm/m depth or 13 percent by volume
silt loam	150 mm/m depth or 15 percent by volume
clay loam	170 mm/m depth or 17 percent by volume

Average root depths and peak rate of moisture use by crops are given in Table 20.3.

Example 20.1. A sprinkler irrigation system is to be designed to irrigate 16.2 ha (40 ac) of pasture on a silt loam soil over compact subsoil near Cleveland, Ohio. The field is flat. Determine the limiting rate of application, the irrigation period, the net depth of water per application, the depth of water pumped per application, and the required system capacity in hectares per day.

Solution.   From Table 20.2 the limiting rate of application is 15 mm/h (0.6 iph). The available moisture-holding capacity of the soil is 150 mm/m (1.8 in./ft) and the depth of the root zone from Table 20.3 is about 0.6 m (2 ft). The total available moisture capacity is thus $(0.6 \times 150) = 90$ mm (3.6 in.). The net depth of application is $(0.45 \times 90) = 40.5$ mm (1.6 in.). Assuming a water application efficiency of 70 percent, the depth of water pumped per application is $(40.5/0.70) = 58$ mm (2.3 in.). From Table 20.3 the peak rate of use by the crop is 5 mm/d (0.2 ipd). The irrigation period is 8.1 (40.5/5) days. To cover this field in 8.1 days the system must be able to pump and discharge 58 mm on 2.0 (16.2/8.1)

Table 20.3   Root Depth and Peak Rate of Moisture Use

		Peak Rate of Moisture Use (mm/d)		
			Climate	
Crop	Root Depth (m)	Cool	Moderate	Hot
Alfalfa	0.9−1.2	5	6	8
Beans	0.3−0.9	3	4	6
Corn	0.6−1.2	5	6	8
Pasture	0.5−0.8	5	6	8
Potatoes	0.3−0.6	4	5	6
Strawberries	0.3−0.5	3	4	6

hectares per day. This information is then used as a guide in the selection of equipment.

20.5. Layouts for Hand-Move, Side-Roll, and End-Pull Systems. The number of possible arrangements for the mains, laterals, and sprinklers is practically unlimited. The arrangement selected should allow a minimal investment in irrigation pipe, have a low labor requirement, and provide for an application of water over the total area in the required period of time. The most suitable layout can be determined only after a careful study of the conditions to be encountered. The choice will depend to a large extent upon the types and capacities of the sprinklers and the pressure required. For medium-pressure systems, the laterals are moved about 18 m at each setting and the sprinklers are spaced about 12 m along each lateral.

Typical layouts for sprinkler irrigation systems are shown in Fig. 20.4. The layout in Fig. 20.4a is suitable where the water supply can be obtained from a stream or canal alongside the field to be irrigated. This arrangement either eliminates the main line or requires a relatively short main, depending on the number of moves for the pump. Less pipe is required for this method than for any of the others. The layout illustrated in Figs. 20.4b and 20.4c is suitable where the water supply is from a well or pit. In Fig. 20.4b the two laterals are started at opposite ends of the field and are moved in opposite directions. Since the farther half of the main supplies a maximum of one lateral at a time, the diameter of this section can be reduced. This arrangement is well suited to day and night operation when the required amount of water can be added in about 6 or 8 hours. The system shown in Fig. 20.4c can be designed for continuous operation. While line A is in operation, the operator moves line B. When the required amount of water has been applied, line B is turned on and then line A is moved. With this procedure the capacity of the pump needs to be adequate to supply only one lateral.

20.6. Layout for Center Pivot Systems. With a center pivot system, sprinklers are spaced along the moving lateral each covering an area equal to $2\pi r S_l$ in one revolution, where r is the distance from the pivot point and S_l is the sum of half the distances to the two adjacent sprinklers as shown in Fig. 20.5. To obtain uniform application of water the sprinkler spacing may be constant with a variable sprinkler discharge or with a combination of spacings and discharges. A large part-circle sprinkler (end gun) is usually placed at the end of the lateral to extend the diameter of coverage as far as possible. The length of the lateral line and the size of the end gun are adjusted so that the effective circular area to be irrigated is adequately covered. Note in Fig. 20.5 that the outer area of the sprinkler pattern of the end gun is outside this area. Some manufacturers provide a fold-back extension arm that attaches to the end of the

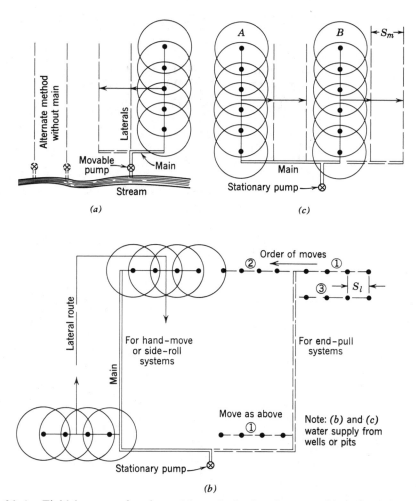

Fig. 20.4. Field layouts of main and laterals for hand-move, side-roll, and end-pull sprinkler systems. (a) Fully portable. (b) Portable or permanent (buried) main and portable laterals. (c) Portable main and laterals.

sprinkler lateral (replaces the end gun) to cover most of the corner areas of a square field. An end gun is then located at the end of the extension arm. These systems will also cover odd-shaped fields. Some center pivot laterals can be pulled to another field by rotating the tower wheels 90 degrees, but during normal sprinkling the lateral rotates about the center pivot as before.

20.7. Capacity of the Sprinkler System. The capacity of a sprinkler system depends on the area to be irrigated, depth of water application at each irriga-

CENTER PIVOT

① SYSTEM $Q = \dfrac{(Area)(PD)}{IF} = 500/min$

② Lateral Design — spacing
        — per nozzle   Area, flow, dist, ∑flow H loss ∑H loss

③ Add flow for end gun + carry through

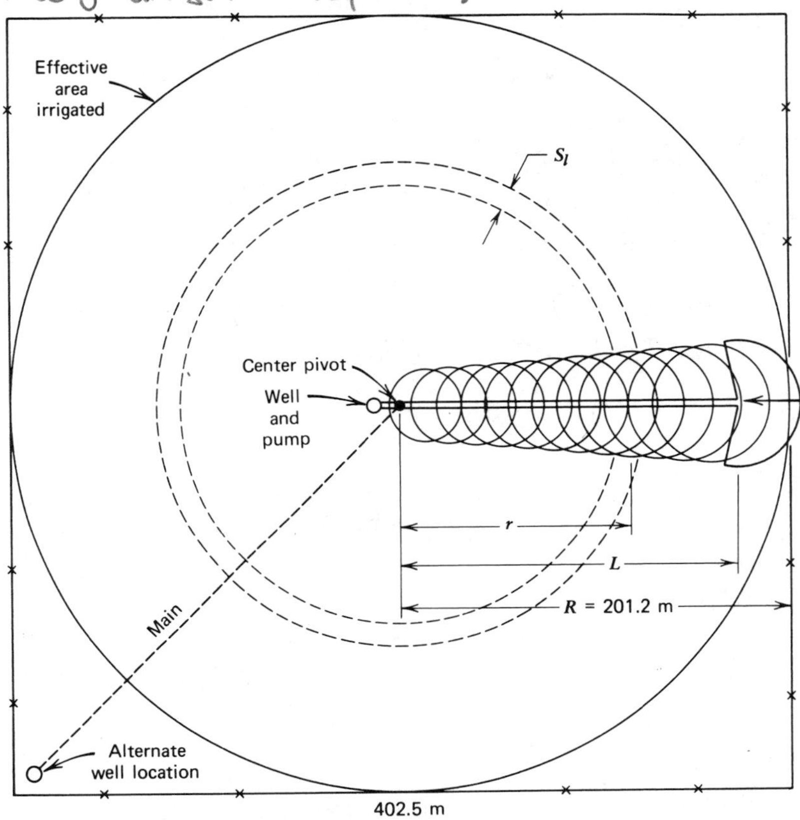

res acre 7 100 in sum circle

Fig. 20.5. Center pivot sprinkler system for a 16.2-ha (40-ac) field. Alternate well location is for convenience of supplying four such adjacent fields.

tion, efficiency of application, and actual operating time for each irrigation. For 100 percent efficiency of application,

$$q_s = da/t = d\pi R^2/t \qquad (20.1)$$

where q_s = capacity of the system (L³/T),
          d = depth of application per revolution (L),
          a = area to be irrigated (L²),
         R = radius of a circular area as in a center pivot system (L),
          t = total time of operation per revolution (T).

For a center pivot system the area covered by one revolution of the lateral by an infinitesimally small sprinkler is $2\pi r\, dr$, where r is the distance from the sprinkler to the pivot point. The sprinkler capacity for this area is

$$q = d(2\pi r)\, dr/t \tag{20.2}$$

where dr/t = derivative of r with time.

As given by Chu and Moe (1972), the integration of Eq. 20.2 from the limits r to R gives the flow rate in the lateral at any point r as

$$q_r = (\pi R^2 d/t)\,(1 - r^2/R^2) \tag{20.3}$$

At the outer end of the lateral at length L the discharge required for the end gun can be obtained by substituting in Eq. 20.3, $r = L$. The discharge of the sprinklers along the lateral is thus the difference between Eq. 20.1 and Eq. 20.3. Discharge from Eq. 20.3 is theoretically correct for an infinite number of sprinklers applying the average water depth uniformly along the lateral. In practice the distribution will not be as ideal because of the limited number of sprinklers. Chu and Moe (1972) found that the theoretical distribution was close to experimental data.

20.8. Sprinkler Capacity. When the rate of application and the spacing of the sprinklers have been determined, the required sprinkler capacity can be computed by the formula

$$q = S_l S_m r \tag{20.4}$$

where q = discharge of each sprinkler (L³/T),
 S_l = sprinkler spacing along the lateral (L),
 S_m = sprinkler spacing between lines or along the main (L),
 r = rate of application (L/T).

For example, a spacing of 12 by 18 m and an application rate of 10 mm/h require a sprinkler with a capacity of 0.6 L/s. The theoretical discharge of a nozzle may be computed from the orifice flow equation (see Chapter 9), which is

$$q = aC\sqrt{2gh} \tag{20.5}$$

For simplification of calculations, this equation is

$$q = 0.00111 C d_n^2 P^{1/2} \tag{20.6}$$

where q = nozzle discharge in L/s,
 C = coefficient of discharge,
 d_n = diameter of the nozzle orifice in mm,
 P = pressure at the nozzle in kPa.

The coefficient of discharge for well-designed, small nozzles varies from about 0.95 to 0.98. Some nozzles have coefficients as low as 0.80. Normally, the larger the nozzle, the lower is the coefficient. Where the sprinkler has two nozzles, the total discharge is the combined capacity of both.

20.9.   Distribution Pattern of Sprinklers.   A typical distribution pattern showing the effect of wind for a single sprinkler is illustrated in Fig. 20.6. Since one sprinkler does not apply water uniformly over the area, the overlapping of sprinkler patterns is relied upon to provide more uniform coverage. The distribution pattern shown in Fig. 20.7 illustrates how the overlapping patterns combine to give a relatively uniform distribution between sprinklers. Although Fig. 20.7 shows relatively uniform distribution over the area, wind will skew the pattern so as to give less uniform distribution.

The factors that influence the distribution pattern of sprinklers are the nozzle pressure, wind velocity, and the speed of rotation. Too low a pressure will result in a "ring-shaped" distribution and a reduction in the area covered, while

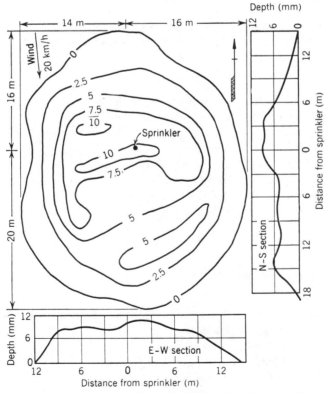

Fig. 20.6.   Distribution pattern for a single sprinkler showing the effect of wind. (Nozzle pressure 207 kPa and discharge 1.23 L/s.) (Redrawn from Christiansen, 1948.)

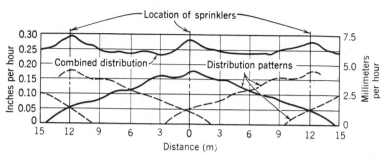

Fig. 20.7. Distribution pattern from a sprinkler showing overlapping to give relatively uniform combined distribution for no wind. (Redrawn from Gray, 1961.)

high pressures produce smaller drops with high application rates near the sprinkler. Wind will cause a variable diameter of coverage and also somewhat higher rates near the sprinkler. A high speed of rotation of the sprinkler greatly reduces the area covered and causes excessive wear of the sprinkler. The variation in speed of rotation is not due to the wind, but to changes in frictional resistance attributed to lack of precision in manufacture. Uniformity of application of a sprinkler system can be expressed by the uniformity coefficient given in Chapter 19. In applying the coefficient to evaluate a sprinkler system, the depths can be the water applied or the depths of soil moisture penetration. Depth of water applied evaluates the sprinkler system alone. Depth of penetration combines the effects of the sprinkler system, the topography, and the soil conditions. Observations at many points are made in the field with cans uniformly spaced in the sprinkler area bounded by four adjacent regularly spaced sprinklers. An estimate of the uniformity coefficient can also be determined graphically if the distribution pattern, as shown in Fig. 20.6, is known. An absolutely uniform application would give a uniformity coefficient of 1.0. Uniformity coefficients of 0.85 or more are acceptable. Distribution in wind may be improved by moving the sprinkler lateral only one half the normal spacing, by irrigating at night when the wind is low, or by using only the range nozzle on two-nozzle sprinklers.

20.10. Sprinkler Selection and Spacing.   The actual selection of the sprinkler is based largely upon design information furnished by manufacturers of the equipment. The choice depends primarily upon the diameter of coverage required, pressure available, and capacity of the sprinkler. The data given in Table 20.4 may serve as a guide in selecting the pressure and spacing desired.

Sprinkler discharge and diameter of coverage for a given sprinkler head design are given in Table 20.5. Each manufacturer will recommend a combination of nozzle sizes and pressures to give the best breakup of the stream and

Table 20.4 Classification of Sprinklers and Their Adaptability

Type of Sprinkler	Very Low Pressure (34–105 kPa)	Low Pressure (100–210 kPa)	Medium Pressure (210–415 kPa)
General characteristics	Special thrust springs or reaction-type arms	Usually single-nozzle oscillating or long-arm dual-nozzle design	Either single or dual nozzle design
Range of wetted diameters	6–15 m	18–24 m	23–37 m
Recommended application rate	10 mm/h	5 mm/h	5 mm/h
Jet characteristics (assuming proper pressure-nozzle size relations)	Waterdrops are large due to low pressure	Waterdrops are fairly well broken	Waterdrops are well broken over entire wetted diameter
Moisture distribution pattern (assuming proper spacing and pressure-nozzle size relation)	Fair	Fair to good at upper limits of pressure range	Very good
Adaptations and limitations	Small acreages confined to soils with intake rates exceeding 13 mm/h and to good ground cover on medium-to-coarse textured soils	Primarily for undertree sprinkling in orchards. Can be used for field crops and vegetables	For all field crops and most irrigable soils. Well adapted[†] to overtree sprinkling in orchards and groves and to tobacco shades

Source: U.S. Soil Conservation Service (1960).

Table 20.4—Continued

High Pressure (350–690 kPa)	Very High Pressure (550–830 kPa)	Undertree Low Angle (70–350 kPa)	Perforated Pipe (30–140 kPa)
Either single or dual nozzle design	One large nozzle with smaller supplemental nozzles to fill in pattern gaps. Small nozzle rotates the sprinkler	Designed to keep stream trajectories below fruit and foliage by lowering the nozzle angle	Portable irrigation pipe with lines of small perforations in upper third of pipe perimeter
34–70 m	60–120 m	12–27 m	Rectangular strips 3–15 m wide
13 mm/h	16 mm/h	8 mm/h	13 mm/h
Waterdrops are well broken over entire wetted diameter	Waterdrops are extremely well broken	Waterdrops are extremely well broken	Waterdrops are large due to low pressure
Good *except* where wind velocities exceed 6 km/h	Acceptable in calm air. Severely distorted by wind	Fairly good. Diamond pattern recommended where laterals are spaced more than one tree interspace	Good, pattern is rectangular
Same as for intermediate pressure sprinklers except where wind is excessive	Adaptable to close-growing crops that provide a good ground cover. For rapid coverage and for odd-shaped areas. Limited to soils with high intake rates	For all orchards or citrus groves. In orchards where wind will distort overtree sprinkler patterns. In orchards where available pressure is not sufficient for operation of overtree sprinklers	For low-growing crops only. Unsuitable for tall crops. Limited to soils with relatively high intake rates. Best adapted to small acreages of high-value crops. Low operating pressure permits use of gravity or municipal supply

Table 20.5 Manufacturer's Sprinkler Characteristics

Nozzle Pressure		Nozzle Diameters [mm (in.)]					
		3.97 × 3.18 (5/32 × 1/8)		4.76 × 3.97 (3/16 × 5/32)		6.35 × 3.97 (1/4 × 5/32)	
(kPa)	(psi)	Dia.[a]	L/s	Dia.	L/s	Dia.	L/s
207	(30)	25	0.37	26	0.52	28	0.76
276	(40)	27	0.43	28	0.61	31	0.90
345	(50)	28	0.47	30	0.68	34	1.00
414	(60)	30	0.52	31	0.74	36	1.10

[a] Diameter of coverage in meters.
Pressures to left and below dashed line recommended for best break up of stream.
Note: 1 L/s = 15.85 gpm and 1 m = 3.28 ft.

distribution pattern for uniform application. Single nozzle sprinklers are generally recommended only for small nozzles with limited diameter of coverage or for part-circle sprinklers. Because the distribution and coverage depend on the angle of the stream from the horizontal and the rate of rotation, sprinklers should be selected from manufacturer's tables.

The minimum sprinkler spacings for satisfactory uniformity of application are given in Table 20.6 for a distribution pattern which is triangular in cross section or for one similar to that shown in Fig. 20.7. These values may be applied in design when uniformity coefficients are not available.

20.11. Size of Laterals and Mains. Laterals and mains should provide the required rate of flow with a reasonable head loss. For laterals the sections at the distant end of the line have less water to carry and may therefore be smaller. However, many authorities advise against "tapering" of pipe diameters in laterals, as it then becomes necessary to keep the various pipe sizes in the same relative position. The system may also be less adaptable to other fields and situations.

Table 20.6 Maximum Spacings for Low or Medium Pressure Sprinklers[a]

Wind Velocity (km/h)	Lateral Spacing in Percent of the Diameter of Coverage	
	S_l, along the lateral	S_m, along the main
0	50%	65%
6 or less	45	60
7–12	40	50
13 or more	30	30

[a] For laterals normal to the wind direction only.

NOMOGRAPH
GIVEN — AVG Q FOR ALL SPRINKLERS ON LATERAL
— AVG OPERATING PRESSURE REQ'D
— NUMBER OF SPRINKLERS ON LATERAL
} TAKE 20% OF
PRESSURE FOR
INITIAL PRESSURE
LOSS

SPRINKLER IRRIGATION 461

— GET LATERAL
SIZE AND GO BACK

ASAE (1978) recommends that the total pressure variation in the laterals, when practicable, should be not be more than ±10 percent of the design pressure. If the lateral runs up or downhill, allowance for this difference in elevation should be made in determining the variation in head. If the water runs uphill, less pressure will be available at the nozzle; if it runs downhill, there will be a tendency to balance the loss of head due to friction.

Scobey's (1930) equation for friction or head loss in pipes may be expressed as

FOR UNIFORM Q — MUST REWORK AS YOU GO OUT

$$H_f = \frac{K_s L Q^{1.9}}{D^{4.9}}(4.10 \times 10^6) \qquad (20.7)$$

.32-.42

$$= \frac{K_s Q^{1.9}(1.45 \times 10^{-8})}{D^{4.9}}$$

$$= f_t/f_t$$

where H_f = total friction loss in line in m,
K_s = Scobey's coefficient of retardation,
L = length of pipe in m,
Q = total discharge in L/s,
D = inside diameter of pipe in mm.

Although this equation was developed for main line or full flow, it may be adapted to lateral pipe with uniformly spaced sprinkler outlets by multiplying the friction loss H_f by a factor F to obtain the actual loss. Suggested values of F are given in Table 20.7. If the head loss for the main line is 70 kPa, the loss for the same length lateral with eight sprinklers is only 29 kPa (0.41 × 70). For center pivot systems Chu and Moe (1972) found $F = 0.54$. This F is higher than those in Table 20.7 because the flow in a center pivot lateral is more than in a regular lateral due to the end gun. The friction loss of the lateral is also adjusted for the size of the end gun by taking the length of the lateral equal to the radius of the irrigated area. See Example 20.3. The size of the lateral is selected so that the friction loss is within the allowable limits adjusted for elevation differences along the line. The selected pipe diameter should be the closest nominal size available commercially.

SCOBEY
F FACTOR

Recommended values of K_s for design purposes are 0.32 for new Transite pipe, 0.40 for steel pipe or portable aluminum pipe and couplers, and 0.42 for portable galvanized steel pipe and couplers.

K_s

The diameter of the main should be adequate to supply the laterals in each of their positions. The rate of flow required for each lateral may be determined by the total capacity of the sprinklers on the lateral. The position of the laterals that gives the highest friction loss in the main should be used for design purposes. The friction loss in the main may be computed by Eq. 20.7. Allowable friction loss in the main varies with the cost of power and the price differential between different diameters of pipe. For small systems with few irrigations per season an approximate maximum friction loss is 4 kPa per 10 m length of pipe. The most economical size should be determined by balancing the increase in

NOMO CONVERSION —

$$\left(psi_{loss} = .2 \, psi\right)$$

TAKE $\frac{20}{space}$ (psi lost)

WHEN LEAVING NOMO
MULTIPLY (psi lost) $\frac{space}{20}$

SPRINKLER IRRIGATION

Table 20.7 Correction Factor F for Friction Losses in Aluminum Pipes with Multiple Outlets[a]

No. of Sprinklers	Correction Factor, F	
	1st Sprinkler One Sprinkler Interval from Main[b]	1st Sprinkler One-Half Sprinkler Interval from Main[c]
1	1.00	1.00
2	0.63	0.51
4	0.48	0.41
6	0.43	0.38
8	0.41	0.37
12	0.39	0.36
16	0.38	0.36
20	0.37	0.35
30 or more	0.36	0.35

[a] Based on Scobey exponent $m = 1.9$ for Q in Eq. 20.7.
[b] Adapted from Christiansen (1948).
[c] Adapted from Jensen and Frantini (1957).

pumping costs against the amortized cost difference of the pipe. A direct economic solution was proposed by Keller (1965).

The design capacity for sprinklers on a lateral with uniform spacing should be based on the average operating pressure. Where the friction loss in the laterals is within 20 percent of the average pressure, the average head for design in a sprinkler line can be expressed approximately by (see Fig. 20.8)

$$H_a = H_o + 0.25H_f + 0.4 H_e \qquad (20.8)$$

where H_a = average pressure at the nozzle (L),
H_o = nozzle pressure at the farthest end of the line (L),
H_f = friction head loss in the lateral (L),
H_e = maximum difference in elevation between the junction with the main and the farthest sprinkler on the lateral (L).

The pressure or head H_n required at the junction of the lateral and the main is

junction $\longrightarrow$ $$H_n = H_o + H_f + H_e + H_{rp} \qquad (20.9)$$

By solving for H_o in Eq. 20.8 and substituting in Eq. 20.9,

$$H_n = H_a + 0.75 H_f \pm 0.6H_e + H_{rp} \qquad (20.10)$$

IF DOWNHILL THEN He NEGATIVE

where H_{rp} = the riser height (L). The constant 0.6 varies from 0.5 to 1.0 for 100

$(PSI)(2.31) = ft$

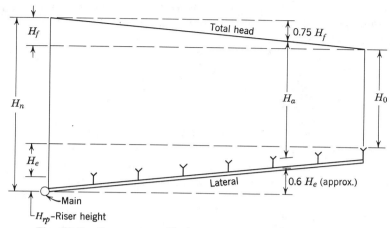

Fig. 20.8. Pressure profile in a sprinkler lateral laid uphill.

and 1 sprinklers, respectively, and is a correction for the percentage of the height at the point where the average pressure exists. Likewise the constant 0.75 for the friction term represents the percentage of the friction loss to this point. If the lateral is located downhill from the main, the elevation term is negative. The relationships in Eqs. 20.8 and 20.10 are shown in Fig. 20.8.

For center pivot systems with an end gun Chu and Moe (1972) derived the pressure distribution along the lateral, from which

$$H_r = H_f \left[1 - 1.875(x - 2x^3/3 + x^5/5)\right] + H_L \qquad (20.11)$$

where H_r = pressure at a distance r from the pivot (L),

$x = r/R$, distance from the pivot divided by the radius of the field to be irrigated,

H_L = pressure at the end of the lateral that is also the nozzle pressure for the end gun (L),

H_f = friction loss in the center pivot lateral (L).

This equation neglects the velocity head and assumes the pressure at R is equal to that at distance L and Scobey's coefficient $m = 2.0$. The theoretical equation is a good approximation of field measurements. From Eq. 20.11 nozzle pressure can be computed for a selected sprinkler location along the lateral. Knowing the pressure, the nozzle size can be selected, and the area covered can be determined.

20.12. General Rules for Sprinkler System Design and Layout. In the absence of experience the following rules may be helpful for design and layout of hand-move or mechanical-move systems other than center pivot. Commercial companies and government agency personnel may also be of great assistance.

In general, sprinkler spacings along the lateral and lateral spacing along the main should be as wide as possible to reduce labor costs. Since greater spacings require higher pressures and thus higher pumping costs, these wide spacings are more easily justified where the power costs are low. Greater spacings also require a higher application rate in which case the infiltration rate may be the limiting factor. The following general rules for layout should be kept in mind: (1) Mains should be laid up and downhill. (2) Laterals should be laid across slope or nearly on the contour. (3) For multiple lateral operation, lateral pipe sizes should be limited to not more than two diameters. (4) If possible, water supply nearest the center of the area should be chosen. (5) For balanced design, lateral operation as shown in Fig. 20.4b should be provided. (6) Layout should facilitate and minimize lateral movement during the season. (7) Difference in the number of sprinklers operating for the various set-ups should be held to a minimum. (8) Booster pumps should be considered where small portions of the field would require a high pressure at the pump. (9) Layout should be modified to apply different rates and amounts of water where soils are greatly different in the design area.

20.13. Pump and Power Units.   In selecting a suitable pump (see Chapter 16), it is necessary to determine the maximum total head against which the pump is working. This head may be determined by

$$H_t = H_n + H_m + H_j + H_s$$ (20.12)

where H_t = total design head against which the pump is working,
H_n = maximum head required at the main to operate the sprinklers on the lateral at the required average pressure, including the riser height (Eq. 20.9),
H_m = maximum friction loss in the main, the suction line, and NPSH of the pump,
H_j = elevation difference between the pump and the junction of the lateral and the main,
H_s = elevation difference between the pump and the water supply after drawdown, all in m or kPa.

The amount of water that will be required is the sum of the capacity of all sprinklers to be operated at the same time. When the total head and rate of pumping are known, the pump may be selected from rating curves or tables furnished by the manufacturer as discussed in Chapter 16.

The following examples illustrate the design of simple sprinkler irrigation systems.

Example 20.2.   Design a side-roll sprinkler irrigation system to irrigate a square 16.2-ha (40-ac) pasture for a maximum rate of application of 15 mm/h

(0.6 iph), for applying 58 mm (2.3 in.) of water in 8.1 days or 2.0 ha (4.9 ac) per day as given in Example 20.1. Assume a wind velocity of 6 km/h (3.7 mph), H_a = 276 kPa (40 psi), H_j = 1.0 m (3.3 ft), H_e = 0.6 m (2.0 ft), H_s = 5.0 m (16.5 ft), H_{rp} = 0.8 m (2.6 ft), NPSH of pump = 2.0 m (6.6 ft), S_l = 12 m (40 ft), S_m = 18 m (60 ft), allowable variation of pressure on the lateral is 20 percent of the average pressure, and the well location is at the center of the field.

Solution.   Location and length of laterals and mains and sprinkler layout are shown in Fig. 20.9. Allowing 12 m from first sprinkler to the main line and 9 m

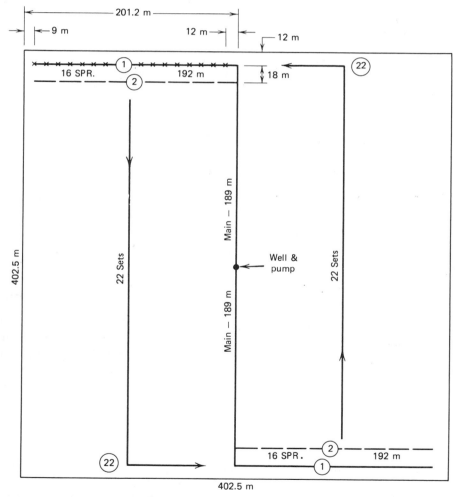

Fig. 20.9. Side-roll sprinkler layout of a 16.2-ha (40-ac) field described in Example 20.2.

from the last sprinkler to the field boundary, 16 sprinklers are required on each lateral (15 spacings).

(1) Number of lateral settings per day

$$\frac{2.0 \text{ ha} \times 10\,000 \text{ m}^2/\text{ha}}{16 \times 12 \text{ m} \times 18 \text{ m}} = 5.8 \text{ or round to } 6$$

Two laterals with three moves per day will be adequate or six laterals if moved once a day. Select two laterals to reduce equipment cost.

(2) Substituting in Eq. 20.4, discharge per sprinkler

$$q = \frac{12 \text{ m} \times 18 \text{ m} \times 15 \text{ mm/h} \times 10\,000 \text{ cm}^2/\text{m}^2}{10 \text{ mm/cm} \times 1000 \text{ cm}^3/\text{L} \times 3600 \text{ s/h}}$$

$$= 0.90 \text{ L/s} \ (14.3 \text{ gpm})$$

(3) Discharge per lateral $= 16 \times 0.9 = 14.4$ L/s (228 gpm). Pump or system capacity, $q_s = 2 \times 14.4 = 28.8$ L/s.

(4) From Eq. 20.6, the theoretical discharge of a sprinkler with 6.35 and 3.97 mm (1/4 and 5/32 in.) diameter nozzles with $C = 0.95$,

$$q = 0.00111 \times 0.95 \times 276^{1/2} \ (6.35^2 + 3.97^2)$$

$$= 0.98 \text{ L/s} \ (15.6 \text{ gpm})$$

Alternative procedure is to select the same nozzle sizes at 276 kPa from manufacturer's values, such as shown in Table 20.5, for which $q = 0.90$ L/s (14.2 gpm).

(5) From Table 20.6 the diameter of coverage required for a 6 km/h wind for sprinkler spacing along the lateral is 27 m (12 m/0.45) and along the main is 30 m (18 m/0.69). Diameter of coverage from Table 20.5 is 31 m, which is adequate.

(6) Diameter of laterals and main. Total allowable variation of pressure in the lateral $= 0.20 \times 276 = 55.2$ kPa (8.0 psi). Allowable variation due to friction only

$$= (55.2/9.8) - H_e = 5.6 - 0.6 = 5.0 \text{ m} \ (7.1 \text{ psi}).$$

Compute the friction loss of a 101.6-mm (4-in.) outside diameter lateral (16 gage with 1.30-mm wall thickness) from Eq. 20.7, using $K_s = 0.40$. Read $F = 0.38$ from Table 20.7 for 16 sprinklers. Compute friction loss for other sizes for the lateral and main as follows:

Tubing Outside diameter		Friction Loss [m (ft)]	
		192-m Lateral	189-m
mm	(in.)	$H_f \times F$	Main
76.2	(3)	13.5 (44)	35.0 (115)
101.6	(4)	3.2[a] (10)	8.2 (27)
127.0	(5)	1.0 (3)	2.7[a] (9)

[a] Select 101.6 mm (4 in.) lateral and 127.0 mm (5 in.) main.

(7)   Determine the pressure or head required at the junction of the lateral and the main for the most remote lateral location and at the highest elevation by substituting in Eq. 20.10,

$$H_n = 28.2 + 0.75(3.2) + 0.6(0.6) + 0.8$$

$$= 31.8 \text{ m (104 ft)}$$

(8)   Determine the pump capacity and total head by substituting in Eq. 20.12,

$$H_t = 31.8 + 2.0 + 2.7 + 1.0 + 5.0 = 42.5 \text{ m (139 ft)}$$

Select the pump from manufacturer's characteristic curves to deliver 28.8 L/s (456 gpm) at a total head of 42.5 m (60 psi) with an efficiency as high as possible. Allow some factor of safety for possible operation of the pump at other conditions and for loss of efficiency with age.

(9)   Determine the size of the power unit. Assuming an efficiency of 70 percent and substituting in Eq. 16.3,

$$kW = 1.34 \, hp = 0.01764 \times 28.8 \text{ L/s} \times 42.5 \text{ m}/0.70$$

$$= 30.7 \text{ kW (22.9 hp)}$$

Select power unit capable of continuously furnishing 30.7 kW, such as a 30-kW electric motor or a (30.7/0.70) = 44-kW (33-hp) water-cooled internal combustion engine.

Example 20.3.   Design a center pivot irrigation system to irrigate the circular area of the square 16.2-ha (40-ac) pasture at a maximum rate of application of 15 mm/h (0.6 iph) with a net depth of 40.5 mm (1.6 in.) of water in 8.1 days as given in Example 20.1. Assuming an application efficiency of 70 percent, the total depth to be applied is 40.5/0.70 = 58 mm (2.3 in.). The field is to be irrigated in two days with two revolutions per day. With a portable unit other fields can be irrigated with the same equipment.

Solution. The maximum area irrigated is about a 400-m (1312-ft) diameter circular area or 125 663 m² (31.0 ac).

The system capacity from Eq. 20.1 using 70 percent application efficiency,

$$q_s = \frac{58 \text{ mm} \times 125 \ 663 \text{ m}^2 \times 1000 \text{ L/m}^3}{4 \text{ rev.} \times 12 \text{ h/rev.} \times 3600 \text{ s/h} \times 1000 \text{ mm/m}}$$

$$= 42.17 \text{ L/s } (668 \text{ gpm})$$

Assuming that the end gun covers an area of about 15 m (49 ft) beyond the end of the lateral (obtain from manufacturer), $L = 200 - 15 = 185$ m (607 ft) and the discharge of the end gun from Eqs. 20.1 and 20.3 for $r = L$ is

$$q_r = 42.17 \left[1 - (185/200)^2 \right] = 6.07 \text{ L/s } (96 \text{ gpm})$$

Select an end gun from manufacturer's catalog to deliver 6.07 L/s at a pressure of 42.25 m (60 psi).

Friction loss in 152-mm (6.0-in.) outside diameter galvanized steel tubing (12 gage, 2.77 mm wall) from Eq. 20.7 with $L = R$, and $F = 0.54$,

$$H_f = \frac{0.4 \times 200 \ (42.17)^{1.9} \times 4.10 \times 10^6 \times 0.54}{(152 - 5.54)^{4.9}}$$

$$= 5.29 \text{ m } (17.4 \text{ ft})$$

Total head at the center pivot including 4.0 m riser height and $H_e = 0.6$ m is

$$H_t = 42.25 + 5.29 + 4.0 + 0.6(0.6) = 51.9 \text{ m } (73.7 \text{ psi})$$

Nozzle size may be determined at any point along the lateral from the pressure head and the required sprinkler capacity. For example, at a distance of 100 m (328 ft) from the center pivot for $x = 0.5$, the nozzle pressure from Eq. 20.11 is

$$H_r = 5.3 \left[1 - 1.875 \ (0.5 - 2(0.5)^3/3 + (0.5)^5/5) \right] + 42.25$$

$$= 43.35 \text{ m } (142 \text{ ft})$$

The required capacity for a 10-m (32.8-ft) spacing of the sprinkler at $r = 100$ m from Eq. 20.1 is

$$q = \frac{58 \times 2 \times 3.14 \times 100 \times 10 \times 1000}{4 \times 12 \times 3600 \times 1000} = 2.11 \text{ L/s (33.4 gpm)}$$

Center pivot systems are usually sold as a complete system from each manufacturer, and thus the example is intended to show only some of the more important design parameters.

20.14. Sprinkler Systems for Environmental Control. Sprinkling has been successful for protecting small plants from wind damage, soil from blowing, and plants from frosts or freezing, and for reducing high air and soil temperatures. Since the entire area usually needs protection at the same time, solid-set systems are required. The rate of application should be as low as possible or just enough to achieve the desired control. Small pipes and low volume sprinklers may be desired to reduce costs, but normal sprinkler systems may be modified for dual use. Because water is applied without regard to irrigation requirements, natural drainage should be adequate or a good drainage system should first be provided.

Especially in organic or sandy soils where onions, carrots, lettuce, and other small seed crops are grown, the soil dries out quickly and the seed may be blown away or covered too deeply for germination. When such plants are small they are also easily damaged by wind-blown soil particles. Protection for such conditions can be provided with a sprinkler system that will apply low rates up to 2.5 mm/h. Operation at night when winds are usually at a minimum will provide more uniform coverage.

Low growing plants can be protected from freezing injury, which is likely either in the early spring or late fall. Sprinkling has been most successful against radiation frosts. Water must be applied continuously at about 2.5 mm/h until the plant is free of ice. Sprinkling should be started before the temperature reaches 0°C at the plant level. Strawberries have been protected from temperatures as low as −6°C. Tomatoes, peppers, cranberries, apples, cherries, and citrus have been successfully protected. Tall plants, such as trees, may suffer limb breakage when ice accumulates, but in some areas low-level undertree sprinkling has provided some control. Rates of application may be reduced by increasing the normal sprinkler spacing. A slightly higher pressure may be desirable to increase the diameter of coverage and to give better breakup of the water droplets.

Sprinkling during the day to reduce plant stress has been successful with many plants, such as lettuce, potatoes, green beans, small fruits, tomatoes, cucumbers, and muskmelons. This practice is sometimes called "misting" or "air conditioning" irrigation. Maximum stress in the plant usually occurs at high temperatures, at low humidity, with rapid air movement, on bright cloudless days, and/or with rapidly growing crops on dry soils. Under these conditions crops at a critical state of growth as during emergence, flowering, or fruit enlargement may benefit greatly from low application of water during the midday. At 27°C water loss was reduced 80 percent with an increase of humidity from 50 to 90 percent. Measured temperature reductions in the plant canopy of about 11°C were attained by misting in an atmosphere of 38 percent relative humidity. Green bean yields were increased 52 percent by midday misting during the bloom and pod development period. Potatoes and corn respond to sprinkling, especially when temperatures exceed 30°C. The tasseling period is a critical time for corn. Small quantities of water applied frequently to strawberries increased the quality and the yield by as much as 55 percent. For low-growing crops the same sprinkler system can be adapted for frost protection. Misting in a greenhouse or under a lath house to reduce transpiration of nursery plants for propagation increases plant growth and root development.

During periods of high incoming radiation soil temperatures may be 20°C greater than the ambient air temperature. Seedlings emerging through soil temperatures as high as 50°C frequently die due to high transpiration. Small applications of water at this critical period often assures emergence and good stands. Another benefit from sprinkling at this stage is to enhance the effectiveness of herbicides applied to control weeds. For further details see Pair et al. (1969) and other current references.

20.15. Sprinkler Systems for Fertilizer, Chemical, or Waste Applications.   Fertilizers, soil amendments, and pesticides may be injected into the sprinkler line as a convenient means of applying these materials to the soil or crop. This method primarily reduces labor costs and in some cases may improve the effectiveness and timeliness of application. Liquid manure and sewage wastes are applied with sprinklers for disposing of unwanted material. A large amount of specialized commercial equipment is on the market. Cannery wastes are usually sprinkled on wooded areas or on land in permanent grass. Good subsurface drainage is required.

Liquid and dry fertilizers have been successfully applied with sprinklers. Dry material must first be dissolved in a supply tank. The liquid may be injected on the suction side of the pump, forced under pressure into the discharge line, or injected into the discharge line by a differential pressure device, such as a venturi section. The material is applied for a short-time during the irrigation set. Sprinkling should be continued for at least 30 minutes after it has been applied to rinse the supply tank, pipes, and the crop. For center pivot or other moving systems the material must be applied continuously or until the field has been completely covered.

For waste disposal systems the solids should be well mixed and small enough so as not to plug the nozzles, which necessitates an effective nonplugging screen on the suction side of the pump. Liquid must be stored in a lagoon or other holding pond. Equipment should be resistent to corrosion from chemicals that may be present in the water. The system should be designed to apply water during subfreezing weather, or sufficient storage should be provided during the nonoperation period. Such systems are usually solid-set and are operated for long periods. Application rates should be lower than the infiltration rate so that the water does not run off the surface causing stream pollution.

REFERENCES

American Society of Agricultural Engineers (1978). "Minimum Requirements for the Design, Installation and Performance of Sprinkler Irrigation Equipment." ASAE Recommendation: R264.2.

Berge, I. O., and M. D. Groskopp (1964). "Irrigation Equipment in Wisconsin." Univ. Wisc. Special Cir. 90.

Christiansen, J. E. (1948). "Irrigation by Sprinkling." California Agr. Expt. Sta. Bull. 670.

Chu, S. T., and D. L. Moe (1972). "Hydraulics of a Center Pivot System." *ASAE Trans.* 15 (5), 894–896.

Dillon, R. C., E. A. Hiler, and G. Vittetoe (1972). "Center-Pivot Sprinkler Design Based on Intake Characteristics." *ASAE Trans.* 15 (5), 996–1001.

Frost, K. R. (1963). "Factors Affecting Evapotranspiration Losses During Sprinkling." *ASAE Trans.* 6 (4), 282–283, 287.

Gray, A. S. (1961). *Sprinkler Irrigation Handbook* (7th ed.). Rain Bird Sprinkler Mfg. Corp., Glendora, Calif.

Hagan, R. M., H. R. Haise, and T. W. Edminster (eds.) (1967). *Irrigation of Agricultural Lands,* monograph 11. Am. Soc. Agron., Madison, Wisc.

Jensen, M. C., and A. M. Frantini (1957). "Adjusted "F" Factors for Sprinkler Lateral Design." *Agr. Eng.* 38, 247.

Jensen, M. E. (ed.) (1981). "Design and Operation of Farm Irrigation Systems," Publ. 1-80. Am. Soc. Agr. Eng., St. Joseph, Mich.

Keller, J. (1965). "Selection of Economical Pipe Sizes for Sprinkler Irrigation Systems." *ASAE Trans.* 8 (2), 186-190.

Pair, C. H. et al. (eds.) (1969). *Sprinkler Irrigation* (3rd ed.) and *Supplement* (1973). Sprinkler Irrigation Association, Silver Springs, Md.

Pillsbury, A. F. (1970). *Sprinkler Irrigation.* Unipub Inc., New York.

Scobey, F. C. (1930). "The Flow of Water in Riveted Steel and Analogous Pipes." U.S. Dept. Agr. Tech. Bull. 150.

U.S. Soil Conservation Service (1949). *Regional Engineering Handbook* (Chapter VI, "Conservation Irrigation, Pacific Region XII"). Portland, Oregon.

——— (1960). *National Engineering Handbook* (Section 15, "Irrigation," Chapter 11). Washington, D.C.

PROBLEMS

20.1. Determine the required capacity of a sprinkler system to apply water at a rate of 13 mm/h (0.5 iph). Two 186-m (620-ft) sprinkler lines with 16 sprinklers each at 12-m (40-ft) spacing on the line and 18-m (60-ft) spacing between lines are required.

20.2. Allowing 1 hour for moving each 186-m (620-ft) sprinkler line described in Problem 20.1, how many hours would be required to apply a 50-mm (2-in.) application of water to a square 16.2-ha (40-ac) field? How many 10-hr days are required?

20.3. Determine the discharge rate for one sprinkler operating at 276 kPa (40 psi) and having two nozzles, 4.0- and 2.8-mm (5/32- and 7/64-in.) diameter nozzles with a discharge coefficient of 0.96.

20.4. Compute the depth rate of application for a 0.95-L/s (15-gpm) sprinkler head if the sprinkler spacing is 18 × 24 m (60 × 80 ft)?

20.5. Compute the total friction loss for a sprinkler system having a 102-mm (4-in.) diameter main 244 m (800 ft) long and one 76-mm (3 in.) lateral 116 m (380 ft) long. The pump delivers 8.5 L/s (135 gpm) with 10 sprinklers on the lateral. Sixteen-gage aluminum pipe (1.3-mm or 0.051-in. wall thickness) and standard couplers are used.

20.6. Design a sprinkler irrigation system for a square 16.2-ha (40-ac) field to irrigate the entire field within a 12-day period. Not more than 16 hours per day are available for moving the pipe and sprinkling. Depth of application is to be 5 mm (2 in.) at each setting at a rate not to exceed 9.0 mm/h (0.35 iph). A well 23 m (75 ft) deep located in the center of the field will provide the following drawdown-discharge characteristics obtained from a well test:

$$12 \text{ m (40 ft)} — 12.6 \text{ L/s (200 gpm)}$$
$$15 \text{ m (50 ft)} — 15.8 \text{ L/s (250 gpm)}$$
$$20 \text{ m (65 ft)} — 18.9 \text{ L/s (300 gpm)}$$

Design for an average pressure of 276 kPa (40 psi) at the nozzle. Highest point in the field is 1.2 m (4 ft) above the well site, and 1 m (3 ft) risers are needed on the sprinklers. Assuming a pump efficiency of 60 percent and assuming that the engine will furnish 70 percent of its rated output for continuous operation, determine the rated output for a watercooled internal combustion engine.

20.7. Determine the total pumping head for a sprinkler irrigation system on level land. The average operating pressure at the nozzle is 276 kPa (40 psi); the friction loss in the main is 6 m (20 ft) and in the lateral 3.7 m (12 ft); the drawdown of the well is 4.3 m (14 ft) at the required discharge of 31.6 L/s (500 gpm); the riser height is 1.5 m (5 ft); and friction loss in all valves is 3.0 m (10 ft). Determine the power requirements for the pump if it operates at 65 percent efficiency.

20.8. Derive the equation for sprinkler nozzle discharge, $q = 0.00111Cd_n^2P^{1/2}$. (*Hint:* Pressure is proportional to velocity head.)

20.9. Assuming a triangular depth-distribution pattern for a rectangular placement of sprinklers and a diameter of coverage of 30 m (100 ft), compute the uniformity coefficient for a 12 × 18 m (40 × 60 ft) sprinkler spacing using 6-m (20-ft) grids for depth measurements. Assume also that the depth of application at the sprinkler is 25 mm (1 in.) and zero at 15 m (50 ft).

20.10. A 12.1-ha (30-ac) field is to be irrigated at a maximum rate of 10 mm/h (0.4 iph) with a sprinkler system. The root zone is 0.9 m (3 ft) deep and the available moisture capacity of the soil is 50 mm (2 in.) per foot of depth. The water application efficiency is 70 percent, and the soil is to be irrigated when 45 percent of the available moisture capacity is depleted. The peak rate of moisture use is 0.5 mm/d (0.2 ipd). Determine the net depth of application per irrigation, depth of water to be pumped, days to cover the field, and area to be irrigated per day.

20.11. If the pressure at opposite ends of a sprinkler lateral is 290 and 262 kPa (42 and 38 psi), what would be the discharge of the distal sprinkler provided the sprinkler at the 290-kPa (42-psi) end discharged 0.76 L/s (12 gpm)?

20.12. If 100 sample cans are uniformly spaced in the area covered by four sprinklers and the average depth caught in a given time is 13 mm (0.5 in.) with the average variation from the mean 1.9 mm (0.075 in.), what is the uniformity coefficient? Assuming that the infiltration rate was not exceeded and the water did not penetrate below the root zone, what is the application efficiency?

CHAPTER 21

Trickle Irrigation

Trickle irrigation is a method of applying water directly to plants through a number of low flow-rate outlets generally placed at short intervals along small tubing. At these outlets specially designed orifices may apply water to individual plants or to a row of plants. Trickle irrigation, sometimes referred to as *drip irrigation,* is similar to watering a plant with a slowly leaking bucket. Unlike sprinkler or surface irrigation only the soil near the plant is watered rather than the entire area.

According to Karmeli and Keller (1975), trickle irrigation research began in Germany about 1860. In the 1940s it was introduced in England especially for watering and fertilizing plants in greenhouses. With the increased availability of plastic pipe and the development of emitters in Israel in the 1950s, it has since become an important method of irrigation in Australia, Europe, Israel, Japan, Mexico, S. Africa, and the United States (California, Hawaii, and Florida). According to the *Irrigation Journal* (1978) survey, California alone had 50 000 ha (124 000 ac) and the total in the United States was over 80 000 ha (198 000 ac).

Trickle irrigation has been accepted mostly in the more arid regions for watering high value crops, such as fruit and nut trees, grapes, and other vine crops, sugar cane, pineapples, strawberries, flowers, and vegetables. Although successfully used on cotton, sorghum, and sweet corn, trickle irrigation is not as well adapted to field crops.

21.1. Advantages and Disadvantages of Trickle Irrigation. With trickle irrigation only the root zone of the plant is supplied with water, and with proper system management deep percolation losses are minimal. Soil evaporation is lower because only a portion of the surface area is wet. Like solid-set sprinkler systems labor requirements are less and the systems can be readily automated. Reduced percolation and evaporation losses result in a greater economy of water use. Weeds are more easily controlled, especially for the soil area that is not irrigated. Bacteria, fungi, and other pests and diseases that depend on a moist environment are reduced as the above-ground plant parts normally are completely dry. Because soil is kept at a high moisture level and the water does not contact the plant, use of more saline water may be possible with less stress and damage to the plant, such as leaf burn. Field edge losses and spray evaporation, such as occurs with sprinklers, are reduced with trickle systems. Low rates of water application at lower pressures are possible so as to eliminate runoff. With some crops, yields and quality are increased probably due to maintenance of a high temporal soil moisture level adequate to meet transpiration demands. Crop yield experiments have shown wide differences varying from little or no difference to 50 percent increase compared to other methods of irrigation. Crop quality may also be improved.

The major disadvantages of trickle irrigation are high cost and clogging of system components, especially emitters by particulate, biological, and chemical matter. Emitters are not well-suited to certain crops and special problems may be caused by salinity. Salt tends to accumulate along the fringes of the wetted surface strip (Fig. 21.1). Since trickle systems normally wet only part of the potential soil-root volume, plant roots may be restricted to the soil volume near each emitter as shown in Fig. 21.1. The dry soil area between emitter lateral lines may result in dust formation from tillage operations and subsequent wind erosion. Compared to surface irrigation systems, more highly skilled labor is required to operate and maintain the filtration equipment and other specialized components.

21.2. Layout and Components of Trickle Systems. Trickle system layouts are similar to sprinkler systems (Chapter 20). As with sprinkler systems many arrangements are possible. The one given in Fig. 21.2 shows split-line operation for the upper left quadrant of the 16.2-ha (40-ac) field described in Fig. 20.9. The well is located in the center of the larger field. The trickle layout would be similar for the other quadrants. Tree rows are parallel to the trickle laterals. Sections 1, 2, 3, or 4 of the 4.05-ha orchard in Fig. 21.2 could be operated independently of each other or in any combination since each section has its own control valve.

As shown in Figs. 21.2 and 21.3, the primary components for a trickle system are an efficient filter, a main and submain, a manifold, and a lateral line to which the emitters are attached. The manifold is a line to which the trickle laterals are connected. Pressure regulators, pressure gages, a water meter, flushing valves, time clocks, and automating control devices are other desirable components. The manifold, submain, and main may be laid on the surface or buried underground. The manifold is usually flexible pipe if laid on the surface or rigid pipe if buried. The main lines may be any type of pipe, such as polyethylene (PE), polyvinylchloride (PVC), galvanized steel, aluminum, or asbestos-cement (A/C). The lateral lines that have emitters are usually flexible PVC or PE tubing. They generally range from 10 to 32 mm ($^3/_8$ to $1^1/_4$ in.) in diameter and have emitters spaced at short intervals appropriate for the crop to be grown.

A filter is the most important component of the trickle system because of

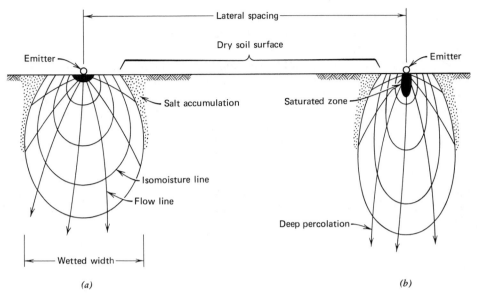

Fig. 21.1. Soil moisture pattern with trickle irrigation. (a) Medium and heavy soils. (b) Sandy soils. (Adapted from Karmeli and Keller, 1975, and United Nations, 1973.)

emitter clogging. Most water should be cleaner than drinking water. Trickle irrigation systems generally require screen, gravel, or graded sand filters. Recommendations of the emitter manufacturer should be followed in selecting the filtration system. In the absence of such recommendations the net opening diameter of the filter shall be smaller than $1/4$ to $1/10$ of the emitter opening diameter. For clean ground water an 80–200 mesh filter may be adequate. This filter will remove soil, sand, and debris, but should not be used with high algae water. For high silt and algae water a sand filter backed up with a screen filter may be required. A sand separator ahead of the filter may be necessary if the water contains considerable sand. In-line strainers with replaceable screens and clean-out plugs may be adequate with small amounts of sand. Secondary filters may be installed at the inlet to each manifold. These are recommended as a safety precaution should accidents during cleaning or filter damage allow particles or unfiltered water to pass into the system. Filters must be cleaned and serviced regularly. Pressure loss through the filter should be monitored as an indication for maintenance.

Lateral lines may be located along the row of trees with several emitters required for each tree as shown in Fig. 21.4. Many laterals have multiple emitters, such as the "spagetti" tubing or "pigtail" lines shown in Fig. 21.4c. One or two laterals per row (Figs. 21.4a or 21.4b) may be provided, depending on the size of the trees. With small trees a single line is adequate.

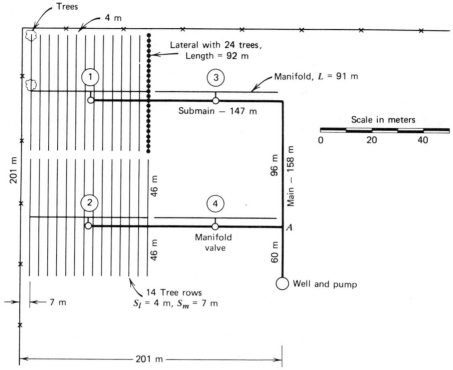

Fig. 21.2. Trickle system layout for a 4.05-ha (10-ac) orchard with a well in the center of a 16.2-ha (40-ac) square field.

Many types and designs of emitters are commercially available, some of which are shown in Fig. 21.5. The emitter controls the flow from the lateral. The pressure is greatly decreased by the emitter; this loss is accomplished by small openings, long passageways, vortex chambers, manual adjustment, or other mechanical devices. Some emitters may be pressure-regulated by changing the length or cross section of passageways or size of orifice. These emitters (Fig. 21.5c) give nearly a constant discharge over a wide range of pressures. Some are selfcleaning and flush automatically. Porous pipe or tubing may have many small openings as shown in Figs. 21.5e–21.5g. The actual size is much smaller than indicated in the drawing. Some holes are barely visible to the naked eye. The double-tube lateral shown in Fig. 21.5g has more openings in the outer channel than in the main flow channel. Such tubes have thin walls and are low in cost. In Hawaii they are often discarded after the crop is harvested and replaced with new lines. Most emitters are placed on the soil surface, but they may be buried at shallow depths for protection.

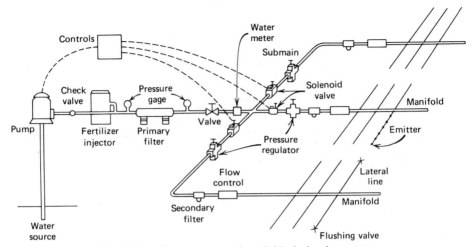

Fig. 21.3. Components of a trickle irrigation system.

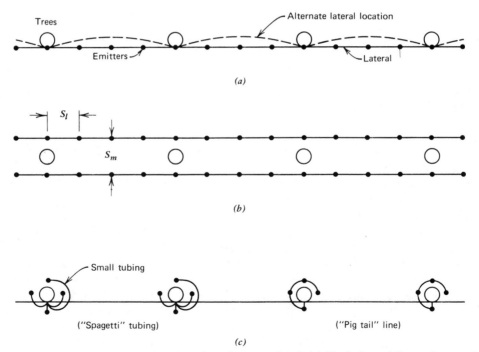

Fig. 21.4. Lateral and emitter locations for an orchard. (a) Single lateral for each row of trees. (b) Two laterals for each row of trees. (c) Multiple-exit emitters.

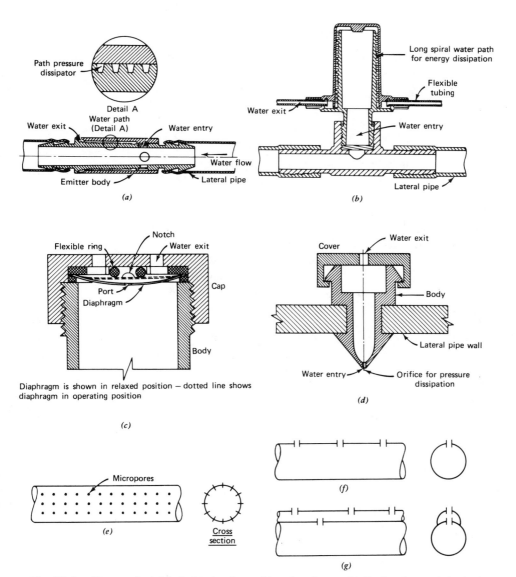

Fig. 21.5. Types of trickle irrigation laterals and emitters. (a) In-line long-path single-exit emitter. (b) In-line long-path multiple-exit emitter. (c) Flushing-type emitter. (d) Orifice-type emitter. (e) Porous tubing lateral. (f) Single-tube emitter lateral. (g) Double-tube emitter lateral. (Figs. 21.5a–21.5d redrawn from Karmeli and Keller, 1975.)

21.3. Emitter Discharge. In an orifice-type emitter (Fig. 21.5*d*) the flow is fully turbulent and the discharge can be determined from the sprinkler nozzle equation (Eq. 20.6). The discharge of any emitter may be expressed by the power-curve equation (Karmeli and Keller, 1975) in which

$$q = Kh^x \qquad\qquad (21.1)$$

where q = emitter discharge (L^3/T),
K = constant for each emitter,
h = pressure head (L),
x = emitter discharge exponent.

The exponent x can be determined by measuring the slope of the log-log plot of head vs discharge. With x known K can be determined from Eq. 21.1. In fully turbulent flow $x = 0.5$ and in laminar flow regime $x = 1.0$. In a fully pressure-compensating emitter K is a constant for a wide range of pressure and $x = 0$. Because of the large number of emitters available, it may be more convenient to determine discharge directly from manufacturer's curves. Some are shown in Fig. 21.6 from which K and x were computed using Eq. 21.1. An average $x = 0.63$ was used to compute K for the small, medium, and large long-path emitters. These were rated at 0.0005, 0.001, and 0.0026 L/s (0.5, 1.0, and 2.5 gph), respectively. For the double-wall tube $K = 0.00046$ and $x = 0.5$. The two Rain Bird emitters have straight lines for which $x = 1.0$, indicating laminar flow. Emitter discharge usually varies from about 0.0003 to 0.0084 L/s (0.3 to 8 gph) and pressures range from about 14 to 276 kPa (2 to 40 psi). Average diameters of openings for emitters range from 0.0025 to 0.25 mm (0.0001 to 0.01 in.) for 147 kPa pressure (Walker, 1979).

Emitters made from thermoplastic material may vary in discharge depending on the temperature. Thus, discharge curves should be corrected for temperature.

21.4. Water Distribution from Emitters. Trickle irrigation was developed to provide a more efficient application of water. An ideal system should provide a uniform discharge from each emitter. Application efficiency depends on the variation of emitter discharge, pressure variation along the lateral, and seepage below the root zone or other losses, such as soil evaporation. Emitter discharge variability is greater than that for sprinkler nozzles because of smaller openings (lower flow) and lower design pressures. Such variability may be due to the design of the emitter, materials, and care in manufacture. Solmon (1974) found that the statistical coefficient of variation may range from 0.02 to 0.40. Nakayama et al. (1979) found that the uniformity for emitters is approximately

$$EU = 1 - (0.8 \, C_v/n^{0.5}) \qquad (21.2)$$

where EU = emission uniformity,
C_v = manufacturer's coefficient of variation,
n = number of emitters per plant.

Application efficiency for trickle irrigation can thus be defined as

$$E_{ea} = EU \times E_a \times 100 \qquad (21.3)$$

where E_{ea} = trickle irrigation efficiency,
E_a = application efficiency as defined for sprinkler or surface irrigation (water stored in the root zone divided by the water delivered).

As indicated by Karmeli and Keller (1975), a reasonable safe design value for E_a is 90 percent.

Example 21.1. Determine the trickle irrigation efficiency of a system using emitters that have a manufacturer's coefficient of variation of 0.125 and an application efficiency of 90 percent.
Solution: Assuming one emitter per plant and substituting in Eq. 21.2,

$$EU = 1 - (0.8 \times 0.125/1^{0.5}) = 0.90$$

Substituting in Eq. 21.3,

$$E_{ea} = 0.90 \times 0.90 \times 100 = 81 \text{ percent}$$

21.5. Trickle System Design. Trickle systems may be designed using the same general rules and procedures outlined in Chapter 20 for sprinkler systems. The primary differences are that the spacing of emitters is much less than that for sprinkler nozzles and that water must be filtered and treated to prevent clogging of the small emitter openings. Another major difference with trickle irrigation, especially for widely spaced tree crops, is that not all of the area will be irrigated. Karmeli and Keller (1975) suggest that a minimum of 33 percent of the potential root volume should be irrigated. For closely spaced plants a much higher percentage may be necessary to assure sufficient water to the plants. In design the water use rate or the area irrigated may be decreased to account for this reduced area. Karmeli and Keller (1975) suggest the following water use rate for trickle irrigation design:

$$ET_t = ET \times P/85 \qquad (21.4)$$

TOTAL HEAD = EMITTER PRESSURE + FRICTION LOSSES
ELEVATION HEADS AND DRAWDOWN

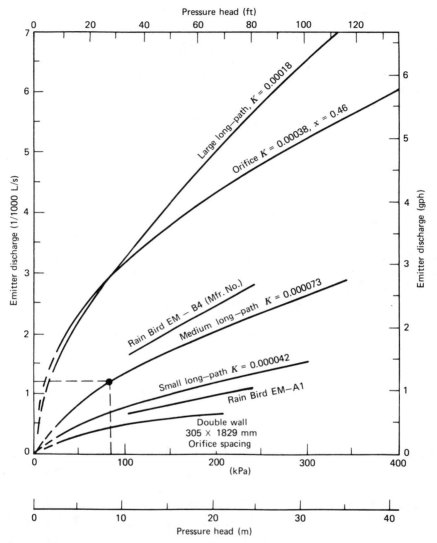

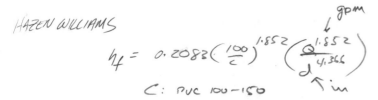

Fig. 21.6. Discharge of various emitters vs pressure head. (Adapted from Karmeli and Keller, 1975, Walker, 1979, and Davis, 1976.)

HAZEN WILLIAMS

$$h_f = 0.2083 \left(\frac{100}{c}\right)^{1.852} \left(\frac{Q^{1.852}}{d^{4.366}}\right)$$

gpm

C: PVC 100-150

d in

where ET_t = average evapotranspiration rate for crops under trickle
                irrigation (L/T),
        P = percent of the total area shaded by the crop,
        ET = conventional ET rate for the crop (L/T).

For example, if a mature orchard shades 70 percent of the area and the conventional ET is 7 mm/d, the trickle irrigation design rate is 5.8 mm/d ($7 \times 70/85$).

The diameter of the lateral or of the manifold should be selected so that the difference in discharge between emitters operating simultaneously will not exceed 10 percent. This allowable variation is the same as for sprinkler irrigation laterals discussed in Chapter 20. To stay within this 10 percent variation in flow, the head difference between emitters should not exceed 10 to 15 percent of the average operating head for long-path emitters or 20 percent for turbulent flow emitters (Karmeli and Keller, 1975). The maximum difference in pressure is the head loss between the control point at the inlet and the pressure at the emitter farthest from the inlet. The inlet is usually at the manifold where the pressure is regulated. In Fig. 21.2 the maximum difference in head loss to the farthest emitter is that for one half the lateral length plus one half the manifold length. Where the manifold is connected to the end of each lateral and the submain is connected to the end of a manifold, the head loss would be computed for their entire length.

For minimum cost, Karmeli and Keller (1975) recommended that on a level area 55 percent of the allowable head loss should be allocated to the lateral and 45 percent to the manifold. As in sprinkler laterals (Chapter 20) allowable head loss should be adjusted for elevation differences along the lateral and along the manifold.

The friction loss for mains and submains can be computed from Scobey's equation (Eq. 20.7), but suitable values for K_s have not been developed. The Hazen–Williams or the Darcy–Weisbach equations are more widely accepted for trickle system design. From Watters and Keller (1978) the Darcy–Weisbach equation for smooth pipes in trickle systems when combined with the Blasius equation for the friction factor is

$$H_f = KLQ^{1.75} D^{-4.75} \qquad (21.5)$$

where H_f = friction loss in m,
        K = a constant = 7.89×10^5 for SI units for water at 20°C,
        L = pipe length in m,
        Q = total pipe flow in L/s,
        D = inside diameter of pipe in mm.

Equation 21.5 applies for continuous sections of plastic pipe. For in-line emitters (Figs. 21.5a and 21.5b) the head loss should be increased. Such losses may be expressed as an equivalent length of lateral pipe.

The friction loss in laterals and manifolds having multiple outlets can also be computed from Eq. 21.5 by multiplying H_f by a suitable F factor described in Chapter 20. For 20 or more outlets on a line, the value of F is nearly constant, and for practical purposes (Table 20.7) $F = 0.35$.

Example 21.2. Design a trickle irrigation system for a fully matured orchard with the layout shown in Fig. 21.2. Assume that the field is level, maximum time for irrigation is 12 hours per day, allowable pressure variation in the emitters is 15 percent, maximum lift at the well is 20 m (66 ft), the evapotranspiration rate corrected for shaded area is 5.8 mm/d (0.23 ipd), and the trickle irrigation efficiency is 80 percent. Sections 1 and 2 in Fig. 21.2 are to be irrigated at the same time and alternated with sections 3 and 4.

Solution. Determine the required discharge for each section of the system and the total pumping head for the most remote lateral location in Section 1 of Fig. 21.2.

(1) The discharge for each tree with a spacing of 4 × 7 m (13 × 23 ft) using Eq. 20.1 is

$$q_s = \frac{5.8 \times 4 \times 7 \times 10\ 000}{10 \times 12 \times 3600 \times 1\ 000} = 0.00376 \text{ L/s (0.060 gpm)}$$

With an application efficiency of 80 percent the required discharge per tree is

$$0.00376/0.80 = 0.0047 \text{ L/s (0.075 gpm)}$$

The discharge per emitter assuming four emitters per tree (pig tail layout) is thus $0.0047/4 = 0.00118$ L/s (0.0187 gpm).

(2) Rounding the emitter discharge to 0.0012 L/s, the discharge of each of the lines is as follows:

Line	Number Trees	Number Emitters	Required Discharge (L/s)	(gpm)
Half lateral	12	48	0.0576	0.914
Half manifold	168	672	0.806	12.80
Submain, A to section 1	336	1 344	1.613	25.60
Main, A to pump	672	2 688	3.226	51.21

(3)   From Fig. 21.6 select the medium long-path emitter with $K = 0.000073$ and $x = 0.63$. Substituting in Eq. 21.1 using the average emitter discharge of 0.0012 L/s,

$$\log h = (\log 0.0012 - \log 0.000073)/0.63$$

from which $h = 87$ kPa (8.9 m or 29 ft). This head is the average operating head, H_a.

(4)   The total allowable pressure loss of 15 percent of H_a in both the lateral and the manifold is $8.9 \times 0.15 = 1.3$ m (4.3 ft) of which (0.55×1.3) 0.7 m (2.4 ft) is allowed for the lateral and (0.45×1.3) 0.6 m (1.9 ft) is for the manifold.

(5)   Compute the friction loss in each of the lines from Eq. 21.5 by selecting a diameter to keep the loss within the allowable limits previously specified.

Line	Q (L/s)	Pipe Dia., D (mm)	(in.)	L (m)	F^a	H_f (m)
Half lateral	0.0576	12.7	1/2	46	0.36	0.51
Half manifold	0.806	31.75	1¼	45.5	0.375	0.68
Submain, A to section 1	1.613	44.45	1¾	243	1	6.59
Main, A to pump	3.226	50.80	2	60	1	2.90

[a] From Table 20.7 for multiple-outlet lines.

(6)   From Eq. 20.10 the pressure head at the inlet to the manifold using values from (3 and 5) above,

$$H_n = 8.9 + 0.75 \,(0.51 + 0.68) + 0 + 0 = 9.79 \text{ m (32 ft)}$$

(7)   The total operating head for the pump assuming NPSH for the pump is 4.0 m and substituting in Eq. 20.12,

$$H_t = 9.79 + (6.59 + 2.90) + 0 + 20 + 4.0$$
$$= 43.28 \text{ m (142 ft or 62 psi) (424 kPa)}$$

From the above the pump must deliver 3.23 L/s (51 gpm) at a head of about 43 m. Allowance must be made for pump wear and additional head must be added to account for losses through filters, pressure regulators, valves, and other devices.

REFERENCES

Baars, I. C. (1976). "Design of Trickle Irrigation Systems." Irrigation and Civil Engineering Dept., Agricultural University, Wageningen, The Netherlands.

Davis, D. D. (1976). *Rain Bird Irrigation Systems Design Handbook*. Rain Bird Sprinkler Mfr. Corp., Glendora, Calif.

Goldberg, D. S., B. Gornat, and D. Rimon (1976). *Drip Irrigation: Principles, Design and Agricultural Practices*. Drip Irrigation Scientific Publ., Kfar Shmaryahu, Israel.

Howell, T. A., and E. A. Hiler (1974). "Trickle Irrigation Lateral Design." *ASAE Trans*. **17** (5), 902–908.

Irrigation Journal (1978). "1978 Irrigation Survey" **28**, 46A–46H.

Karmeli, D., and J. Keller (1975). *Trickle Irrigation Design*. Rain Bird Sprinkler Mfr. Corp., Glendora, Calif.

Marsh, A. W. (1974). *Proceedings Second International Drip Irrigation Congress*. Riverside Printers, Riverside, Calif.

Nakayama, F. S., D. A. Bucks, and A. T. Clemmens (1979). "Assessing Trickle Emitter Application Uniformity." *ASAE Trans*. **22** (4), 816–821.

Solmon, K. (1979). "Manufacturing Variation of Trickle Emitters." *ASAE Trans*. **22** (5), 1034–1038, 1043.

United Nations, Food and Agriculture Organization (1973). *Trickle Irrigation*. FAO, Rome, Italy.

Walker, W. R. (1979). *Sprinkler and Trickle Irrigation*, 3rd ed. Mimeo. Dept. Agr. and Chem. Eng., Colorado State Univ.

Watters, G. Z., and J. Keller (1978). "Trickle Irrigation Tubing Hydraulics." ASAE Paper 78-2015.

Wu, I. P., and D. D. Fangmeier (1974). "Hydraulic Design of Twin-Chamber Trickle Irrigation Laterals." Arizona Agr. Exp. Sta., Tech. Bull. 216.

Wu, I. P., and H. M. Gitlin (1974). "Design of Drip Irrigation Lines." Hawaiian Agr. Expt. Sta. Tech. Bull. 96.

——— (1977). "Drip Irrigation System Design in Metric Units." Hawaiian Coop. Ext. Ser., Univ. of Hawaii, Misc. Publ. 144.

PROBLEMS

21.1. Determine the trickle irrigation efficiency for emitters that have a manufacturer's coefficient of variation of 0.20. Assume four emitters per plant and 46 mm (1.8 in.) of water is stored in the root zone of the 51 mm (2.0 in.) delivered to the field.

21.2. If the conventional *ET* is 7.6 mm/d (0.3 ipd) and 75 percent of the area is shaded by trees in an orchard, determine the *ET* rate for the design of a trickle irrigation system. What is the application rate in L/s for each tree if the tree spacing is 3 × 6 m (10 × 20 ft)?

21.3. From data in Example 21.2 determine the diameter of a trickle lateral and the friction loss if the manifold is connected to the end of the 92-m line (24 trees) rather than at the midpoint as in Example 21.2. The allowable friction loss in the lateral is 0.7 m (2.4 ft).

21.4. From data in Example 21.2 determine the friction loss in the 12.7-mm diameter half lateral (46-m length with 12 trees) if the 48 emitters were uniformly spaced along the line as shown in Fig. 21.4a. If the 48 in-line emitters increased the friction loss by 10 percent, what is the total friction loss?

21.5. From data in Example 21.2 determine the pump capacity if all four Sections 1, 2, 3, and 4 are to be irrigated at the same time, but in a 6-hour rather than a 12-hour period. Determine the diameter and friction loss in the 60-m length main if the maximum allowable loss is 4 kPa per 10 m of length.

APPENDIX A

Runoff Determination

Table A.1 Time of Concentration for Small Watersheds

Maximum Length of Flow		Time of Concentration[a] (min)					
		Watershed Gradient (%)					
m	(ft)	0.05	0.1	0.5	1.0	2.0	5.0
152	(500)	18	13	7	6	4	3
305	(1000)	30	23	11	9	7	5
610	(2000)	51	39	20	16	12	9
1220	(4000)	86	66	33	27	21	15
1830	(6000)	119	91	46	37	29	20
2440	(8000)	149	114	57	47	36	25
3050	(10 000)	175	134	67	55	42	30
6100	(20 000)	306	234	117	97	74	52

[a] Computed from Eq. 4.2.

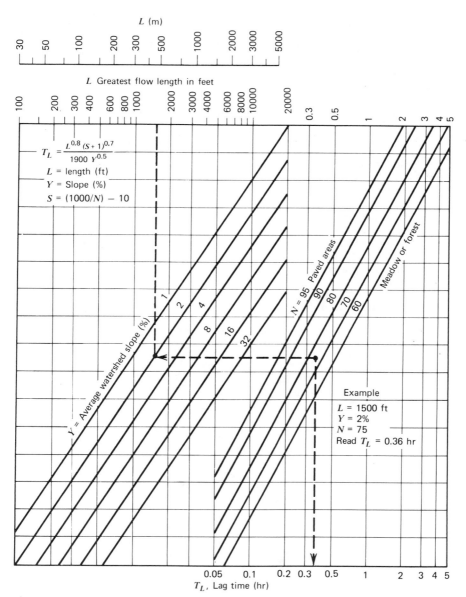

Fig. A.1. Nomograph for estimating watershed lag and time of concentration. (U.S. SCS, Hydrology, National Eng. Hbk., 1972.)

Table A.2 Runoff Coefficients for Urban Areas

Type of Drainage Area	Runoff Coefficient, C
Business	
Downtown areas	0.70−0.95
Neighborhood areas	0.50−0.70
Residential	
Single-family areas	0.30−0.50
Multi units, detached	0.40−0.60
Multi units, attached	0.60−0.75
Suburban	0.25−0.40
Apartment dwelling areas	0.50−0.70
Industrial	
Light areas	0.50−0.80
Heavy areas	0.60−0.90
Parks, cemeteries	0.10−0.25
Playgrounds	0.20−0.35
Railroad yard areas	0.20−0.40
Unimproved areas	0.10−0.30
Streets	0.70−0.95
Brick	
Drives and walks	0.75−0.85
Roofs	0.75−0.95

Source: Ill. Eng. Expt. Sta. Bull. 462, 1962.

APPENDIX B

Manning Velocity Formula

Table B.1 Roughness Coefficient *n* for Manning Formula

Line No.	Type and Description of Conduits	n Values[a]		
		Min.	Design	Max.
	Channels, Lined			
1	Asphaltic concrete, machine placed		0.014	
2	Asphalt, exposed prefabricated		0.015	
3	Concrete	0.012	0.015	0.018
4	Concrete, rubble	0.017		0.030
5	Metal, smooth (flumes)	0.011		0.015
6	Metal, corrugated	0.021	0.024	0.026
7	Plastic	0.012		0.014
8	Shotcrete	0.016		0.017
9	Wood, planed (flumes)	0.010	0.012	0.015
10	Wood, unplaned (flumes)	0.011	0.013	0.015
	Channels, Earth			
11	Earth bottom, rubble sides	0.028	0.032	0.035
	Drainage ditches, large, no vegetation			
12	(*a*) <0.8 m, hydraulic radius	0.040		0.045
13	(*b*) 0.8−1.2 m, hydraulic radius	0.035		0.040
14	(*c*) 1.2−1.5 m, hydraulic radius	0.030		0.035
15	(*d*) >1.5 m, hydraulic radius	0.025		0.030
16	Small drainage ditches	0.035	0.040	0.040
17	Stony bed, weeds on bank	0.025	0.035	0.040
18	Straight and uniform	0.017	0.0225	0.025
19	Winding, sluggish	0.0225	0.025	0.030
	Channels, Vegetated (grassed waterways) (See Chapter 7)			
	Dense, uniform stands of green vegetation more than 250 mm long			
20	(*a*) Bermuda grass	0.04		0.20
21	(*b*) Kudzu	0.07		0.23
22	(*c*) Lespedeza, common	0.047		0.095
	Dense, uniform stands of green vegetation cut to a length less than 60 mm			

Table B.1   Roughness Coefficient n for Manning Formula

Line No.	Type and Description of Conduits	Min.	Design	Max.
			n Values[a]	
23	(a) Bermuda grass, short	0.034		0.11
24	(b) Kudzu	0.045		0.16
25	(c) Lespedeza	0.023		0.05
26	Sorghum, 1-m rows	0.04		0.15
27	Wheat, mature poor	0.08		0.15
	Natural Streams			
28	(a) Clean, straight bank, full stage, no rifts or deep pools	0.025		0.033
29	(b) Same as (a) but some weeds and stones	0.030		0.040
30	(c) Winding, some pools and shoals, clean	0.035		0.050
31	(d) Same as (c), lower stages, more ineffective slopes and sections	0.040		0.055
32	(e) Same as (c), some weeds and stones	0.033		0.045
33	(f) Same as (d), stony sections	0.045		0.060
34	(g) sluggish river reaches, rather weedy or with very deep pools	0.050		0.080
35	(h) Very weedy reaches	0.075		0.150
	Pipe			
36	Asbestos cement		0.009	
37	Cast iron, coated or uncoated	0.011	0.013	0.015
38	Clay or concrete drain tile (102-305 mm dia.)	0.011	0.013	0.020
39	Concrete or clay vitrified sewer pipe	0.01	0.014	0.017
40	Corrugated plastic tubing	0.014	0.016	0.018
41	Metal, corrugated, ring	0.021	0.025	0.026
42	Metal, corrugated, helical	0.013	0.015	
43	Steel, riveted and spiral	0.013	0.016	0.017
44	Wood stave	0.010	0.013	
45	Wrought iron, black	0.012		0.015
46	Wrought iron, galv.	0.013	0.016	0.017

[a] Selected from numerous sources.

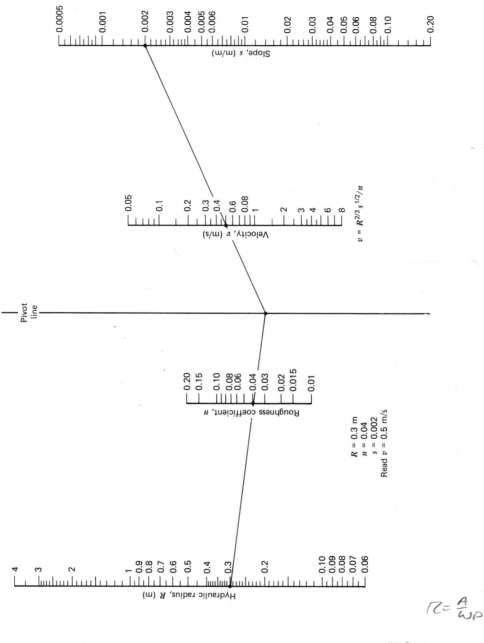

Slope, s (m/m)

Velocity, v (m/s)

$v = R^{2/3} s^{1/2} / n$

Pivot line

Roughness coefficient, n

$R = 0.3$ m
$n = 0.04$
$s = 0.002$
Read $v = 0.5$ m/s

Hydraulic radius, R (m)

$$R = \frac{A}{WP}$$

$$V = R^{2/3} S^{1/2} / n = \frac{1.486}{n} R^{2/3} S^{1/2}$$

$\underline{\text{METRIC}}$ $\underline{\text{ENGLISH}}$

TRAPEZOIDAL;

$$A = bd + zd^2$$

$$WP = b + 2d\sqrt{z^2 + 1}$$

$$t = b + 2dz$$

APPENDIX C

Pipe and Conduit Flow

Table C.1 Friction Loss Coefficients for Circular or Square Pipe at Bends

$\dfrac{R}{D} = \dfrac{\textit{Bend Radius to Pipe Center Line}}{\textit{Pipe Diameter}}$	Bend Coefficient, K_b	
	45° Bend	90° Bend
0.5	0.7	1.0
1	0.4	0.6
2	0.3	0.4
5	0.2	0.3

Source: U.S. Soil Conservation Service, Engineering Handbook, Hydraulics Section 5, 1951.

Table C.2   Head Loss Coefficients for Circular Pipe Flowing Full (SI units)

$$K_c = \frac{1\ 244\ 522\ n^2}{d^{4/3}}, \text{ where } d = \text{diameter (mm)}$$

Pipe Inside Dia.		Flow area	Manning Coefficient of Roughness, n				
mm	(in.)	(sq. mm)	0.010	0.013	0.016	0.020	0.025
13	(0.5)	133	4.071	6.881	10.423	16.286	25.447
25	(1)	491	1.702	2.877	4.358	6.810	10.641
51	(2)	2 043	0.658	1.112	1.685	2.632	4.113
76	(3)	4 536	0.387	0.653	0.990	1.546	2.416
102	(4)	8 171	0.261	0.441	0.669	1.045	1.632
127	(5)	12 668	0.195	0.329	0.499	0.780	1.218
152	(6)	18 146	0.153	0.259	0.393	0.614	0.959
203	(8)	32 365	0.104	0.176	0.267	0.417	0.652
254	(10)	50 671	0.0774	0.131	0.198	0.309	0.484
305	(12)	73 062	0.0606	0.102	0.155	0.242	0.379
381	(15)	114 009	0.0451	0.0761	0.115	0.180	0.282
457	(18)	164 030	0.0354	0.0598	0.0905	0.141	0.221
533	(21)	223 123	0.0288	0.0487	0.0737	0.115	0.180
610	(24)	292 247	0.0241	0.0407	0.0616	0.0962	0.150
762	(30)	456 037	0.0179	0.0302	0.0458	0.0715	0.112
914	(36)	656 119	0.0140	0.0237	0.0359	0.0561	0.0877
1219	(48)	1 167 071	0.00956	0.0162	0.0245	0.0382	0.0597
1524	(60)	1 824 147	0.00710	0.0120	0.0182	0.0284	0.0444

Note: K_c (English units) = K_c (SI units)/3.28, d = (in).

Table C.3   Head Loss Coefficients for Square Conduits Flowing Full

$$K_c = \frac{19.60\ n^2}{R^{4/3}}, \text{ where } R = \text{hydraulic radius (m)}$$

Conduit Size		Flow Area	Manning Coefficient of Roughness, n			
m × m	(ft × ft)	(sq. m)	0.012	0.014	0.016	0.020
0.61 × 0.61	(2 × 2)	0.372	0.0347	0.0472	0.0616	0.0963
0.91 × 0.91	(3 × 3)	0.828	0.0203	0.0277	0.0361	0.0564
1.22 × 1.22	(4 × 4)	1.488	0.0138	0.0187	0.0245	0.0382
1.52 × 1.52	(5 × 5)	2.310	0.0103	0.0140	0.0182	0.0285
1.83 × 1.83	(6 × 6)	3.349	0.00800	0.0109	0.0142	0.0222
2.13 × 2.13	(7 × 7)	4.537	0.00653	0.00889	0.0116	0.0181
2.44 × 2.44	(8 × 8)	5.954	0.00545	0.00742	0.00970	0.0152
2.74 × 2.74	(9 × 9)	7.508	0.00467	0.00636	0.00831	0.0130
3.05 × 3.05	(10 × 10)	9.303	0.00405	0.00551	0.00720	0.0113

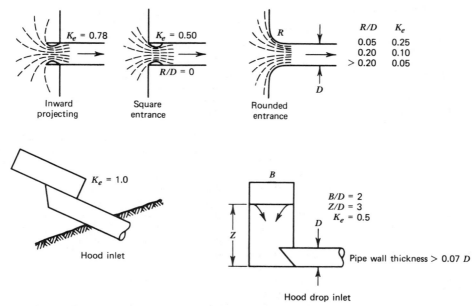

Fig. C.1. Entrance loss coefficients of pipe conduits. (From U.S. SCS, Engineering Handbook, Hyd. Sect. 5, 1951; Mavis, Hydraulics of Culverts, Penn. Eng. Expt. Sta. Bull., 56, 1943; Blaisdell and Donnelly, Agr. Eng. Jour., 1956, and Yalamanchili and Blaisdell, Hood Drop Inlet, ARS-NC-23, 1975.)

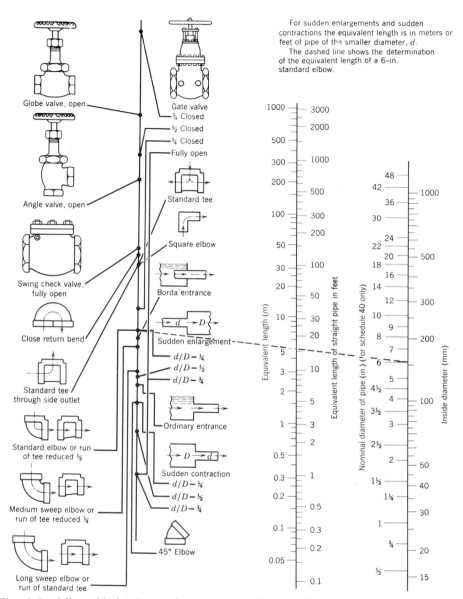

Fig. C.2. Minor friction losses for valves and fittings. (Revised from Wolfe, Oregon Agr. Expt. Sta. Bull. 181, 1950.)

APPENDIX D

Drain Tile and Pipe Specifications

D.1. Clay Drain Tile. The test requirements given in Tables D.1 and D.2 are condensed from ASTM C4-62, *Tentative Specifications for Clay Drain Tile,* but the most current standard should be checked for possible changes.

Table D.1   Physical Test Requirements for Clay Drain Tile

	Standard		Extra-Quality		Heavy-Duty	
Internal Dia. (in.)	Avg.[a] Minimum Strength (lbs/ft)	Avg.[b] Maximum Absorption (%)	Avg.[a] Minimum Strength (lbs/ft)	Avg.[b] Maximum Absorption (%)	Avg.[a] Minimum Strength (lbs/ft)	Avg.[b] Maximum Absorption (%)
4,5,6	800	13	1100	11	1400	11
8	800	13	1100	11	1500	11
10	800	13	1100	11	1550	11
12	800	13	1100	11	1700	11
15	870	13	1150	11	1980	11
18			1300	11	2340	11
21			1450	11	2680	11
24			1600	11	3000	11

[a] Average of 5 tile using the 3-edge bearing method.
[b] Average of 5 tile using the 5-hour boiling test.
Note: Tile diameters 14, 16, 27, and 30 are omitted for brevity.
Dia. in mm = Dia. in in. × 25.4.
N/m = lbs/ft × 14.6.
kg/m = lbs/ft × 1.488.

Drain tile subject to these specifications may be made from clay, shale, fire clay, or mixtures thereof, and burned. The quality of tile selected should be such that the strength exceeds the soil load by a suitable margin. Where subjected to extreme freezing and thawing, extra-quality or heavy-duty tile are recommended.

Size and Minimum Lengths.   The nominal sizes of clay drain tile shall be designated by their inside diameter. Tile less than 12 in. in diameter shall be not less than 1 ft in length; 12- to 30-in. tile, not less than their diameter; and tile larger than 30 in., not less than 30 in. in length.

Other Physical Properties.   Some of the general physical requirements for the three classes of clay drain tile are given in Table D.2. Drain tile, while dry, shall give a clear ring when stood on end and tapped with a light hammer. They shall also be reasonably straight and smooth on the inside. Drain tile shall be free from cracks and checks extending into the tile in such a manner as to decrease its strength appreciably. They shall be neither chipped nor broken so as to decrease their strength materially or to admit soil into the drain.

Table D.2   Distinctive General Physical Properties of Clay Drain Tile

Physical Properties Specified	*Standard*	*Extra-Quality and Heavy-Duty*
Number of freezings and thawings (reversals)	36	48
Permissible variation of average diameter below specified diameter, percent	3	3
Permissible variation between maximum and minimum diameters of same tile, percentage of thickness of wall	75	65
Permissible variation of average length below specified length, percent	3	3
Permissible variation from straightness, percentage of length	3	3
Permissible thickness of exterior blisters, lumps, and flakes which do not weaken tile and are few in number, percentage of thickness of wall	20	15
Permissible diameters of above blisters, lumps, and flakes, percentage of inside diameter	15	10
General inspection	Rigid	Very rigid

D.2.   Concrete Drain Tile.   The specifications given in Table D.3 are condensed from ASTM C412-63. Standard and extra-quality concrete tile are intended for ordinary soils, while special-quality tile are for soils or drainage waters that are markedly acid (pH of 6.0 or lower) or where they contain unusual quantities of soil sulfates, chiefly sodium or magnesium, singly or in combination (assumed to be 3000 ppm or more).

Table D.3 Physical Test Requirements for Concrete Drain Tile

| Internal Diameter (in.) | Standard | | Extra-Quality | | |
	Avg. Min.[a] Strength (lbs/ft)	Avg. Max.[b] Absorption (%)	Wall[c] Thickness (in.)	Avg. Min.[a] Strength (lbs/ft)	Avg. Max.[b] Absorption (%)
4	800	10	$1/2$	1100	9
5	800	10	$9/16$	1100	9
6	800	10	$5/8$	1100	9
8	800	10	$3/4$	1100	9
10	800	10	$7/8$	1100	9
12	800	10	1	1100	9
15			$1 1/4$	1100	9
18			$1 1/2$	1200	9
21			$1 3/4$	1400	9
24			2	1600	9

[a] Average of 5 tile using the 3-edge bearing method.
[b] Average of 5 tile using the 5-hour boiling test.
[c] Minimum diameters shall not be less than the nominal diameters by more than $1/4$ in. for 4- and 5-in. tile, $3/8$ in. for 6- and 8-in. tile, $1/2$ in. for 10- to 14-in. tile, $5/8$ in. for 15- to 18-in. tile, and $3/4$ in. for sizes of 20- to 24-in. No wall thickness is specified for standard quality.
Note. Tile diameters 14, 16, and 20 are omitted for brevity.
Dia. in mm = Dia. in in. × 25.4.
N/m = lbs/ft × 14.6.
kg/m = lbs/ft × 1.488.

Special-quality tile should have the same wall thickness and strength as extra-quality tile. Average maximum absorption is 8 percent, and closer tolerances are required for wall thicknesses than for extra-quality tile. The 10-minute, room-temperature maximum soaking absorption shall be 3 percent for individual tile. The hydrostatic pressure test may be made in lieu of the above 10-minute test. For sulfate exposures, sulfate-resistant cement shall be specified.

Size and Minimum Lengths.   Concrete tile less than 12 in. in diameter shall not be less than 1 ft in length, and 12- to 24-in. diameter tile shall have nominal lengths not less than their diameter.

D.3. Corrugated Plastic Tubing.   Specifications for tubing and fittings are given in ASTM F405, F667, and D2412. Pipe stiffness is the slope of the load-deflection curve (kPa or psi) at a specified percent deflection based on the original inside diameter. The load is applied between two parallel plates at a constant deflection rate of 12.7 mm/min. at a test temperature of 23°C. The test specimen shall be 305 mm long. Minimum tubing stiffness and maximum elongation are given in Table D.4. Heavy duty tubing is required for leach beds.

Tubing Size and Perforations. Nominal diameters range from 76 to 203 mm in 25.4-mm (1-in.) increments. Perforations shall be cleanly cut and uniformly spaced along the length and circumference of the tubing in a size, shape, and pattern to suit the needs of the user.

Elongation. A 1.27-m length specimen shall be tested with the axis vertical using a test load of 5D lbs, where D is the nominal inside tubing diameter in inches.

Table D.4 Physical Test Requirements for Corrugated Plastic Tubing (76 to 203 mm dia.) ASTM F405-76b

Physical property	Standard		Heavy Duty[a]	
	MPa	*(psi)*	*MPa*	*(psi)*
Pipe stiffness at 5 percent deflection, minimum	0.17	(24)	0.21	(30)
Pipe stiffness at 10 percent deflection, minimum	0.13	(19)	0.175	(25)
Elongation, maximum percent	10		5	

[a] Pipe stiffnesses for 254, 305, and 381 mm diameters as per ASTM F667-80.

Table D.5 Specifications for Tile and Pipe

Type of Drain and Specification Title	Specification Number	
Clay drain tile	ASTM[a]	C4-Yr[b]
Perforated clay drain tile	ASTM	C498-Yr
Standard strength clay sewer pipe	ASTM	C13-Yr
Extra strength clay pipe	ASTM	C200-Yr
Standard and extra strength perforated clay pipe	ASTM	C211-Yr
Concrete drain tile	ASTM	C412-Yr
Concrete sewer pipe	ASTM	C14-Yr
Reinforced concrete culvert, storm drain, and sewer pipe	ASTM	C76-Yr
Concrete pipe for irrigation or drainage	ASTM	C118-Yr
Manufacture and placement of reinforced concrete casings for irrigation wells	ASAE	(see Yearbook, current issue)
Asbestos cement nonpressure sewer pipe	ASTM	C428-Yr
Bituminized fiber drain and sewer pipe, homogeneous	ASTM	D1861-Yr
laminated wall	ASTM	D1862-Yr
Minimum standards for irrigation equipment (aluminum tubing)	ASAE	(see Yearbook, current issue)
Physical properties of concrete tile or pipe, Determining	ASTM	C497-Yr
Corrugated pipe,	Interim	
Aluminum alloy	Federal Spec	WW-P-00402[c]
Iron or steel, zinc coated	Federal Spec	WW-P-00405[c]
Plastic drain and sewer pipe	Commercial Std	CS-228-Yr[c]
External Loading properties of plastic pipe by parallel-plate loading	ASTM	D2412-Yr
Corrugated polyethylene tubing and fittings	ASTM	F405-Yr
	ASTM	F667-Yr
Subsurface installation of corrugated thermoplastic tubing for agricultural drainage or water table control	ASTM	F449-Yr
Installation of thermoplastic pipe and corrugated tubing in septic tank leach fields	ASTM	F481-Yr

[a] Am. Soc. for Testing Materials, 1916 Race St., Philadelphia, Pa.

[b] Current specifications include last two digits of calendar year, i.e., C4-59T, where T means tentative.

[c] U.S. Government Printing Office, Washington, D.C. 20250.

APPENDIX E

Loads on Underground Conduits

Underground conduits should be installed such that the load does not cause failure. For concrete and clay tile and corrugated plastic tubing the required strength and stiffness are set forth in current ASTM Specifications. Tile should meet the required minimum crushing strength and tubing should have the minimum stiffness at the specified vertical deflection.

Methods of calculating static soil loads are given in Chapter 15. Loads from wheeled vehicles or from super concentrated loads can be estimated from Fig. E.1 using the procedure given in Examples E.1 and E.2.

Example E.1. Determine the average load per lineal length and the total load transmitted to a 610-mm (24-in.) drain tile installed at a depth of 1.65 m (5.4 ft) from a static concentrated load of 454 kg (1000 lbs) directly over the center of the tile.

Solution. Depth to the top of conduit = depth to bottom of tile − B_c = 1.65 − 0.73 = 0.92 m (3.0 ft). Read I_c = 1.0 for static superload from Fig. E.1. Read $(100 \times C_t)$ = 20 percent from curve.

$$W_t = (1/0.61) \times 1.0 \times 0.20 \times 454 = 149 \text{ kg/m (100 lb/ft)}$$

$$\text{Total load} = LW_t = 0.61 \times 149 = 91 \text{ kg (200 lbs)}$$

Example E.2. Determine the average load per 0.3-m (1-ft) length on a 305-mm (12-in.) tile installed at a depth of 1 m (3.3 ft) if a 454-kg (1000-lb) concentrated load is moving at 32 km/h (20 mph).

Solution. From Fig. E.1 for a depth of 0.6 m (1.0 − 0.4), read $(100 \times C_t)$ = 13 percent, and select I_c = 2.0 for moving load.

$$W_t = (1/0.305) \times 2.0 \times 0.13 \times 454 = 387 \text{ kg/m (260 lb/lin ft)}$$

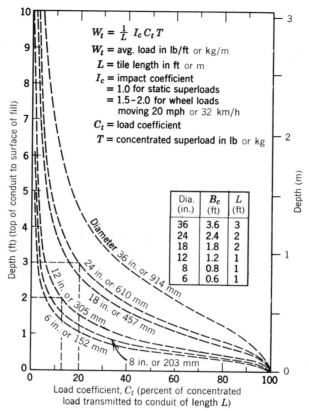

Fig. E.1.   Concentrated surface load coefficients. (Revised from Spangler et al., Iowa Eng. Expt. Sta. Bull. 79, 1926 and Marston, Iowa Eng. Expt. Sta. Bull. 96, 1930.)

Recommended maximum depths for corrugated plastic tubing from static soil loads are given in Table E.1. For ditch conditions the soil load was computed using the equation for flexible conduits, and for wide trenches the projecting load equation for rigid pipe was applied. These loads were substituted in the deflection equation described in Chapter 15.

Table E.1 Recommended Maximum Depths for Tubing Buried in Loose, Fine-Grained Soils in Meters

Nominal Tubing Diameter		Tubing	Trench width at top of tubing (m)					
mm	*(in.)*	*Quality*	0.2	0.3	0.4	0.6	0.8	1 m or more
102	(4)	Standard	[a]	3.9	2.1	1.7	1.6	1.6
		Heavy Duty	[a]	[a]	3.0	2.1	1.9	1.9
152	(6)	Standard	[a]	3.1	2.1	1.7	1.6	1.6
		Heavy Duty	[a]	[a]	2.9	2.0	1.9	1.9
203	(8)	Standard	[a]	3.1	2.2	1.7	1.6	1.6
		Heavy Duty		[a]	3.0	2.1	1.9	1.9
254	(10)			[a]	2.8	2.0	1.9	1.9
305	(12)				2.7	2.0	1.9	1.9
381	(15)					2.1	1.9	1.9

[a] Any depth is permissible at this width or less. Minimum side clearance between tubing and trench should be about 0.08 m.

Assumptions: Soil modulus of reaction, $E' = 345$ kN/m^2; deflection lag factor, $D = 3.4$; vertical deflection 20 percent; bedding angle factor, $K = 0.096$, and soil density of 1.75 gm/cc. See Chapter 15.

Source: Fenemor et al. *Trans. Am. Soc. Agr. Eng.* 22(6): 1338–1342, 1979.

APPENDIX F

Conversion Constants[1]

Table F.1 Conversion of Drainage Coefficients

Drainage Coefficient		Cfs per acre	Gpm per acre	L/s per hectare	m³/d per hectare
mm	(in.)				
1.0	($^{1}/_{25}$)	0.0017	0.75	0.118	10.15
1.6	($^{1}/_{16}$)	0.0026	1.18	0.185	15.85
3.2	($^{1}/_{8}$)	0.0052	2.36	0.367	31.70
6.4	($^{1}/_{4}$)	0.0105	4.71	0.735	63.41
8.5	($^{1}/_{3}$)	0.0142	6.29	1.000	85.76
9.5	($^{3}/_{8}$)	0.0157	7.07	1.102	95.12
12.7	($^{1}/_{2}$)	0.0210	9.43	1.469	126.83
15.9	($^{5}/_{8}$)	0.0262	11.79	1.836	158.53
19.1	($^{3}/_{4}$)	0.0315	14.14	2.204	190.23
22.2	($^{7}/_{8}$)	0.0367	16.50	2.571	222.17
25.4	(1)	0.0420	18.86	2.939	253.65

1 cfs/acre = 69.96 $Ls^{-1}ha^{-1}$ = 6 044 $m^3d^{-1}ha^{-1}$
1 L/s = 15.85 gpm

[1] See inside of back cover.

Table F.2 Miscellaneous Conversion Constants

Pressure and Force

1 in. Hg = 3386.4 Pa	1 Pa = 1 N/m²
= 2.04 psi	1 atm = 1013 mb
1 mm Hg = 133.3 Pa	= 101.3 kPa
1 mm water = 9.8 Pa	= 760 mm Hg
1 psi = 51.7 mm Hg	= 33.93 ft water
1 psi = 6.895 kPa	1 mb = 100 Pa
1 lb f = 4.45 N	1 mb = 10.2 mm water
1 lb f/ft² = 47.88 Pa	1 ft water = 2.985 kPa
1 kg f = 9.807 N	1 kPa = 0.335 ft water

Power and Energy

1 Btu = 1055 joules	1 kW = 1.341 hp
1 cal = 4.19 joules	1 kWh = 3.6 × 10⁶ joules
1 hp = 550 ft-lb/sec	
= 746 watts	

Volume and Weight

1 U.S. gal = 8.34 lb	1 oz = 28.35 g
1 Imp. gal = 10.02 lb	1 lb = 453.6 g
1 ft³ water = 62.4 lb	1 lb/ft³ = 16.02 kg/m³
1 yd³ = 0.7656 m³	1 short ton = 907.18 kg
1 ft³ = 7.48 gal	1 metric tonne = 1000 kg (Mg)
	= 2205 lb
	1 t/a = 2.24 Mg/ha (metric tonne)
	1 Kg = 2.205 lbs

Velocity

1 fps = 0.305 m/s	1 mph = 1.61 km/h = 0.447 m/s
= 1097 m/h	1 km/h = 0.621 mph

APPENDIX G

Useful Formulas and Procedures

VOLUME FORMULAS

The average end area formula for computing the volume of storage in a reservoir, earth fill in a dam, or ditch excavation, is

$$V = \frac{d}{2} (A_1 + A_2) \tag{G.1}$$

where V = volume of storage (L^3)
  d = vertical distance between end areas (L),
  A_1 and A_2 = end area (L^2).

The prismoidal formula is

$$V = \frac{d}{6} (A_1 + 4A_m + A_2) \tag{G.2}$$

where A_m = middle area halfway between the end areas (L^2).

Where preliminary surveys are made by taking slopes in the reservoir area, the storage may be estimated from the approximate formula for a frustrum of a cone,[1]

$$V = A_0 d + \frac{177d^2 A_0^{1/2}}{S} \tag{G.3}$$

where A_0 = area at spillway crest (L^2),
  d = depth of water above spillway crest (L),
  S = average slope of reservoir banks, through range of d, in percent.

---

[1] M. M. Culp. "The Effect of Spillway Storage on the Design of Upstream Reservoirs." *Agr, Eng.* **29**, 344–346 (1948).

LAYOUT OF CIRCULAR CURVES

The procedure to be followed in laying out a circular curve is as follows: The transit is first set up at the point of intersection (P.I.) as indicated in Fig. G.1 and the angle I is measured. Next calculate the tangent distance by the formula

$$T = R \tan \frac{I}{2} \tag{G.4}$$

where T = the tangent distance (L),
          I = the intersection angle in degrees,
          R = radius of curvature (L).

The point of curvature P.C. and the point of tangency P.T. are located by measuring the computed distance T from the P.I. Set up the transit at P.C. and locate stations on the curve by chaining and measuring off deflection angles as computed by the equation

$$e = \frac{c}{100} \cdot \frac{D}{2} = \frac{cD}{200} \tag{G.5}$$

where e = deflection angle in degrees,
          c = chord length in feet,
         D = degree of curve.

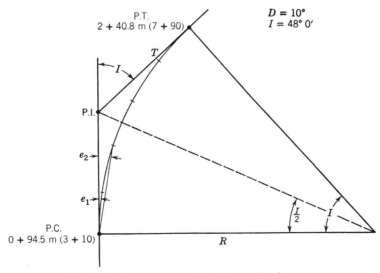

Fig. G.1. Layout procedure for a circular curve.

From this equation 100-foot stations require deflection angles of $1/2D$, 50-foot stations $1/4D$, and so on. After the first station beyond the P.C. is located, the deflection angle for each succeeding station is the summation of the deflection angles for all previous chord distances. Since most curves are rather flat, the arc distance is nearly equal to the chord length for 50- or 100-foot stations. The total length of the curve is

$$L = 30.48 \, I/D \text{ (meters)} \tag{G.6}$$

The design of a circular curve is illustrated by the following problem. Because the degree of curvature was defined in English units originally, computation is simplier than in SI units.

Example G.1.   Design a 10-degree curve for Fig. G.1 if the angle I is 48 deg. 0 min.

Solution.   From definition of degree of curvature given in Chapter 13,

$$R = 15.24/\sin 5 \text{ deg.} = 174.85 \text{ m (573.6 ft)}$$

and from Eq. G.4,

$$T = 174.85 \text{ m} \times \tan 24 \text{ deg.} = 77.85 \text{ m (255.4 ft)}$$

From Eq. G.5, the deflection angle to Sta. $1 + 21.92$ m ($4 + 00$ ft) is

$$e_1 = (27.43 \times 10)/60.96 = 4.5 \text{ deg.}$$

and similarly e_2 for Sta. $1 + 52.4$ m ($5 + 00$) is $(4.5 + 5) = 9.5$ deg., and so on.

From Eq. G.6,

$$L = (30.48 \times 48)/10 = 146.30 \text{ m (480 ft)}$$

and $94.5 + 146.3$ m ($310 + 480$) $= 240.80$ m (790 ft) or P.T. Sta. is $2 + 40.80$ m ($7 + 90$).

SETTING SLOPE STAKES

In making the location survey prior to construction of a dam or a ditch, center line stakes and slope stakes are set at each station or at more frequent intervals to guide the operator. On level or nearly level topography the offset of the slope stakes from the center line can be easily computed by adding one-half the top or bottom width plus the side slope ratio (z) times the depth. On irregular land the slope stakes are set by trial and error. Although the following procedure, illustrated in Fig. G.2, applies to ditch location, the same method is applicable to earth dam construction. First, the offset distance from the center line is estimated and an elevation for the point determined. If the depth from this point to the bottom of the ditch corresponds to the computed distance to the center line, the slope stake has been set correctly. For example, in Fig. G.2 the slope stake is at an elevation of 52.5 m and the bottom of the ditch is 45.0 m. The computed distance is

$$d = \Delta E z + w/2 = 52.5 - 45.0 + 6/2 = 10.5 \text{ m}$$

where d = distance from center line to edge of fill or ditch (L),
          ΔE = difference in elevation from the stake to the bottom of ditch or top of fill (L),
          z = side slope ratio (horizontal to vertical),
          w = bottom width of ditch or top width of dam (L).

If this distance is not 10.5, a new trial point must be selected, the elevation determined, and the distance from the center line again compared to the computed distance.

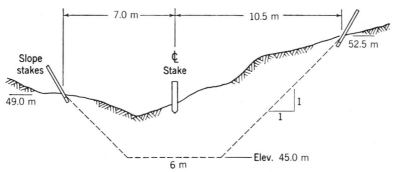

Fig. G.2.   Setting slope stakes for a ditch on uneven ground.

APPENDIX H

Filter Design Criteria

Considerable experimentation with the design of filters has been performed by U.S. Corps of Engineers, U.S. Bureau of Reclamation, and many others. The U.S. Bureau of Reclamation (1973)[1] recommends the following limits which will satisfy filter stability criteria and provide ample increase in permeability from the base to the filter material. These criteria are satisfactory for natural sand and gravel or crushed rock.

$$\frac{D_{15} \text{ (filter)}}{D_{15} \text{ (base)}} = 5 \text{ to } 40 \qquad (H.1)$$

provided the filter does not contain more than 5 percent finer than 0.074 mm,

$$\frac{D_{15} \text{ (filter)}}{D_{85} \text{ (base)}} = 5 \text{ or less} \qquad (H.2)$$

$$\frac{D_{85} \text{ (filter)}}{\text{Maximum size openings in pipe drain}} = 2 \text{ or more} \qquad (H.3)$$

and the distribution size curve of the filter should be roughly parallel to that of the base material. In the above criteria D_{15} and D_{85} are the particle diameters at which 15 and 85 percent, respectively, of the total soil particles are smaller on a weight basis. In addition to the above criteria the maximum particle size is 76.2 mm (3 in.).

The following example illustrates a typical design for a thin filter around a perforated pipe drain:

Example H.1. Determine the filter size limits for the conditions given in Fig. H.1 if the pipe drain openings are 12.7 mm (1/2 in.) in diameter and the gradation of the foundation soil is that shown in Fig. H.2.

---

[1] U.S. Bureau Reclamation (1973). *Design of Small Dams*. U.S. Government Printing Office, Washington, D.C.

512

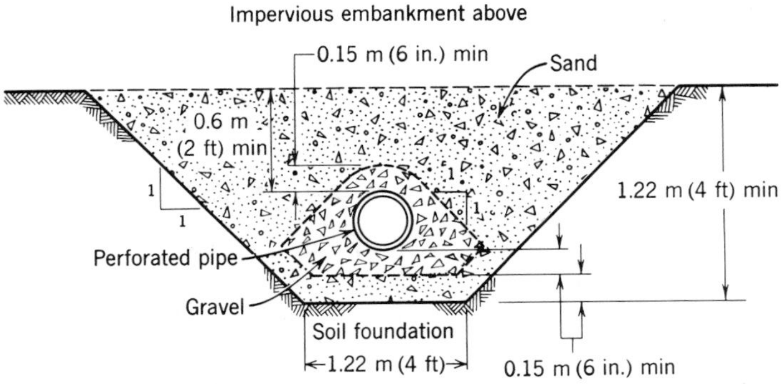

Impervious embankment above

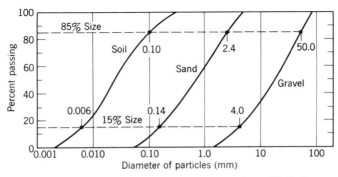

Fig. H.1. Typical filter for a toe drain of a dam. (Redrawn and revised from U.S. Bureau Reclamation, Design of Small Dams, 1973.)

Fig. H.2. Filter distribution curves for dams. (Redrawn from U.S. Bureau Reclamation, Design of Small Dams, 1973.)

Solution. From available materials the sand and gravel with size distribution curves shown in Fig. H.2 were selected, based on the following calculated values:

| Layer | Layer Sizes for | | D_{15} Size Requirements for the Next Coarser Layer (Filter) | | Max. Opening in Pipe Drain |
	D_{15} (mm)	D_{85} (mm)	From Eq. (H.1) (mm)	From Eq. (H.2) (mm)	(H.3) (mm)
Soil	0.006	0.1	0.03 to 0.24	0.5 or less	0.05
Sand	0.14[a]	2.4[a]	0.70 to 5.6	12.0 or less	1.2
Gravel	4.0[a]	50.0[a]	20.0 to —[b]	[b]	25.0

[a] These materials gave distribution curves nearly parallel to that for the soil.
[b] Maximum permissible size is 76.2 mm at D_{100} for any filter material.

Since the size of openings in the drain pipe is 12.7 mm ($1/2 \times 25.4$), the sand will not meet the criteria in Eq. H.3, but the gravel with a permissible opening of 25.0 mm is satisfactory. Thus, it was necessary to provide the gravel layer next to the perforated drain. Theoretically, each layer of the filter could be very thin, but practically a reasonable thickness is necessary to make sure that some slight readjustment during or after construction does not disrupt the layer. For earth dams the U.S. Bureau Reclamation (1973) recommends a minimum filter thickness of 0.9 m (3 ft), but the thickness of individual layers should be a minimum of 0.15 m (6 in.). Other minimum criteria are shown in Fig. H.1.

INDEX